LIBRARY
THE NORTH HIGHLAND COLLEGE
ORMLIE ROAD
THURSO
CAITHNESS KW14 7EE

Pump Characteristics and Applications

Second Edition

MECHANICAL ENGINEERING
A Series of Textbooks and Reference Books

Founding Editor

L. L. Faulkner

Columbus Division, Battelle Memorial Institute
and Department of Mechanical Engineering
The Ohio State University
Columbus, Ohio

1. *Spring Designer's Handbook*, Harold Carlson
2. *Computer-Aided Graphics and Design*, Daniel L. Ryan
3. *Lubrication Fundamentals*, J. George Wills
4. *Solar Engineering for Domestic Buildings*, William A. Himmelman
5. *Applied Engineering Mechanics: Statics and Dynamics*, G. Boothroyd and C. Poli
6. *Centrifugal Pump Clinic*, Igor J. Karassik
7. *Computer-Aided Kinetics for Machine Design*, Daniel L. Ryan
8. *Plastics Products Design Handbook, Part A: Materials and Components; Part B: Processes and Design for Processes*, edited by Edward Miller
9. *Turbomachinery: Basic Theory and Applications*, Earl Logan, Jr.
10. *Vibrations of Shells and Plates*, Werner Soedel
11. *Flat and Corrugated Diaphragm Design Handbook*, Mario Di Giovanni
12. *Practical Stress Analysis in Engineering Design*, Alexander Blake
13. *An Introduction to the Design and Behavior of Bolted Joints*, John H. Bickford
14. *Optimal Engineering Design: Principles and Applications*, James N. Siddall
15. *Spring Manufacturing Handbook*, Harold Carlson
16. *Industrial Noise Control: Fundamentals and Applications*, edited by Lewis H. Bell
17. *Gears and Their Vibration: A Basic Approach to Understanding Gear Noise*, J. Derek Smith
18. *Chains for Power Transmission and Material Handling: Design and Applications Handbook*, American Chain Association
19. *Corrosion and Corrosion Protection Handbook*, edited by Philip A. Schweitzer
20. *Gear Drive Systems: Design and Application*, Peter Lynwander
21. *Controlling In-Plant Airborne Contaminants: Systems Design and Calculations*, John D. Constance
22. *CAD/CAM Systems Planning and Implementation*, Charles S. Knox
23. *Probabilistic Engineering Design: Principles and Applications*, James N. Siddall

24. *Traction Drives: Selection and Application*, Frederick W. Heilich III and Eugene E. Shube
25. *Finite Element Methods: An Introduction*, Ronald L. Huston and Chris E. Passerello
26. *Mechanical Fastening of Plastics: An Engineering Handbook*, Brayton Lincoln, Kenneth J. Gomes, and James F. Braden
27. *Lubrication in Practice: Second Edition*, edited by W. S. Robertson
28. *Principles of Automated Drafting*, Daniel L. Ryan
29. *Practical Seal Design*, edited by Leonard J. Martini
30. *Engineering Documentation for CAD/CAM Applications*, Charles S. Knox
31. *Design Dimensioning with Computer Graphics Applications*, Jerome C. Lange
32. *Mechanism Analysis: Simplified Graphical and Analytical Techniques*, Lyndon O. Barton
33. *CAD/CAM Systems: Justification, Implementation, Productivity Measurement*, Edward J. Preston, George W. Crawford, and Mark E. Coticchia
34. *Steam Plant Calculations Manual*, V. Ganapathy
35. *Design Assurance for Engineers and Managers*, John A. Burgess
36. *Heat Transfer Fluids and Systems for Process and Energy Applications*, Jasbir Singh
37. *Potential Flows: Computer Graphic Solutions*, Robert H. Kirchhoff
38. *Computer-Aided Graphics and Design: Second Edition*, Daniel L. Ryan
39. *Electronically Controlled Proportional Valves: Selection and Application*, Michael J. Tonyan, edited by Tobi Goldoftas
40. *Pressure Gauge Handbook*, AMETEK, U.S. Gauge Division, edited by Philip W. Harland
41. *Fabric Filtration for Combustion Sources: Fundamentals and Basic Technology*, R. P. Donovan
42. *Design of Mechanical Joints*, Alexander Blake
43. *CAD/CAM Dictionary*, Edward J. Preston, George W. Crawford, and Mark E. Coticchia
44. *Machinery Adhesives for Locking, Retaining, and Sealing*, Girard S. Haviland
45. *Couplings and Joints: Design, Selection, and Application*, Jon R. Mancuso
46. *Shaft Alignment Handbook*, John Piotrowski
47. *BASIC Programs for Steam Plant Engineers: Boilers, Combustion, Fluid Flow, and Heat Transfer*, V. Ganapathy
48. *Solving Mechanical Design Problems with Computer Graphics*, Jerome C. Lange
49. *Plastics Gearing: Selection and Application*, Clifford E. Adams
50. *Clutches and Brakes: Design and Selection*, William C. Orthwein
51. *Transducers in Mechanical and Electronic Design*, Harry L. Trietley
52. *Metallurgical Applications of Shock-Wave and High-Strain-Rate Phenomena*, edited by Lawrence E. Murr, Karl P. Staudhammer, and Marc A. Meyers
53. *Magnesium Products Design*, Robert S. Busk
54. *How to Integrate CAD/CAM Systems: Management and Technology*, William D. Engelke

55. *Cam Design and Manufacture: Second Edition; with cam design software for the IBM PC and compatibles*, disk included, Preben W. Jensen
56. *Solid-State AC Motor Controls: Selection and Application*, Sylvester Campbell
57. *Fundamentals of Robotics*, David D. Ardayfio
58. *Belt Selection and Application for Engineers*, edited by Wallace D. Erickson
59. *Developing Three-Dimensional CAD Software with the IBM PC*, C. Stan Wei
60. *Organizing Data for CIM Applications*, Charles S. Knox, with contributions by Thomas C. Boos, Ross S. Culverhouse, and Paul F. Muchnicki
61. *Computer-Aided Simulation in Railway Dynamics*, by Rao V. Dukkipati and Joseph R. Amyot
62. *Fiber-Reinforced Composites: Materials, Manufacturing, and Design*, P. K. Mallick
63. *Photoelectric Sensors and Controls: Selection and Application*, Scott M. Juds
64. *Finite Element Analysis with Personal Computers*, Edward R. Champion, Jr. and J. Michael Ensminger
65. *Ultrasonics: Fundamentals, Technology, Applications: Second Edition, Revised and Expanded*, Dale Ensminger
66. *Applied Finite Element Modeling: Practical Problem Solving for Engineers*, Jeffrey M. Steele
67. *Measurement and Instrumentation in Engineering: Principles and Basic Laboratory Experiments*, Francis S. Tse and Ivan E. Morse
68. *Centrifugal Pump Clinic: Second Edition, Revised and Expanded*, Igor J. Karassik
69. *Practical Stress Analysis in Engineering Design: Second Edition, Revised and Expanded*, Alexander Blake
70. *An Introduction to the Design and Behavior of Bolted Joints: Second Edition, Revised and Expanded*, John H. Bickford
71. *High Vacuum Technology: A Practical Guide*, Marsbed H. Hablanian
72. *Pressure Sensors: Selection and Application*, Duane Tandeske
73. *Zinc Handbook: Properties, Processing, and Use in Design*, Frank Porter
74. *Thermal Fatigue of Metals*, Andrzej Weronski and Tadeusz Hejwowski
75. *Classical and Modern Mechanisms for Engineers and Inventors*, Preben W. Jensen
76. *Handbook of Electronic Package Design*, edited by Michael Pecht
77. *Shock-Wave and High-Strain-Rate Phenomena in Materials*, edited by Marc A. Meyers, Lawrence E. Murr, and Karl P. Staudhammer
78. *Industrial Refrigeration: Principles, Design and Applications*, P. C. Koelet
79. *Applied Combustion*, Eugene L. Keating
80. *Engine Oils and Automotive Lubrication*, edited by Wilfried J. Bartz
81. *Mechanism Analysis: Simplified and Graphical Techniques, Second Edition, Revised and Expanded*, Lyndon O. Barton
82. *Fundamental Fluid Mechanics for the Practicing Engineer*, James W. Murdock
83. *Fiber-Reinforced Composites: Materials, Manufacturing, and Design, Second Edition, Revised and Expanded*, P. K. Mallick
84. *Numerical Methods for Engineering Applications*, Edward R. Champion, Jr.

85. *Turbomachinery: Basic Theory and Applications, Second Edition, Revised and Expanded*, Earl Logan, Jr.
86. *Vibrations of Shells and Plates: Second Edition, Revised and Expanded*, Werner Soedel
87. *Steam Plant Calculations Manual: Second Edition, Revised and Expanded*, V. Ganapathy
88. *Industrial Noise Control: Fundamentals and Applications, Second Edition, Revised and Expanded*, Lewis H. Bell and Douglas H. Bell
89. *Finite Elements: Their Design and Performance*, Richard H. MacNeal
90. *Mechanical Properties of Polymers and Composites: Second Edition, Revised and Expanded*, Lawrence E. Nielsen and Robert F. Landel
91. *Mechanical Wear Prediction and Prevention*, Raymond G. Bayer
92. *Mechanical Power Transmission Components*, edited by David W. South and Jon R. Mancuso
93. *Handbook of Turbomachinery*, edited by Earl Logan, Jr.
94. *Engineering Documentation Control Practices and Procedures*, Ray E. Monahan
95. *Refractory Linings Thermomechanical Design and Applications*, Charles A. Schacht
96. *Geometric Dimensioning and Tolerancing: Applications and Techniques for Use in Design, Manufacturing, and Inspection*, James D. Meadows
97. *An Introduction to the Design and Behavior of Bolted Joints: Third Edition, Revised and Expanded*, John H. Bickford
98. *Shaft Alignment Handbook: Second Edition, Revised and Expanded*, John Piotrowski
99. *Computer-Aided Design of Polymer-Matrix Composite Structures*, edited by Suong Van Hoa
100. *Friction Science and Technology*, Peter J. Blau
101. *Introduction to Plastics and Composites: Mechanical Properties and Engineering Applications*, Edward Miller
102. *Practical Fracture Mechanics in Design*, Alexander Blake
103. *Pump Characteristics and Applications*, Michael W. Volk
104. *Optical Principles and Technology for Engineers*, James E. Stewart
105. *Optimizing the Shape of Mechanical Elements and Structures*, A. A. Seireg and Jorge Rodriguez
106. *Kinematics and Dynamics of Machinery*, Vladimír Stejskal and Michael Valásek
107. *Shaft Seals for Dynamic Applications*, Les Horve
108. *Reliability-Based Mechanical Design*, edited by Thomas A. Cruse
109. *Mechanical Fastening, Joining, and Assembly*, James A. Speck
110. *Turbomachinery Fluid Dynamics and Heat Transfer*, edited by Chunill Hah
111. *High-Vacuum Technology: A Practical Guide, Second Edition, Revised and Expanded*, Marsbed H. Hablanian
112. *Geometric Dimensioning and Tolerancing: Workbook and Answerbook*, James D. Meadows
113. *Handbook of Materials Selection for Engineering Applications*, edited by G. T. Murray

114. *Handbook of Thermoplastic Piping System Design*, Thomas Sixsmith and Reinhard Hanselka
115. *Practical Guide to Finite Elements: A Solid Mechanics* Approach, Steven M. Lepi
116. *Applied Computational Fluid Dynamics*, edited by Vijay K. Garg
117. *Fluid Sealing Technology*, Heinz K. Muller and Bernard S. Nau
118. *Friction and Lubrication in Mechanical Design*, A. A. Seireg
119. *Influence Functions and Matrices*, Yuri A. Melnikov
120. *Mechanical Analysis of Electronic Packaging Systems*, Stephen A. McKeown
121. *Couplings and Joints: Design, Selection, and Application, Second Edition, Revised and Expanded*, Jon R. Mancuso
122. *Thermodynamics: Processes and Applications*, Earl Logan, Jr.
123. *Gear Noise and Vibration*, J. Derek Smith
124. *Practical Fluid Mechanics for Engineering Applications*, John J. Bloomer
125. *Handbook of Hydraulic Fluid Technology*, edited by George E. Totten
126. *Heat Exchanger Design Handbook*, T. Kuppan
127. *Designing for Product Sound Quality*, Richard H. Lyon
128. *Probability Applications in Mechanical Design*, Franklin E. Fisher and Joy R. Fisher
129. *Nickel Alloys*, edited by Ulrich Heubner
130. *Rotating Machinery Vibration: Problem Analysis and Troubleshooting*, Maurice L. Adams, Jr.
131. *Formulas for Dynamic Analysis*, Ronald L. Huston and C. Q. Liu
132. *Handbook of Machinery Dynamics*, Lynn L. Faulkner and Earl Logan, Jr.
133. *Rapid Prototyping Technology: Selection and Application*, Kenneth G. Cooper
134. *Reciprocating Machinery Dynamics: Design and Analysis*, Abdulla S. Rangwala
135. *Maintenance Excellence: Optimizing Equipment Life-Cycle Decisions*, edited by John D. Campbell and Andrew K. S. Jardine
136. *Practical Guide to Industrial Boiler Systems*, Ralph L. Vandagriff
137. *Lubrication Fundamentals: Second Edition, Revised and Expanded*, D. M. Pirro and A. A. Wessol
138. *Mechanical Life Cycle Handbook: Good Environmental Design and Manufacturing*, edited by Mahendra S. Hundal
139. *Micromachining of Engineering Materials*, edited by Joseph McGeough
140. *Control Strategies for Dynamic Systems: Design and Implementation*, John H. Lumkes, Jr.
141. *Practical Guide to Pressure Vessel Manufacturing*, Sunil Pullarcot
142. *Nondestructive Evaluation: Theory, Techniques, and Applications*, edited by Peter J. Shull
143. *Diesel Engine Engineering: Thermodynamics, Dynamics, Design, and Control*, Andrei Makartchouk
144. *Handbook of Machine Tool Analysis*, Ioan D. Marinescu, Constantin Ispas, and Dan Boboc

145. *Implementing Concurrent Engineering in Small Companies*, Susan Carlson Skalak
146. *Practical Guide to the Packaging of Electronics: Thermal and Mechanical Design and Analysis*, Ali Jamnia
147. *Bearing Design in Machinery: Engineering Tribology and Lubrication*, Avraham Harnoy
148. *Mechanical Reliability Improvement: Probability and Statistics for Experimental Testing*, R. E. Little
149. *Industrial Boilers and Heat Recovery Steam Generators: Design, Applications, and Calculations*, V. Ganapathy
150. *The CAD Guidebook: A Basic Manual for Understanding and Improving Computer-Aided Design*, Stephen J. Schoonmaker
151. *Industrial Noise Control and Acoustics*, Randall F. Barron
152. *Mechanical Properties of Engineered Materials*, Wolé Soboyejo
153. *Reliability Verification, Testing, and Analysis in Engineering Design*, Gary S. Wasserman
154. *Fundamental Mechanics of Fluids: Third Edition*, I. G. Currie
155. *Intermediate Heat Transfer*, Kau-Fui Vincent Wong
156. *HVAC Water Chillers and Cooling Towers: Fundamentals, Application, and Operation*, Herbert W. Stanford III
157. *Gear Noise and Vibration: Second Edition, Revised and Expanded*, J. Derek Smith
158. *Handbook of Turbomachinery: Second Edition, Revised and Expanded*, edited by Earl Logan, Jr. and Ramendra Roy
159. *Piping and Pipeline Engineering: Design, Construction, Maintenance, Integrity, and Repair*, George A. Antaki
160. *Turbomachinery: Design and Theory*, Rama S. R. Gorla and Aijaz Ahmed Khan
161. *Target Costing: Market-Driven Product Design*, M. Bradford Clifton, Henry M. B. Bird, Robert E. Albano, and Wesley P. Townsend
162. *Fluidized Bed Combustion*, Simeon N. Oka
163. *Theory of Dimensioning: An Introduction to Parameterizing Geometric Models*, Vijay Srinivasan
164. *Handbook of Mechanical Alloy Design*, edited by George E. Totten, Lin Xie, and Kiyoshi Funatani
165. *Structural Analysis of Polymeric Composite Materials*, Mark E. Tuttle
166. *Modeling and Simulation for Material Selection and Mechanical Design*, edited by George E. Totten, Lin Xie, and Kiyoshi Funatani
167. *Handbook of Pneumatic Conveying Engineering*, David Mills, Mark G. Jones, and Vijay K. Agarwal
168. *Clutches and Brakes: Design and Selection, Second Edition*, William C. Orthwein
169. *Fundamentals of Fluid Film Lubrication: Second Edition*, Bernard J. Hamrock, Steven R. Schmid, and Bo O. Jacobson
170. *Handbook of Lead-Free Solder Technology for Microelectronic Assemblies*, edited by Karl J. Puttlitz and Kathleen A. Stalter
171. *Vehicle Stability*, Dean Karnopp

172. *Mechanical Wear Fundamentals and Testing: Second Edition, Revised and Expanded*, Raymond G. Bayer
173. *Liquid Pipeline Hydraulics*, E. Shashi Menon
174. *Solid Fuels Combustion and Gasification*, Marcio L. de Souza-Santos
175. *Mechanical Tolerance Stackup and Analysis*, Bryan R. Fischer
176. *Engineering Design for Wear,* Raymond G. Bayer
177. *Vibrations of Shells and Plates: Third Edition, Revised and Expanded*, Werner Soedel
178. *Refractories Handbook*, edited by Charles A. Schacht
179. *Practical Engineering Failure Analysis*, Hani M. Tawancy, Anwar Ul-Hamid, and Nureddin M. Abbas
180. *Mechanical Alloying and Milling*, C. Suryanarayana
181. *Mechanical Vibration: Analysis, Uncertainties, and Control, Second Edition, Revised and Expanded*, Haym Benaroya
182. *Design of Automatic Machinery*, Stephen J. Derby
183. *Practical Fracture Mechanics in Design: Second Edition, Revised and Expanded*, Arun Shukla
184. *Practical Guide to Designed Experiments*, Paul D. Funkenbusch
185. *Gigacycle Fatigue in Mechanical Practive*, Claude Bathias and Paul C. Paris
186. *Selection of Engineering Materials and Adhesives*, Lawrence W. Fisher
187. *Boundary Methods: Elements, Contours, and Nodes*, Subrata Mukherjee and Yu Xie Mukherjee
188. *Rotordynamics*, Agnieszka Muszynska
189. *Pump Characteristics and Applications: Second Edition*, Michael Volk

Pump Characteristics and Applications

LIBRARY
THE NORTH HIGHLAND COLLEGE
ORMLIE ROAD
THURSO
CAITHNESS KW14 7EE

621.
69
VOL

Published in 2005 by
CRC Press
Taylor & Francis Group
6000 Broken Sound Parkway NW, Suite 300
Boca Raton, FL 33487-2742

© 2005 by Taylor & Francis Group, LLC
CRC Press is an imprint of Taylor & Francis Group

No claim to original U.S. Government works
Printed in the United States of America on acid-free paper
10 9 8 7 6 5 4 3 2 1

International Standard Book Number-10: 0-82472-755-X (Hardcover)
International Standard Book Number-13: 978-0-82472-755-0 (Hardcover)

This book contains information obtained from authentic and highly regarded sources. Reprinted material is quoted with permission, and sources are indicated. A wide variety of references are listed. Reasonable efforts have been made to publish reliable data and information, but the author and the publisher cannot assume responsibility for the validity of all materials or for the consequences of their use.

No part of this book may be reprinted, reproduced, transmitted, or utilized in any form by any electronic, mechanical, or other means, now known or hereafter invented, including photocopying, microfilming, and recording, or in any information storage or retrieval system, without written permission from the publishers.

For permission to photocopy or use material electronically from this work, please access www.copyright.com (http://www.copyright.com/) or contact the Copyright Clearance Center, Inc. (CCC) 222 Rosewood Drive, Danvers, MA 01923, 978-750-8400. CCC is a not-for-profit organization that provides licenses and registration for a variety of users. For organizations that have been granted a photocopy license by the CCC, a separate system of payment has been arranged.

Trademark Notice: Product or corporate names may be trademarks or registered trademarks, and are used only for identification and explanation without intent to infringe.

Library of Congress Cataloging-in-Publication Data

Catalog record is available from the Library of Congress

Taylor & Francis Group
is the Academic Division of T&F Informa plc.

Visit the Taylor & Francis Web site at
http://www.taylorandfrancis.com

and the CRC Press Web site at
http://www.crcpress.com

Preface to the Second Edition

Thankfully, the laws of physics have not changed since the first edition of this book was written in 1996. Therefore, virtually everything about pump selection, sizing, system analysis, and other aspects of pump hydraulics remains unchanged from the first edition. There have, however, been a number of innovations in the world of pumps, which are introduced in this second edition. This edition also expands the material on many components of typical pump installations that were only briefly covered in the first edition, if at all. Some of the most important new or expanded topics covered in this second edition include:

- *Chapter 1* — Several new types of positive displacement (P.D.) pumps are introduced, while the information on other types of P.D. pumps has been expanded.
- *Chapter 2* — Important new topics in this chapter include NPSH analysis for closed systems, expansion of the discussion on NPSH margin, and system head curve development for existing systems and for parallel pumping systems.
- *Chapter 3* — In the world of software, 9 years is an eternity, and so the entire section of this chapter covering software used to design and analyze pump piping systems has been completely rewritten. A new CD is included with the second edition of the book, demonstrating one such software tool, including solving some of the problems covered in the book.

- *Chapter 4* — Entire new sections of this chapter have been added to provide in-depth coverage of two very important and relevant topics: pump couplings and electric motors. Additionally, several types of centrifugal pumps that were not included in the first edition are covered in this chapter.
- *Chapter 5* — This chapter has an entire new section on O-rings used in pumps, as well as additional information about sealless pumps.
- *Chapter 6* — Two major additions to the book are included in this chapter. The first is an in-depth discussion of variable-frequency drives. Second, this chapter includes a section covering pump life-cycle cost, an innovative approach to the study of the cost of pumping equipment that looks way beyond the capital cost of the pump.
- *Chapter 7* — This chapter has added in-depth discussion of metallic corrosion in pumps, as well as discourse on elastomers commonly used in pumps for sealing components.
- *Chapter 8* — New topics covered in this chapter include ten methods to prevent low flow damage in pumps, and a much more detailed discussion of vibration, including a detailed vibration troubleshooting chart.

Acknowledgments

Thanks to my colleagues in the pump field who provided input for this second edition, or who reviewed particular sections of it. Finally, I wish to thank my daughter Sarah, who typed major portions of the new material for this edition.

Preface to the First Edition

This book is a practical introduction to the characteristics and applications of pumps, with a primary focus on centrifugal pumps. Pumps are among the oldest machines still in use and, after electric motors, are probably the most widely used machines today in commercial and industrial activities. Despite the broad use of pumps, this subject is covered only briefly in many engineering curricula. Furthermore, companies which use pumps are often unable to provide their engineers, operators, mechanics, and supervisors the kind of training in pump application, selection, and operation that this vital equipment merits.

The purpose of this book is to give engineers and technicians a general understanding of pumps, and to provide the tools to allow them to properly select, size, operate, and maintain pumps. There are numerous books on the market aout pumps, but most of them are very, very technical, and are mainly design oriented, or else are directed to a specific niche market. I have attempted to provide practical information on pumps and systems to readers with with all levels of experience, without getting so immersed in design details as to overwhelm the reader.

This book begins with the basics of pump and system hydraulics, working gradually to more complex concepts. The topics are covered in a clear and concise manner, and are accompanied by examples along the way. Anyone reading the material, regardless of education and experience with pumps, will be able to achieve a better understanding of pump characteristics and applications.

While it is not possible to cover pump hydraulics without getting into some mathematics, this book covers the subject without resorting to differential equations and other high level matchematics that most people forgot right after school. For the reader who is interested in a more complex or sophisticated approach to particular topics, or who wants additional information in a given area, references are made to other sources which provide a more analytical approach.

A theme that is repeated throughout this book is that all aspects of pumps — from system design, to pump selection, to piping design, to installation, to operation — are interrelated. Lack of attention to the sizing of a pump or improper design of the piping system can cause future problems with pump maintenance and operation. Even the most precisely sized pump will not perform properly if its installation and maintenance are not performed carefully. A better understanding of how these issues are related will help to solve problems or to prevent them from occurring in the first place.

In addition to a thorough treatment of the fundamentals, this book also provides information on the current state of the art of various technologies in the pump field. Variable speed pumping systems, sealless pumps, gas lubricating noncontacting mechanical seals, and nonmetallic pumps are examples of recent technological trends in the pump industry which are introduced in this book. Computer software for system design and pump selection is previewed in Chapter 3, and a demonstration CD is included with this book. This is another example of a powerful new technology related to pumps that is covered in this book.

Because the book focuses on pump applications and characteristics, rather than on design, it is intended for a broader audience than typical books about pumps. The readership for this book includes the following:

- *Engineers* — This book has broad appeal to mechanical, civil, chemical, industrial, and electrical engineers. Any engineer whose job it is to design or modify systems; select, specify, purchase, or sell pumps; or oversee operation, testing, or maintenance of pumping equipment will find this book very helpful.

- *Engineering Supervisors* — Because they have broad responsibility for overseeing the design and operation of pumps and pump systems, engineering supervisors will benefit from the integrated systems approach provided in this book.
- *Plant Operators* —Employees of plants which utilize pumps are required to oversee the operation of the pumps, and often their maintenance, troubleshooting, and repair. A better understanding of hydraulics and applications will help these people do a better job of operating their pumps most efficiently while reducing maintenance costs and downtime.
- *Maintenance Technicians* — Maintenance personel and their supervisors can do a much better job of installing, maintaining, troubleshooting, and repairing pumps if they have a better understanding of how pumps are applied and operated in a system.
- *Engineering Students* — The "real world" problems which are presented in this book demonstrate to students that a pump is more than a "black box." Many university engineering departments are expanding their technology program to better prepare students for jobs in industry. This book can make an important contribution to a program in industrial machinery.

Formulae used in this book will generally be stated in United States Customary System (USCS) units, the system most widely used by the pump industry in the United States. Appendix B at the end of this book provides simple conversion formulae from USCS to SI (metric) units. The most common terms mentioned in this book will be stated in both untis.

I wish to thank my colleagues in the pump field who reviewed various sections of this book, or who assisted in obtaining materials and illustrations. I'm especially grateful to my friends Jim Johnston, Paul Lahr, and Buster League, who reviewed the entire manuscript and provided me with valuable feedback. Final thanks go to my wife, Jody Lerner, for her word processing and editorial skills, as well as for her patience and encouragement.

About the Author

Michael W. Volk, P.E., is President of Volk & Associates, Inc., Oakland, California, www.volkassociates.com, a consulting company specializing in pumps and pump systems. Volk's services include pump training seminars; pump equipment evaluation, troubleshooting, and field testing; expert witness for pump litigation; witnessing of pump shop tests; pump market research; and acquisition and divestiture consultation and brokerage. A member of the American Society of Mechanical Engineers (ASME), and a registered professional engineer, Volk received the B.S. degree (1973) in mechanical engineering from the University of Illinois, Urbana, and the M.S. degree (1976) in mechanical engineering and the M.S. degree (1980) in management science from the University of Southern California, Los Angeles. He may be contacted at mike@volkassociates.com.

Contents

1 Introduction to Pumps 1
    I. What Is a Pump? 1
    II. Why Increase a Liquid's Pressure? 2
    III. Pressure and Head 3
    IV. Classification of Pumps 5
        A. Principle of Energy Addition 5
            1. Kinetic 5
            2. Positive Displacement 5
        B. How Energy Addition Is Accomplished 7
        C. Geometry Used 7
    V. How Centrifugal Pumps Work 7
    VI. Positive Displacement Pumps 14
        A. General 14
        B. When to Choose a P.D. Pump 15
        C. Major Types of P.D. Pumps 22
            1. Sliding Vane Pump 24
            2. Sinusoidal Rotor Pump 25
            3. Flexible Impeller Pump 25
            4. Flexible Tube (Peristaltic) Pump 26
            5. Progressing Cavity Pump 27
            6. External Gear Pump 29
            7. Internal Gear Pump 33
            8. Rotary Lobe Pump 33
            9. Circumferential Piston and Bi-Wing
               Lobe Pumps 35
           10. Multiple-Screw Pump 36
           11. Piston Pump 38

12. Plunger Pump 40
13. Diaphragm Pump 41
14. Miniature Positive Displacement Pumps 47

2 Hydraulics, Selection, and Curves 51

I. Overview 51
II. Pump Capacity 54
III. Head 54
   A. Static Head 56
   B. Friction Head 58
   C. Pressure Head 66
   D. Velocity Head 70
IV. Performance Curve 71
V. Horsepower and Efficiency 80
   A. Hydraulic Losses 82
   B. Volumetric Losses 82
   C. Mechanical Losses 83
   D. Disk Friction Losses 83
VI. NPSH and Cavitation 89
   A. Cavitation and NPSH Defined 89
      1. $NPSH_a$ 98
      2. $NPSH_r$ 99
   B. Calculating $NPSH_a$: Examples 101
   C. Remedies for Cavitation 102
   D. More $NPSH_a$ Examples 106
   E. Safe Margin $NPSH_a$ vs. $NPSH_r$ 109
   F. NPSH for Reciprocating Pumps 114
VII. Specific Speed and Suction Specific Speed 116
VIII. Affinity Laws 122
IX. System Head Curves 127
X. Parallel Operation 139
XI. Series Operation 146
XII. Oversizing Pumps 152
XIII. Pump Speed Selection 155
   A. Suction Specific Speed 156
   B. Shape of Pump Performance Curves 156
   C. Maximum Attainable Efficiency 157

Contents

    D. Speeds Offered by Manufacturers 158
    E. Prior Experience 159

3 Special Hydraulic Considerations 161

I. Overview 161
II. Viscosity 162
III. Software to Size Pumps and Systems 185
    A. General 185
    B. Value of Piping Design Software 186
    C. Evaluating Fluid Flow Software 186
    D. Building the System Model 187
        1. Copy Command 189
        2. Customize Symbols 190
        3. CAD Drawing Features 190
        4. Naming Items 190
        5. Displaying Results 190
        6. The Look of the Piping Schematic 191
    E. Calculating the System Operation 191
        1. Sizing Pipe Lines 192
        2. Calculating Speed 192
        3. Showing Problem Areas 192
        4. Equipment Selection 192
        5. Alternate System Operational Modes 193
    F. Communicating the Results 193
        1. Viewing Results within the Program 193
        2. Incorporating User-Defined Limits 194
        3. Selecting the Results to Display 194
        4. Plotting the Piping Schematic 194
        5. Exporting the Results 194
        6. Sharing Results with Others 195
        7. Sharing Results Using a Viewer Program ... 195
    G. Conclusion 195
    H. List of Software Vendors 196
IV. Piping Layout 196
V. Sump Design 200
VI. Field Testing 203
    A. General 203
    B. Measuring Flow 205

1. Magnetic Flowmeter 205
2. Mass Flowmeter 205
3. Nozzle 205
4. Orifice Plate 206
5. Paddle Wheel 206
6. Pitot Tube 206
7. Segmental Wedge 207
8. Turbine Meter 207
9. Ultrasonic Flowmeter 207
10. Venturi 208
11. Volumetric Measurement 208
12. Vortex Flowmeter 208
    C. Measuring TH 209
    D. Measuring Power 211
    E. Measuring NPSH 212

4 Centrifugal Pump Types and Applications 213
  I. Overview 213
 II. Impellers 215
     A. Open vs. Closed Impellers 215
     B. Single vs. Double Suction 223
     C. Suction Specific Speed 225
     D. Axial Thrust and Thrust Balancing 227
     E. Filing Impeller Vane Tips 230
     F. Solids Handling Impellers 232
III. End Suction Pumps 233
     A. Close-Coupled Pumps 233
     B. Frame-Mounted Pumps 237
 IV. Inline Pumps 240
  V. Self-Priming Centrifugal Pumps 242
 VI. Split Case Double Suction Pumps 245
VII. Multi-Stage Pumps 250
     A. General 250
     B. Axially Split Case Pumps 250
     C. Radially Split Case Pumps 254
VIII. Vertical Column Pumps 256
 IX. Submersible Pumps 260
  X. Slurry Pumps 264

Contents

XI.	Vertical Turbine Pumps	268
XII.	Axial Flow Pumps	277
XIII.	Regenerative Turbine Pumps	278
XIV.	Pump Specifications and Standards	279
	A. General	279
	1. Liquid Properties	280
	2. Hydraulic Conditions	280
	3. Installation Details	281
	B. ANSI	282
	C. API	284
	D. ISO	286
XV.	Couplings	287
XVI.	Electric Motors	291
	A. Glossary of Frequently Occurring Motor Terms	294
	1. Amps	294
	2. Code Letter	295
	3. Design Letter	295
	4. Efficiency	296
	5. Frame Size	296
	6. Frequency	296
	7. Full Load Speed	297
	8. High Inertial Load	297
	9. Insulation Class	297
	10. Load Types	297
	11. Phase	298
	12. Poles	298
	13. Power Factor	298
	14. Service Factor	298
	15. Slip	299
	16. Synchronous Speed	299
	17. Temperature	299
	18. Time Rating	300
	19. Voltage	300
	B. Motor Enclosures	300
	1. Open Drip Proof	300
	2. Totally Enclosed Fan Cooled	301
	3. Totally Enclosed Air Over	301

    4. Totally Enclosed Non-Ventilated 301
    5. Hazardous Location 302
  C. Service Factor 302
  D. Insulation Classes 303
  E. Motor Frame Size 303
    1. Historical Perspective 303
    2. Rerating and Temperature 307
    3. Motor Frame Dimensions 307
    4. Fractional Horsepower Motors 307
    5. Integral Horsepower Motors 312
    6. Frame Designation Variations 312
  F. Single Phase Motors 314
  G. Motors Operating on Variable Frequency
    Drives 319
  H. NEMA Locked Rotor Code 321
  I. Amps, Watts, Power Factor, and Efficiency 322
    1. Introduction 322
    2. Power Factor 322
    3. Efficiency 323
    4. Amperes 325
    5. Summary 325

5 Sealing Systems and Sealless Pumps 327

 I. Overview 327
 II. O-Rings 328
  A. What Is an O-Ring? 328
  B. Basic Principals of the O-Ring Seal 329
  C. The Function of the O-Ring 329
  D. Static and Dynamic O-Ring Sealing
    Applications 330
  E. Other Common O-Ring Seal Configurations 330
  F. Limitations of O-Ring Use 333
 III. Stuffing Box and Packing Assembly 333
  A. Stuffing Box 334
  B. Stuffing Box Bushing 334
  C. Packing Rings 335
  D. Packing Gland 336
  E. Lantern Ring 337

Contents   xxvii

   IV. Mechanical Seals 338
      A. Mechanical Seal Advantages 338
          1. Lower Mechanical Losses 338
          2. Less Sleeve Wear 338
          3. Zero or Minimal Leakage 338
          4. Reduced Maintenance 339
          5. Seal Higher Pressures 339
      B. How Mechanical Seals Work 339
      C. Types of Mechanical Seals 343
          1. Single, Inside Seals 343
          2. Single, Outside Seals 345
          3. Single, Balanced Seals 346
          4. Double Seals 347
          5. Tandem Seals 349
          6. Gas Lubricated Non-Contacting Seals 351
    V. Sealless Pumps 352
      A. General 352
      B. Magnetic Drive Pumps 354
          1. Bearings in the Pumped Liquid 357
          2. Dry Running 358
          3. Inefficiency 358
          4. Temperature 358
          5. Viscosity 359
      C. Canned Motor Pumps 359
          1. Fewer Bearings 360
          2. More Compact 361
          3. Double Containment 361
          4. Lower First Cost 361

6 Energy Conservation and Life-Cycle Costs 363

     I. Overview 363
   II. Choosing the Most Efficient Pump 364
  III. Operating with Minimal Energy 372
   IV. Variable-Speed Pumping Systems 373
    V. Pump Life-Cycle Costs 395
      A. Improving Pump System Performance:
         An Overlooked Opportunity? 395
      B. What Is Life-Cycle Cost? 397

C. Why Should Organizations Care about Life-Cycle Cost? 397
D. Getting Started 399
E. Life Cycle Cost Analysis 399
   1. C_{ic} — Initial Investment Costs 401
   2. C_{in} — Installation and Commissioning (Start-up) Costs 402
   3. C_e — Energy Costs 403
   4. C_o — Operation Costs 404
   5. C_m — Maintenance and Repair Costs 404
   6. C_s — Downtime and Loss of Production Costs 406
   7. C_{env} — Environmental Costs, Including Disposal of Parts and Contamination from Pumped Liquid 407
   8. C_d — Decommissioning/Disposal Costs, Including Restoration of the Local Environment 407
F. Total Life-Cycle Costs 408
G. Pumping System Design 408
H. Methods for Analyzing Existing Pumping Systems 413
I. Example: Pumping System with a Problem Control Valve 414
J. For More Information 419
   1. About the Hydraulic Institute 419
   2. About Europump 419
   3. About the U.S. Department of Energy's Office of Industrial Technologies 421

7 Special Pump-Related Topics 423

I. Overview 423
II. Variable-Speed Systems 424
III. Sealless Pumps 425
IV. Corrosion 426
   1. Galvanic, or Two-Metal Corrosion 428
   2. Uniform, or General Corrosion 429
   3. Pitting Corrosion 430

		4. Intergranular Corrosion 430

- 4. Intergranular Corrosion 430
- 5. Erosion Corrosion 431
- 6. Stress Corrosion 431
- 7. Crevice Corrosion 432
- 8. Graphitization or Dezincification Corrosion 432

V. Nonmetallic Pumps 432
VI. Materials Used for O-Rings in Pumps 435
   A. General 435
   B. Eight Basic O-Ring Elastomers 437
      1. Nitrile (Buna N) 437
      2. Neoprene 437
      3. Ethylene Propylene 438
      4. Fluorocarbon (Viton) 438
      5. Butyl 439
      6. Polyacrylate 439
      7. Silicone 439
      8. Fluorosilicone 440
VII. High-Speed Pumps 441
VIII. Bearings and Bearing Lubrication 446
IX. Precision Alignment Techniques 447
X. Software to Size Pumps and Systems 449

8 Installation, Operation, and Maintenance 451

I. Overview 451
II. Installation, Alignment, and Start-Up 452
   A. General 452
   B. Installation Checklist 453
      1. Tag and Lock Out 453
      2. Check Impeller Setting 453
      3. Install Packing or Seal 453
      4. Mount Bedplate, Pump, and Motor 454
      5. Check Rough Alignment 454
      6. Place Grout in Bedplate 454
      7. Check Alignment 456
      8. Flush System Piping 457
      9. Connect Piping to Pump 457
      10. Check Alignment 459

          11. Turn Pump by Hand 459
          12. Wire and Jog Motor................................. 459
          13. Connect Coupling 459
          14. Check Shaft Runout 460
          15. Check Valve and Vent Positions 460
          16. Check Lubrication/Cooling Systems........... 460
          17. Prime Pump if Necessary 460
          18. Check Alignment 461
          19. Check System Components Downstream 461
          20. Start and Run Pump................................ 462
          21. Stop Pump and Check Alignment 462
          22. Drill and Dowel Pump to Base.................. 462
          23. Run Benchmark Tests 462
   III. Operation 462
       A. General.. 462
       B. Minimum Flow 463
          1. Temperature Rise 464
          2. Radial Bearing Loads................................ 465
          3. Axial Thrust... 465
          4. Prerotation... 465
          5. Recirculation and Separation 466
          6. Settling of Solids...................................... 468
          7. Noise and Vibration.................................. 468
          8. Power Savings, Motor Load 468
       C. Ten Ways to Prevent Low Flow Damage
          in Pumps 468
          1. Continuous Bypass 470
          2. Multi-Component Control Valve System 471
          3. Variable Frequency Drive 472
          4. Automatic Recirculation Valve 473
          5. Relief Valve 473
          6. Pressure Sensor 475
          7. Ammeter... 475
          8. Power Monitor 475
          9. Vibration Sensor 476
          10. Temperature Sensor 476
   IV. Maintenance 477
       A. Regular Maintenance 477
          1. Lubrication 477

Contents

    2. Packing 478
    3. Seals 479
   B. Preventive Maintenance 479
    1. Regular Lubrication 480
    2. Rechecking Alignment 480
    3. Rebalance Rotating Element 480
    4. Monitoring Benchmarks 480
   C. Benchmarks 480
    1. Hydraulic Performance 480
    2. Temperature 481
    3. Vibration 482
 V. Troubleshooting 489
 VI. Repair 489
   A. General 489
   B. Repair Tips 492
    1. Document the Disassembly 492
    2. Analyze Disassembled Pump 492
    3. Bearing Replacement 493
    4. Wear Ring Replacement 494
    5. Guidelines for Fits and Clearances 495
    6. Always Replace Consumables 495
    7. Balance Impellers and Couplings 495
    8. Check Runout of Assembled Pump 496
    9. Tag Lubrication Status 497
    10. Cover Openings Prior to Shipment 497

Appendix A: Major Suppliers of Pumps in the United States by Product Type 499

Appendix B: Conversion Formulae 511

References 525

Index 527

1

Introduction to Pumps

I. WHAT IS A PUMP?

Simply stated, a pump is a machine used to move liquid through a piping system and to raise the pressure of the liquid. A pump can be further defined as a machine that uses several energy transformations to increase the pressure of a liquid. The centrifugal pump shown in Figure 1.1 illustrates this definition. The energy input into the pump is typically the energy source used to power the driver. Most commonly, this is electricity used to power an electric motor. Alternative forms of energy used to power the driver include high-pressure steam to drive a steam turbine, fuel oil to power a diesel engine, high-pressure hydraulic fluid to power a hydraulic motor, and compressed air to drive an air motor. Regardless of the driver type for a centrifugal pump, the input energy is converted in the driver to a rotating mechanical energy, consisting of the driver output shaft, operating at a certain speed, and transmitting a certain torque, or horsepower.

The remaining energy transformations take place inside the pump itself. The rotating pump shaft is attached to the pump impeller (see Figure 1.4). The rotating impeller causes the liquid that has entered the pump to increase in velocity. This is the second energy transformation in the pump, where the input power is used to raise the kinetic energy of the liquid. Kinetic energy is a function of mass and velocity. Raising a liquid's velocity increases its kinetic energy.

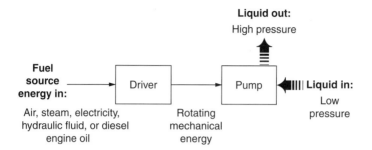

Figure 1.1 A centrifugal pump uses several energy transformations to raise the pressure of a liquid.

After the liquid leaves the impeller, but before exiting the pump, the final transformation of energy occurs in a *diffusion* process. An expansion of the flow area causes the liquid's velocity to decrease to more than when it entered the pump, but well below its maximum velocity at the impeller tip. This diffusion transforms some of the velocity energy to pressure energy.

II. WHY INCREASE A LIQUID'S PRESSURE?

There are actually three distinct reasons for raising the pressure of a liquid with a pump, plus another related factor:

1. *Static elevation.* A liquid's pressure must be increased to raise the liquid from one elevation to a higher elevation. This might be necessary, for example, to move liquid from one floor of a building to a higher floor, or to pump liquid up a hill.
2. *Friction.* It is necessary to increase the pressure of a liquid to move the liquid through a piping system and overcome frictional losses. Liquid moving through a system of pipes, valves, and fittings experiences frictional losses along the way. These losses vary with the geometry and material of the pipe, valves, and fittings, with the viscosity and density of the liquid, and with the flow rate.

Introduction to Pumps 3

3. *Pressure.* In some systems it is necessary to increase the pressure of the liquid for process reasons. In addition to moving the liquid over changes in elevation and through a piping system, the pressure of a liquid must often be increased to move the liquid into a pressurized vessel, such as a boiler or fractionating tower, or into a pressurized pipeline. Or, it may be necessary to overcome a vacuum in the supply vessel.
4. *Velocity.* There is another factor to be considered here, namely that not all of the velocity energy in a pump is converted to potential or pressure energy. The outlet or *discharge* connection of most pumps is smaller than the inlet or *suction* connection. Because liquids are, practically speaking, incompressible, the velocity of the liquid leaving the pump is higher than that entering the pump. This *velocity head* may need to be taken into account (depending on the point of reference) when computing pump *total head* to meet system requirements. This is discussed further in Chapter 2, Section III.

III. PRESSURE AND HEAD

It is important to understand the relationship between *pressure* and *head*. Most plant engineers and those involved in operations tend to speak of the pressure of the liquid at various points in the process. Pressure is measured in *psi* (pounds per square inch, sometimes simply called "pounds") if United States Customary System (USCS) units are used. In SI (metric) units, the equivalent units for pressure are kilopascal (kPa), bar, or kilograms per square centimeter (kg/cm^2), while the equivalent units for head are meters (m). Most readers should be familiar with the difference between gauge pressure and absolute pressure. Absolute pressure is gauge pressure plus atmospheric pressure. Atmospheric pressure is 14.7 psi (1 bar) at sea level.

In the study of pump hydraulics, it is important to realize that any pressure expressed in psi (kPa) is equivalent to a

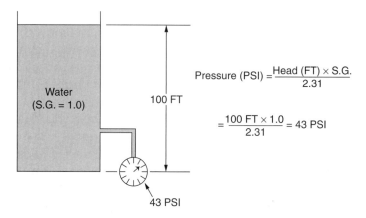

Figure 1.2 Pressure (in psi) is equivalent to a vertical column of liquid with a certain specific gravity.

static column of liquid expressed in feet (meters) of head. This is not meant to imply that pressure and head are interchangeable terms, because conceptually head is a specific energy term and pressure is a force applied to an area. However, the units used in hydrodynamics for specific energy are ft-lb/lbf, which, in the gravitational field of the earth where the acceleration of gravity is 32.2 ft/sec^2, can be numerically reduced to (feet of) head. With this understanding (and because most pumping applications occur under the earth's gravitational influence), the terms will be used interchangeably in this book.

The equivalence between pressure and head is illustrated in Figure 1.2, where the pressure in psi read on a gauge located at the bottom of a column of liquid is related to the height of the column in feet and to the *specific gravity* of the liquid by the following formula (in USCS units):

$$\text{psi} = \frac{\text{feet} \times \text{s.g.}}{2.31} \tag{1.1}$$

where:
psi = pounds per square inch
s.g. = specific gravity

Introduction to Pumps

Specific gravity is the weight of a given volume of liquid compared to the same volume of water. When applying centrifugal pumps, pressures should be expressed in units of feet (meters) of head rather than in psi (kPa).

As an example of the equivalency between psi and feet of head demonstrated by Equation 1.1, atmospheric pressure at sea level can be expressed as 14.7 psi, or as 34 feet of water.

For positive displacement pumps (discussed in Section VI to follow), the conversion to feet of head is not made, and pressures are expressed in psi (kPa).

IV. CLASSIFICATION OF PUMPS

There are many ways to classify pumps: according to their function, their conditions of service, materials of construction, etc. The pump industry trade association, the Hydraulic Institute, has classified pumps as shown in Figure 1.3. This classification divides pumps as follows:

A. Principle of Energy Addition

The first classification is according to the principle by which energy is added to the liquid. There are two broad classes of pumps, defined below.

1. Kinetic

In a *kinetic* pump, energy is continuously added to the liquid to increase its velocity. When the liquid velocity is subsequently reduced, this produces a pressure increase. Although there are several special types of pumps that fall into this classification, for the most part this classification consists of centrifugal pumps.

2. Positive Displacement

In a *positive displacement* pump, energy is periodically added to the liquid by the direct application of a force to one or more movable volumes of liquid. This causes an increase in pressure up to the value required to move the liquid through ports in

6 Pump Characteristics and Applications

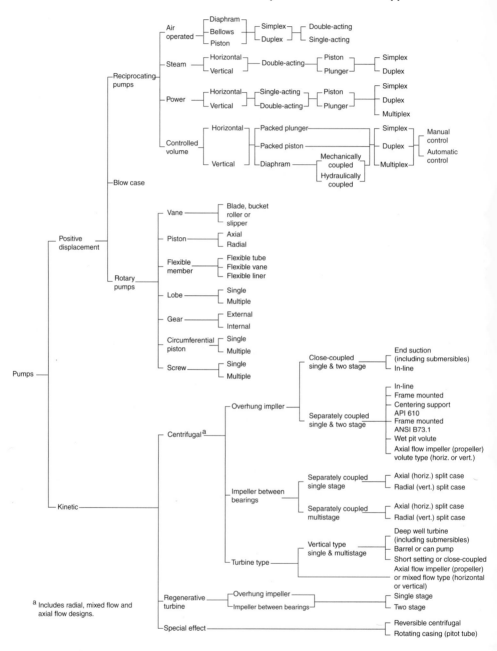

Figure 1.3 Classification of pumps. (Courtesy of Hydraulic Institute, Parsippany, NJ; www.pumps.org and www.pumplearning.org)

Introduction to Pumps

the discharge line. The important points here are that the energy addition is periodic (i.e., not continuous) and that there is a direct application of force to the liquid. This is most easily visualized through the example of a reciprocating piston or plunger pump (see Figure 1.23). As the piston or plunger moves back and forth in the cylinder, it exerts a force directly on the liquid, which causes an increase in the liquid pressure.

B. How Energy Addition Is Accomplished

The second level of pump classification has to do with the means by which the energy addition is implemented. In the kinetic category, the most common arrangement is the centrifugal pump. Other arrangements include *regenerative turbines* (also called *peripheral* pumps), and special pumps such as *jet* pumps that employ an eductor to bring water out of a well.

In the positive displacement category, the two most common sub-categories are *reciprocating* and *rotary* pumps.

C. Geometry Used

The remaining levels of pump classification shown in Figure 1.3 deal with the specific geometry used. With centrifugal pumps, the geometry variations have to do with the support of the impeller (*overhung impeller* vs. *impeller between bearings*), rotor orientation, the number of impellers or stages, how the pump is coupled to the motor, the pump bearing system, how the pump casing is configured, and pump mounting arrangements.

With positive displacement pumps, as is discussed in more detail in Section VI, there are many different types of rotary and reciprocating pumps, each with a unique geometry.

V. HOW CENTRIFUGAL PUMPS WORK

Stripped of all nonessential details, a centrifugal pump (Figure 1.4) consists of an *impeller* attached to and rotating with the shaft, and a *casing* that encloses the impeller. In a centrifugal pump, liquid is forced into the inlet side of the pump casing by atmospheric pressure or some upstream pressure. As

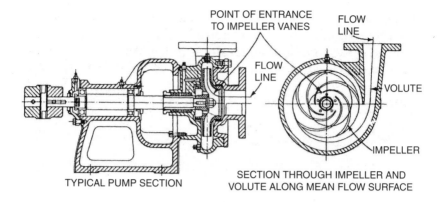

Figure 1.4 Centrifugal pump with single volute casing. (From *Pump Handbook*, I.J. Karassik et al., 1986. Reproduced with permission of McGraw-Hill, Inc., New York, NY.)

the impeller rotates, liquid moves toward the discharge side of the pump. This creates a void or reduced pressure area at the impeller inlet. The pressure at the pump casing inlet, which is higher than this reduced pressure at the impeller inlet, forces additional liquid into the impeller to fill the void.

If the pipeline leading to the pump inlet contains a noncondensable gas such as air, then the pressure reduction at the impeller inlet merely causes the gas to expand, and suction pressure does not force liquid into the impeller inlet. Consequently, no pumping action can occur unless this noncondensable gas is first eliminated, a process known as *priming* the pump.

With the exception of a particular type of centrifugal pump called a *self-priming centrifugal* pump, centrifugal pumps are not inherently self-priming if they are physically located higher than the level of the liquid to be pumped. That is, the suction piping and inlet side of centrifugal pumps that are not self-priming must be filled with noncompressible liquid and vented of air and other noncondensable gases before the pump can be started. Self-priming pumps are designed to first remove the air or other gas in the suction line, and to then pump in a conventional manner.

Introduction to Pumps

If vapors of the liquid being pumped are present on the suction side of the pump, this results in *cavitation*, which can cause serious damage to the pump. Discussed in greater detail in Chapter 2, Section VI, cavitation may also cause the pump to lose prime.

Once it reaches the rotating impeller, the liquid entering the pump moves along the impeller vanes, increasing in velocity as it progresses. The vanes in a centrifugal pump are usually curved backward to the direction of rotation. Some special types of pump impellers (Chapter 7, Section VII) have vanes which are straight rather than curved. The degree of curvature of the vanes and number of vanes, along with other factors, determines the shape and characteristics of the pump *performance curve*, which is described in Chapter 2, Section IV.

When the liquid leaves the impeller vane outlet tip, it is at its maximum velocity. Figure 1.5 illustrates typical velocity

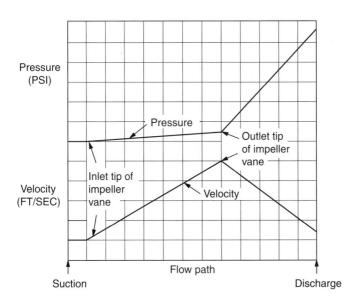

Figure 1.5 Velocity and pressure levels vary as the fluid moves along the flow path in a centrifugal pump.

and pressure changes in a centrifugal pump as the liquid moves through the flow path of the pump. After the liquid leaves the impeller tip, it enters the casing, where an expansion of cross-sectional area occurs. The casing design ensures that the cross-sectional area of the flow passages increases as the liquid moves through the casing. Because the area is increasing as the liquid moves through the path of the casing, a diffusion process occurs, causing the liquid's velocity to decrease, as Figure 1.5 illustrates. By the Bernoulli equation (see Ref. [1] at the end of this book), the decreased kinetic energy is transformed into increased potential energy, causing the pressure of the liquid to increase as the velocity decreases. The increase of pressure while velocity is decreasing is also illustrated in Figure 1.5.

A centrifugal pump operating at a fixed speed and with a fixed impeller diameter produces a differential pressure, or differential *head*. Head is usually expressed in feet or meters, and abbreviated *TH* (total head). The amount of head produced varies with the flow rate, or *capacity* delivered by the pump, as illustrated by the characteristic *head–capacity (H–Q)* curve shown in Figure 1.6. As the head of the pump decreases, the capacity increases. Alternatively, as the pump head increases, the flow decreases. Pump capacity is usually expressed in gallons per minute (gpm) or, for larger pumps, in cubic feet per second (cfs). Metric equivalents, depending on the size of the pump, are cubic meters per second, liters per second, or cubic meters per hour.

The centrifugal pump casing is one of several types. A *single volute* casing is illustrated in Figure 1.4. Note that the single volute casing has a single *cutwater* where the flow is separated. As the flow leaves the impeller and moves around the volute casing, the pressure increases. This increasing pressure as the liquid moves around the casing produces an increasing radial force at each point on the periphery of the impeller, due to the pressure acting on the projected area of the impeller. Summing all of these radial forces produces a net radial force that must be carried by the shaft and radial bearing system in the pump. The radial bearing must also support the load created by the weight of the shaft and impeller.

Introduction to Pumps

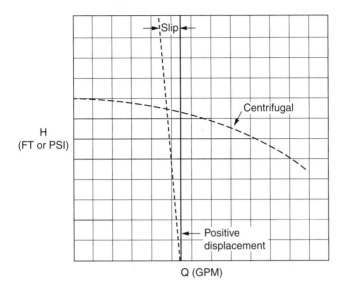

Figure 1.6 Typical head–capacity relationship for centrifugal and positive displacement pumps.

The radial bearing loads generated by a pump also vary as the pump operates at different points on the pump performance curve, with the minimum radial force being developed at the *best efficiency point (BEP)* of the pump (Figure 1.7). See Chapter 2, Section V, for a discussion of BEP. Operation at points on the pump curve to the right or to the left of the BEP produce higher radial loads than are produced when operating at the BEP. This is especially true of single volute casing pumps, as Figure 1.7 illustrates.

Symptoms of excessive radial loads include excessive shaft deflection and premature mechanical seal and bearing failure. Continuous operation of the pump at too low a minimum flow is one of the most common causes of this type of failure. Because for rolling element bearings, bearing life is inversely proportional to the cube of the bearing load, operating well away from the pump best efficiency point can cause a reduction in bearing life by several orders of magnitude.

A *diffuser* casing (Figure 1.8) is a more complex casing arrangement, consisting of multiple flow paths around the

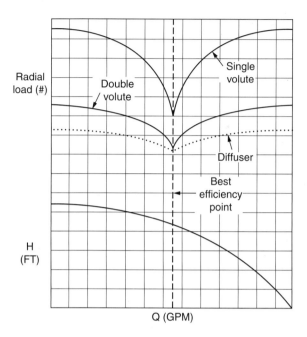

Figure 1.7 Typical radial loads produced by single volute, double volute, and diffuser casings.

periphery of the impeller discharge. The liquid that leaves the impeller vanes, rather than having to move completely around the casing periphery as it does with the single volute casing, merely enters the nearest flow channel in the diffuser casing. The diffuser casing has multiple cutwaters, evenly spaced around the impeller, as opposed to the one cutwater found in a single volute casing. The main advantage of the diffuser casing design is that this results in a near balancing of radial forces (Figure 1.7), thus reducing shaft deflection and eliminating the need for a heavy-duty radial bearing system. The dead weight of the rotating element must still be carried by the radial bearing, but overall the diffuser design minimizes radial bearing loads compared with other casing types.

Because the diffuser design produces minimal radial bearing loads, one might wonder why all pumps do not have diffusers rather than volute type casings. The reason is partially due to

Introduction to Pumps

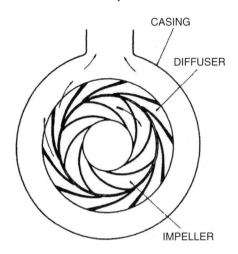

Figure 1.8 Diffuser casing minimizes radial loads in a centrifugal pump. (From *Pump Handbook*, I.J. Karassik et al., 1986. Reproduced with permission of McGraw-Hill, Inc., New York, NY.)

economics, as a pump with a diffuser casing generally has more parts or more complex parts to manufacture than a pump with a volute casing. Depending on pump size and materials of construction, economics often do not justify the use of diffuser casings except where significant savings can be achieved in the size of shaft or radial bearing that is used in the pump. This is usually only found to be the case in multi-stage, high-pressure pumps. However, with multi-stage pumps there are other considerations as well. Volute designs in multi-stage pumps allow, by the use of a cross-over, some of the impellers to be oriented in the opposite direction, providing balancing of axial thrust loads. (Refer to Chapter 4, Section II.D and Section VII.) The leading manufacturers of multi-stage pumps are themselves not in agreement on this subject.

Vertical turbine pumps (Chapter 4, Section XI) usually have diffuser casings. Because the bearings for these vertical pumps are submerged in the liquid being pumped, it is not practical to have a ball or roller type radial bearing for this type of pump. Rather, the radial bearing loads must be accommodated by a sleeve type bearing, which is not an ideal bearing

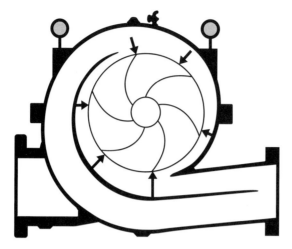

Figure 1.9 Double volute casings are used in larger centrifugal pumps to reduce radial loads. (Courtesy of Goulds Pumps, Inc., ITT Industries, Seneca Falls, NY.)

system in this type of arrangement. Therefore, to minimize radial bearing loads, diffuser type casings are used in this type of pump.

A hybrid between a single volute casing and a diffuser casing is a *double volute* casing (Figure 1.9). With this casing design, the volute is divided, which creates a second cutwater, located 180° from the first cutwater. This design results in much lower radial loads than are present with single volute designs (Figure 1.7). Double volute casings are usually used by pump designers for larger, greater flow pumps (usually for flows higher than about 1500 gpm) to allow the use of smaller shafts and radial bearings.

VI. POSITIVE DISPLACEMENT PUMPS

A. General

This book is primarily about centrifugal pumps. However, as Figure 1.3 illustrates, there is an entire other class of pumps known as *positive displacement (P.D.)* pumps that deserves some attention. One of the earliest decisions that must be

Introduction to Pumps

made in designing a system and applying a pump is the selection of the type of pump to be used. The first issue is the general decision whether the pump should be of the centrifugal or the positive displacement type. Surveys of equipment engineers and pump users indicate that the majority of them have a strong preference for centrifugal pumps over positive displacement pumps (if the hydraulic conditions are such that either type can be considered). Many reasons are given for this preference for centrifugals, but most are related to the belief that centrifugal pumps are more reliable and result in lower maintenance expense. Centrifugal pumps usually have fewer moving parts, have no check valves associated with the pumps (as reciprocating positive displacement pumps do), produce minimal pressure pulsations, do not have rubbing contact with the pump rotor, and are not subject to the fatigue loading of bearings and seals that the periodic aspect of many positive displacement pumps produce. Centrifugals should be considered first when applying a pump, but they are not always suited to the application.

B. When to Choose a P.D. Pump

This preference for centrifugal over P.D. pumps is certainly not always the case, and, in fact, there are certain application criteria that demand the use of a P.D. pump. The following are some key application criteria that would lead to the selection of a P.D. pump over a centrifugal pump:

- High viscosity
- Self-priming
- High pressure
- Low flow
- High efficiency
- Low velocity
- Low shear
- Fragile solids handling capability
- Sealless pumping
- Accurate, repeatable flow measurement
- Constant flow/variable system pressure
- Two-phase flow

The ability to pump viscous liquids is one of the most import attributes of P.D. pumps. It is possible to handle low-viscosity liquids with centrifugal pumps. However, efficiency degrades rapidly as viscosity increases and there is an upper limit of viscosity above which it becomes impractical to consider centrifugal pumps due to the excessive waste of energy. Highly viscous liquids absolutely cannot be pumped with a centrifugal pump. (See Chapter 3, Section II, for more discussion of viscosity and its effect on centrifugal pump performance.) For these liquids, some type of positive displacement pump may be the only practical solution.

Most positive displacement pump types are inherently self-priming, meaning they can be located above the surface of the liquid being pumped without the necessity of the suction line being filled with liquid and the noncondensable gases in the suction line being removed before starting the pump. Therefore, these pump types can be conveniently mounted on top of transfer tanks with no special external priming devices. (Refer to further discussion of the dry self-priming capability of P.D. pumps in Section VI.C and Table 1.1 to follow.)

The high pressure and low flow criteria above must be considered together. How high is high pressure, and how low is low flow? It is possible to find, for instance, centrifugal pumps that produce pressures of several thousand psi. And certainly one can find very small centrifugal pumps whose capacity is only a couple of gallons per minute. But what if one has an application for 5 gpm at 2000 psi? In that case, a positive displacement pump is about the only solution. Figure 1.10 shows in very broad terms the head and flow range of centrifugal, rotary, and reciprocating pumps. While there is a portion of this coverage chart that can be met by all three pump types, the one area that stands out as being able to be met only by a positive displacement pump is low flows in combination with very high pressures.

If energy efficiency were the only consideration in selecting pumps, more positive displacement pumps would be considered, since some positive displacement pumps are quite energy efficient. Energy is not the only consideration though,

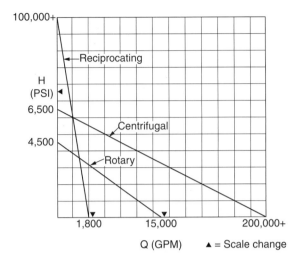

Figure 1.10 Head vs. flow for centrifugal, rotary, and reciprocating pumps.

and other factors such as installed cost and maintenance expense often outweigh the energy savings.

The criteria of low velocity, low shear, and fragile solids handling capability often go hand-in-hand. Centrifugal pumps, because of the high velocities present at the impeller discharge, and because of the close clearances inside the pump, often subject the pumped liquid to high shear stresses. Many liquids cannot tolerate these high velocities and high shear stresses. A good example of this is fruits and vegetables such as cherries and peas that are pumped in food processing plants. If these products were pumped using centrifugal pumps, they would produce cherry juice and pea juice! There is an entire class of centrifugal pump impellers of the *nonclog* type, whose function is to pump sewage and other waste liquids containing solids (Chapter 4, Section II.F). These centrifugal pumps, however, are not concerned with maintaining the integrity of the solids and often shred the solids as they pump them. Another centrifugal impeller type, known as a *recessed*, or *vortex* impeller (Chapter 4, Section II.F), is capable of pumping large solids with minimal degradation, but the downside is that this impeller type is very, very inefficient.

Because of their unique design, several types of P.D. pumps, such as *peristaltic* (Section VI.C.4) and *diaphragm* (Section VI.C.13) pumps, are inherently sealless, requiring no shaft seals and having zero product leakage. While there are available today several types of sealless centrifugal pumps as Chapter 5, Section V, discusses, these centrifugal options also have their limitations and shortcomings. The inherent sealless nature of some P.D. pumps may make them a simpler solution.

With positive displacement pumps, capacity varies directly with speed and is independent of differential pressure or head. Figure 1.6 illustrates that pump capacity is independent of differential pressure for P.D. pumps, while being dependent on differential pressure (head) for centrifugal pumps.

Most positive displacement pumps exhibit *slip*, that is, leakage from the high pressure to the low pressure side of the pump. As shown in Figure 1.6, slip causes the pump to deliver a lower flow rate at higher differential pressures. The amount of slip varies widely from one positive displacement pump to another, as well as varying with pump differential pressure and with the liquid viscosity. Most positive displacement pumps are not nearly as subject to increased leakage back to suction because of wear as are centrifugal pumps. Some types exhibit very little slip. These factors make some types of P.D. pumps ideal for metering applications, where an accurate, controllable flow rate of (usually) an expensive chemical must be dispensed, for example, to treat water with chlorine. Note that centrifugal pumps could also be used to accurately control flow but they would have to rely on a control loop that measures flow and then adjusts a system control valve. Reciprocating pumps are the most common type of P.D. pump used for metering pumps, although other types such as peristaltic pumps are also used for metering.

The requirement for a constant process flow rate where system pressure varies widely can be met with a centrifugal pump. However, this usually requires a feedback control system,

Introduction to Pumps

including flow measurement and the use of an automatic control valve to maintain constant flow. A positive displacement pump may be a much simpler solution because once the liquid characteristics and the speed and size of a P.D. pump are fixed, the pump delivers a nearly constant flow rate regardless of the system pressure against which the pump is pumping.

Finally, many P.D. pump types are much more suited to pumping liquids containing gases than are centrifugal pumps.

One of the hydraulic characteristics of P.D. pumps is that the pump continues to deliver a constant flow rate (if pumping at a constant speed) while building up pressure at the discharge (Figure 1.6). This means that if a valve downstream of the pump is inadvertently closed, the pump continues to build up pressure until something gives. Usually, one of several things happens to prevent damage to the pump. At constant flow and ever-increasing pressure, the required horsepower may continue to build until the motor is overloaded and trips off. Or, a high pressure or high temperature limit switch may trip the motor. If the motor does not trip or the excessive pressure is not relieved, the pump could eventually build up pressure until it over-pressurizes the pump casing or some component downstream of the pump, causing potentially serious damage. Usually, it is recommended that P.D. pumps incorporate a pressure relief valve, built into the pump or supplied just downstream of the pump, to protect against over-pressurization.

Internal relief valves built into the pump relieve pressure via an internal loop that connects discharge to suction. External valves must be piped to an external source, generally the supply reservoir. The external approach can provide a visual indicating that the valve is open.

Relief valve settings are sometimes adjustable and can be tricky. The user should be certain to understand the manufacturer's definition of the valve setting and the valve characteristics. These valves will wear very quickly and generally are not stable if used as a pressure-regulating device. It is best to obtain a pressure-regulating valve for this purpose.

Table 1.1 Key Application Data of Positive Displacement Pumps[a]

Pump Type	Max. Capacity (gpm)	Max. Pressure (psi)	Max. Viscosity (million SSU)	Max. Solid Size (in.)	Dry Self-Priming (Y/N)	Max. Suction Lift (ft H_2O)
Sliding vane	2,500	200	0.5	1/32	N	28
Sinusoidal rotor	300	200	18.0	2	N	30
Flexible impeller	150	60	0.1	1	Y	24
Flexible tube (peristaltic)	200	220	0.2	1	Y	30
Progressing cavity	2,400	2,000	5.0	2	Y	30
External gear	1,200	2,500	2.0	—[b]	N	20
Internal gear	1,500	200	2.0	—[b]	N	20
Rotary lobe	3,000	450	5.0	4	N	20
Circumferential piston	600	200	5.0	2	N	20
Two-screw	15,000	1,500	4.5	—[b]	N	31
Three-screw	4,500	4,500	1.0	—[b]	N	28
Piston	700	5,000	0.05	1/2	Y	25
Plunger	1,200	100,000	0.05	1/2	Y	20
Diaphragm	1,800	17,500	1.0	1	Y	14
Air-operated diaphragm	300	125	0.75	2	Y	25
Wobble plate	50	1,500	0.025	1/8	Y	8

Introduction to Pumps

Pump Type	Able to Run Dry (Y/N)	Abrasive Handling Rating[c]	Fragile Solids/Shear Sensitive Liquids[c]	Pulsations[c]	Metering Ability[c]	Sanitary Designs Avail. (Y/N)
Sliding vane	Y	3	3	3	3	N
Sinusoidal rotor	N	4	1	1	3	Y
Flexible impeller	N	2	2	2	5	Y
Flexible tube (peristaltic)	Y	1	1	4	2	Y
Progressing cavity	N	1	1	1	2	Y
External gear	N	5	4	1	3	N
Internal gear	N	5	4	1	3	Y
Rotary lobe	Y	2	1	3	3	Y
Circumferential piston	Y	2	1	3	3	Y
Two-screw	Y	3	4	1	4	N
Three-screw	N	4	5	1	4	N
Piston	N	2	3	5	1	N
Plunger	N	4	3	5	1	Y
Diaphragm	Y	1	2	5	1	Y
Air-operated diaphragm	Y	1	2	5	5	Y
Wobble plate	Y	1	3	3	1	N

[a] Refer to guidelines in text for the use of this table.
[b] Friable solids only, or rotors must be hardened and clearances opened.
[c] Rating of 1 is best, 3 is medium, 5 is worst.

C. Major Types of P.D. Pumps

The following pages provide a brief description of the most common types of P.D. pumps. The first five types (*sliding vane, sinusoidal rotor, flexible impeller, flexible tube,* and *progressing cavity*) are single-rotor rotary pumps. The next five types (*external gear, internal gear, rotary lobe, circumferential piston,* and *multiple screw*) are multiple-rotor rotary pumps. The last three (*piston, plunger,* and *diaphragm*) are reciprocating pumps.

Table 1.1 summarizes the primary application criteria for P.D. pumps. In using this table, the following guidelines should be observed.

- The information given is approximate only, and is meant to provide assistance in selecting appropriate pump types to consider for a particular application. The parameters of performance for any single pump type vary widely from manufacturer to manufacturer and from one pump design to another. In general, Table 1.1 shows the limits of performance for the major pump types. In many cases, there are only one or two manufacturers that make pumps with the indicated maximum value of flow, pressure, or other performance parameter. For any P.D. pump application, the best advice is to obtain literature and performance data from several manufacturers of the types of pump being considered.
- The parameters given for a particular pump type do not necessarily apply concurrently. As examples of this, the maximum flow listed is likely not available with the maximum pressure listed for that pump type. The maximum solids size listed is probably only able to be handled by the largest pumps of that type, and may be restricted below the maximum size shown in the table for solids of a particular shape, material, or concentration.
- Virtually all P.D. pumps are self-priming to a certain degree. Only the few pump types so indicated in Table 1.1 are truly dry self-priming, that is, will prime on a lift when completely dry.

Introduction to Pumps 23

- The columns of Table 1.1 that rate the various pump types in categories such as ability to handle abrasives and fragile solids have been prepared using the author's best judgment. The ratings may change for particular applications, and arguments could easily be made to move any pump up or down in the relative rankings.
- Some pumps are used for sanitary food-grade applications (manufactured in stainless steel, have food-grade elastomers, etc.), but do not have USDA or 3A certification. Pumps without such certification would be used in the upstream end of the food processing system, prior to the sterilization process, or would otherwise be used in food applications where such certification is not required.
- No attempt has been made in Table 1.1 to present any rating of the pump types relative to cost. The reasons for this include the fact that features and relative costs are so application specific; that many of these pump types were never intended to compete against each other; and the fact that any good cost comparison must include the costs associated with installation, space requirements, maintenance and repair parts, and energy.

For the reader who is interested in finding further information about a specific type of pump, Appendix A at the end of this book contains a listing of most of the major pump suppliers in the United States, segmented according to the pump types they offer. This list is by no means all-inclusive, containing only those manufacturers with which the author is familiar. Readers interested in locating information on a particular pump type should be able to use Appendix A as a guide, and should be able to locate manufacturers via an Internet search or in a *Thomas Register* or similar index of manufacturers. Note that Appendix A lists the manufacturers alphabetically, and that it includes both centrifugal and P.D. pump suppliers.

The major types of P.D. pumps are described below. For each type of pump, the primary application characteristics

Figure 1.11 Sliding vane pump. (Courtesy of Blackmer, Grand Rapids, MI.)

are noted, as well as the benefits and shortcomings of each type.

1. Sliding Vane Pump

In a sliding vane style of rotary pump, shown in Figure 1.11, vanes cooperate with a cam to draw liquid into and force it from the pump chamber. In this pump style, vanes fit into slots cut lengthwise in the rotor, and the rotor is inside an eccentrically shaped casing that acts like a cam. When the rotor is turning at operating speed, the vanes are forced by centrifugal force outward until they come in contact with the casing wall. Some types of vane pumps also rely on springs to force the vanes outward, so that contact between the vanes and the casing walls is maintained even when the pump is operating at slow speeds. Other types use liquid pressure from the pump discharge acting beneath the vanes to force the vanes outward.

Advantages of vane pumps include their rather simple construction, the fact that they are self-compensating for wear on the vanes, and that they operate well with thinner, low-viscosity liquids. They can operate with mildly erosive liquids. Disadvantages of this pump type include their inability to pump highly viscous liquids and the fact that they cannot handle fragile solids.

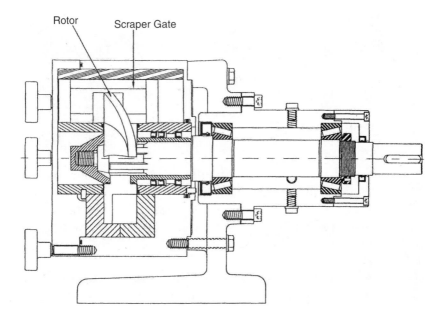

Figure 1.12 Sinusoidal rotor pump. (Courtesy of Sundyne Corporation, Arvada, CO.)

2. Sinusoidal Rotor Pump

Figure 1.12 shows a sinusoidal rotor pump. In this pump type, a rotor having the shape of two complete sine curves turns in a housing, creating four separate, symmetrical pumping compartments. A sliding scraper gate covers part of the rotor, oscillates as the rotor turns, and prevents return of product past the discharge and back to the suction side of the pump.

The major benefits of the sinusoidal rotor pump are its low shear and its gentle handling of fragile solids and highly viscous liquids. Its principal shortcoming is that it has limited ability to handle highly abrasive liquids.

3. Flexible Impeller Pump

In the flexible impeller pump (Figure 1.13), sometimes called a flexible vane pump, the rotor is made of an elastomeric

Figure 1.13 Flexible impeller pump. (Courtesy of ITT Industries, Jabsco, Foothill Ranch, CA.)

material such as rubber. The blades of this "impeller" continuously deflect and straighten as they pass across a cam between the inlet and discharge ports. The flexing of the blades produces a vacuum that causes liquid to flow into the space between the two blades and then moves the liquid through the pump.

Advantages of the flexible impeller pump include the fact that it is dry self-priming; can handle liquids with solids, abrasives, or entrained air; and is relatively inexpensive. Disadvantages include the upper limits of flow (about 150 gpm) and pressure (about 60 psi), and the fact that the pump cannot run dry longer than a couple of minutes without doing damage to the rubber impeller.

4. Flexible Tube (Peristaltic) Pump

The flexible tube pump (Figure 1.14) is also called a peristaltic pump, or simply a hose pump. In this pump type, a flexible tube made of rubber or other material is located inside a circular housing. Rollers or cams attached to the rotor squeeze the tube as they pass across it, drawing the liquid through the pump. This action is similar to what happens when a person swallows, a process called *peristalsis*, which is how the pump gets its name.

Peristaltic pump advantages include the fact that they are sealless (require no packing assembly or mechanical seal), can handle quite corrosive liquids (as long as the tube material is compatible with the liquid being pumped), are dry self-priming, and are relatively inexpensive. They are frequently

Introduction to Pumps 27

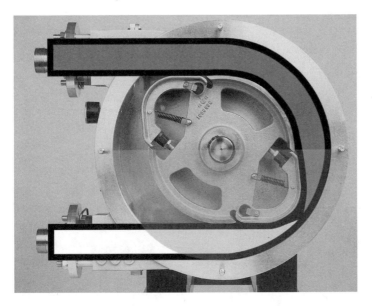

Figure 1.14 Peristaltic (hose) pump. (Courtesy of Crane Pumps & Systems, Inc., Piqua, OH.)

used as low-cost metering pumps, for applications such as chlorine metering in commercial swimming pools.

Disadvantages include relatively low flow and pressure capability for most models (although several manufacturers offer pressures to several hundred psi), and hoses usually require changing about every 3 months. The selection of the proper hose material for the application is the most critical aspect of applying the hose pump. Finally, peristaltic pumps are not as accurate for metering as reciprocating style pumps.

5. Progressing Cavity Pump

The progressing cavity (PC) pump in its most common design has a single-threaded screw or rotor, turning inside a double-threaded stator (Figure 1.15). The stator is made of an elastomeric material, and the rotor has an interference fit inside the stator. As the rotor rotates inside the stator, cavities form at the suction end of the stator, with one cavity closing as the

Figure 1.15 Progressing cavity pump. (Courtesy of Roper Pump Company, Commerce, GA.)

other opens. The cavities progress axially from one end of the stator to the other as the rotor turns, moving the liquid through the pump.

Advantages of PC pumps include their ability to pump highly viscous liquids (as well as very low viscosity liquids), shear sensitive liquids, fragile solids, and abrasives. Also, the pump produces very little pulsation, and is self-priming even when dry. The maximum pressure is limited to about 75 psi, but the PC can be set up with many stages in series using the same driver, so it can achieve upper pressure limits well over 1000 psi. Another advantage is that the packing or seal sees suction pressure rather than discharge pressure.

Disadvantages of the PC pump include the relatively higher cost of replacement parts, its large floor space requirement, the fact that the pump cannot run dry for an extended time period, and upper temperature limits of about 300°F. (The

Introduction to Pumps

material of the elastomeric stator usually sets the upper temperature limit.) Starting torques are quite high, so the pump motor may need to be much larger than would be the case with other types of P.D. pumps.

Note that as the rotor turns, the centerline of the rotor orbits about the centerline of the stator. There are several methods to connect the PC pump's drive shaft to its rotor, and to account for the elliptical motion of the rotor shaft. Two common configurations are the pin drive connecting rod and the crowned gear drive connecting rod (the latter is shown in Figure 1.15). While both configurations are universally accepted, they differ in significant ways. The pin drive configuration is the most common and least expensive of the two types. The crowned gear drive configuration is not as common and is more expensive than the pin drive; however, it is significantly more heavy-duty. The crowning of the gears permits this coupling to accept the elliptical motion of the rotor. The crowned gears also provide a large surface area to transfer the torque load required to turn the rotor.

Variations on progressing cavity designs use more threads on the rotor and stator, or special stator materials.

6. External Gear Pump

External gear pumps (Figure 1.16) have two meshing gears, which may be of the *spur, helical,* or *herringbone* type. These three gear types are illustrated in Figure 1.17. Liquid is carried between the gear teeth and displaced as the teeth mesh. Close clearances between the gear teeth and between the teeth and the casing walls minimize slippage of liquid from the high pressure side to the low pressure side.

Spur gears are simple, and generally cost less to manufacture than helical or herringbone gears. They have good characteristics but can be noisy and inefficient. The trapped liquid between the teeth where the teeth mesh has no place to exit. As the trapped liquid squeezes by the tight clearances, it can make a loud screaming sound. Some manufactures provide a relief slot in this area to give the fluid a place to escape. This minimizes noise and increases efficiency, especially with

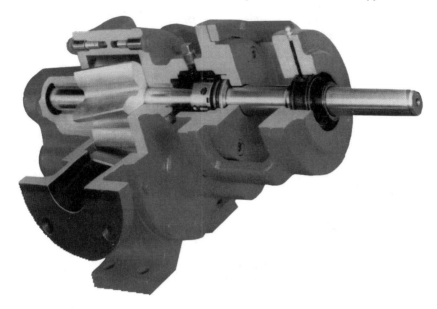

Figure 1.16 External gear pump. (Courtesy of Roper Pump Company, Commerce, GA.)

viscous liquids. Spur gears have the added advantage of minimal axial thrust.

Helical gears (e.g., Figure 1.16) are another means of giving the trapped liquid a path to escape. The helix shape gives the liquid a place to exit. Helical gear pumps are generally efficient and quiet. They do have one disadvantage: they are forced axially against the pump thrust bushing. Axial wear in a gear pump decreases performance much faster than radial wear, so it is important to maintain tight clearances in the axial direction.

Herringbone gears are the most expensive to manufacture, but they are quiet, efficient, and do not exhibit axial thrust. They are very difficult to machine, so sometimes they are made up with two helical gears butted together with the helix of the two gears oriented opposite to each other.

Most external gear pumps have the first gear (which is coupled to the driver) driving the second gear. An alternative

Introduction to Pumps

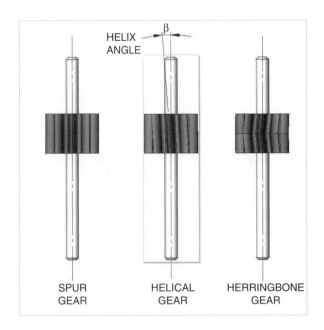

Figure 1.17 Spur, helical, and herringbone gears. (Courtesy of Diener Precision Pumps, Lodi, CA.)

design uses timing gears, which are a separate set of gears, normally isolated from the pumped liquid, which cause the two main gears to remain in mesh without requiring direct contact. This use of timing gears, common on many types of double-rotor rotary pumps, allows larger clearances between the pumping gears, which means that the pump can more easily accommodate some abrasives without doing damage to the gears.

Advantages of external gear pumps include the fact that they operate at relatively high speeds, producing relatively high pressures. The gears are usually supported by bearings on both sides, so there are no overhung loads. The pumps are relatively quiet (particularly if helical or herringbone gears are used), largely free of pulsations, and economical. Disadvantages include the fact that there are usually four bushings in the liquid area, and, unless timing gears are used, these pumps cannot tolerate solids or abrasives.

The weakest element in most gear pumps (external gear or internal gear type) is the driven gear bearing. One might think that the driving gear carries 100% of the load, so the driving bearings should wear the most, but this is not the case.

Radial bearing loads are a result of pump differential pressure and the forces of the driving gear pushing the driven gear. This produces higher loading on the driven gear bearing than on the driving gear bearing. This problem is usually exacerbated by the fact that the driven gear bearing is wetted by the fluid being pumped.

So when examining a gear pump for wear, look carefully at the driven gear bearings first. As they wear, the gear is no longer supported and the teeth will begin to wear against the pump wall or wear plate. The mesh between gears will increase, causing slip and reduced pump performance, especially at high pressures.

Considering the materials of construction of gears, it should be considered that most gear pumps pump lubricating fluids. For these pumps, the gears are made of a variety of metals, but mostly steel. Hardness runs from that of mild steel to very hard treated steels to produce high pressures.

For nonlubricating fluids, bronze and a variety of nonmetallic materials such as plastic composites, carbon, Teflon®, and ceramics are used. The choice is often dictated by chemical resistance, mechanical properties, and cost. For example, Teflon® has excellent chemical resistance but its physical properties often result in excessive wear, low pressure capability, and fluid temperature is limited (high thermal expansion).

Due to the tight clearances required in gear pumps, there is no allowance for abrasive wear. Most gear pumps will not handle abrasives without premature wear failure. Some, however, are made of wear-resistant materials and can perform quite well for some applications. These pumps use hardened shafts, ceramic gears and/or bushings, or highly abrasion-resistant plastic composites. Running the pump very slowly is another means of getting more life out of a pump under all conditions, but is purposely done for abrasive applications.

Figure 1.18 Internal gear pump. (Courtesy of Viking Pump, Inc., Cedar Falls, IA.)

7. Internal Gear Pump

Internal gear pumps (Figure 1.18), like external gear pumps, move and pressurize liquid by the meshing and un-meshing of gear teeth. With an internal gear pump, a rotor having internally cut teeth meshes with and drives an idler gear having externally cut teeth. Pumps of this type usually have a crescent-shaped partition that moves the liquid through the pump with minimal slip.

Internal gear pumps' advantages include their few moving parts, relatively low cost, and the fact that they have only one seal. They can usually operate well in either direction, and reversing rotation causes a reversal in the direction of flow. Disadvantages include the fact that there is one bearing in the pumped liquid, that the one bearing must support an overhung load, and these pumps generally will not work with abrasives or solids.

8. Rotary Lobe Pump

A lobe pump (Figure 1.19) is similar to an external gear pump, in that the liquid is carried between the rotor lobe surfaces

Figure 1.19 Lobe pump. (Courtesy of Viking Pump, Inc., Cedar Falls, IA.)

which cooperate with each other as they rotate to provide continuous sealing, as do the teeth of a gear pump. Unlike a gear pump, though, one lobe cannot drive the other, so this type of pump must have timing gears (described in Section VI.C.6) to allow the lobes to remain in synch with each other. Each rotor may have one lobe or several, with three lobes being most common.

The wide spaces between the lobes and the slow speeds at which these pumps operate make this style of pump ideal for handling food products containing fragile solids. Shear is low with this pump. Also, the fact that the pump has timing gears means that the lobes do not actually touch each other as they rotate, a necessary condition for many sanitary food-grade applications. Disadvantages include the fact that these pumps are subject to pressure pulsations, and they have fairly high amounts of slip with low-viscosity liquids. Finally, they must have timing gears.

Note that some manufacturers make these pumps with elastomeric lobes, while others make them with metal lobes.

Introduction to Pumps

Figure 1.20 Bi-wing lobe pump. (Courtesy of Viking Pump, Inc., Cedar Falls, IA.)

The different lobe materials affect maximum pressure and temperature, as well as the price of the pump.

9. Circumferential Piston and Bi-Wing Lobe Pumps

The circumferential piston and bi-wing lobe pumps are very similar to the traditional lobe pump, both in the way they operate and in their applications. Figure 1.20 shows a bi-wing lobe pump. Instead of traditional lobes, the rotors have arc shaped "pistons," or rotor wings, traveling in annular-shaped "cylinders" machined in the pump body. As with traditional lobe pumps, the rotors are not in direct contact with each other and require the use of timing gears.

Circumferential piston and bi-wing lobe pumps have less slip than do comparably sized traditional lobe pumps. This is because the rounded lobes of a traditional lobe pump only come in close contact with the casing at a single point on the outer surface of each lobe, whereas the circumferential piston and bi-wing lobe pumps have a close clearance between the rotor and the casing over the entire length of the arc. The lower slippage means that these pumps are more energy efficient than traditional lobe pumps. With liquid viscosities greater than about 2000 SSU (Chapter 3, Section II), this advantage disappears.

Advantages and disadvantages of circumferential piston and bi-wing lobe pumps are pretty much the same as for traditional lobe pumps, and again, food processing is the most common application. Circumferential piston and bi-wing lobe pumps normally do not handle abrasives as well as traditional lobe pumps, and the arc-shaped rotor may make contact with the casing at times in higher pressure applications.

10. Multiple-Screw Pump

This pump is called a multiple-screw pump to distinguish it from a single-screw (progressing cavity) pump. In a multiple-screw pump, liquid is carried between rotor screw threads and is displaced axially as the screw threads mesh. Multiple-screw pumps can be either two-screw (Figure 1.21) or three-screw (Figure 1.22). Two-screw pumps usually have timing gears, and because the screws are not in contact with each other, this style can handle more abrasive liquids than three-screw pumps. Three-screw pumps are usually untimed, with the central screw driving the other two screws. Because the screws are in contact with each other, this style of pump cannot tolerate severe abrasives, though these pumps are used to pump crude oil containing sand. The screws on a two-screw pump are opposed, so that they balance out all axial loads. Three-screw pumps, on the other hand, generate some axial loads, although this is usually partially or fully balanced out.

The three-screw pump is usually the most economical choice if the liquid is not overly abrasive, because it has fewer

Introduction to Pumps 37

Figure 1.21 Two-screw pump. (Courtesy of Warren Pumps, Inc., a member of the Colfax Pump Group, Monroe, NC.)

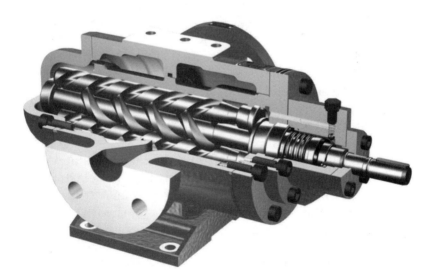

Figure 1.22 Three-screw pump. (Courtesy of Roper Pump Company, Commerce, GA.)

parts (no timing gears). Either type of screw pump has an extremely high upper limit of viscosity, and screw pumps in general produce higher flow rates than any other type of rotary positive displacement pump. Screw pumps operate at high speeds and can produce very high pressures with almost no pulsations and with relatively quiet operation. Two-screw pumps are usually limited to about 1500 psi, while three-screw pumps can go up to about 4500 psi.

11. Piston Pump

Piston pumps are reciprocating pumps consisting of a power end and a liquid end. The power end of the pump takes the driver power and converts the rotary motion to reciprocating motion. The driver can be a motor, an engine, or a turbine, whose output must then be converted by means of crankshafts, connecting rods, and crossheads, to reciprocating motion. An alternative, older design called *direct-acting* has a steam cylinder for the power end, with inlet and outlet valves to allow steam to directly drive the liquid end piston.

The liquid end of a reciprocating piston pump consists of liquid inlet and outlet ports, with most designs having check valves in both the inlet and outlet ports. When the reciprocating piston strokes in one direction, the inlet check valve opens (while the outlet check valve remains closed), directing liquid into the liquid end of the pump. When the piston strokes in the opposite direction, the inlet check valve closes and the outlet check valve opens, allowing liquid to move into the discharge port and out into the system.

Piston pumps have the pistons operating inside cylinders, with the sealing taking place by means of elastomeric packing or O-rings on the outside of the piston. Alternatively, the piston itself may be made of an elastomer with an interference fit. Piston pumps can be *single-acting* or *double-acting*, although most are double-acting. With a double-acting piston pump, liquid is discharged during both the forward and return motions of the piston (requiring four check valves for each piston). A single-acting piston pump has liquid discharging only during the forward stroke of the piston. Double-acting designs are generally associated with slower speeds and medium pressures, while single-acting designs are more generally associated with higher speeds and higher pressures.

Note that high inlet pressures are a very important consideration in the selection of a piston pump. Double-acting piston pumps are quite a bit more efficient than single-acting pumps when the inlet pressures are quite high. Double-acting piston pumps directly utilize the inlet pressure to reduce the power end load. Also, high inlet pressures can affect the selection of a prime mover. Double-acting piston pumps can use the pressure differential in the calculation of the horsepower requirement while single-acting pumps can only use a fraction of the inlet pressure to offset discharge pressure.

Reciprocating pumps can have a single reciprocating piston, plunger, or diaphragm (called *simplex* construction), or there may be multiple reciprocating components. A pump having two reciprocating members is called *duplex* construction; one having three is called *triplex*; etc.

Piston pumps are used in applications where their high pressure capability makes them one of the only alternatives to consider. Relatively high pressure service with an abrasive liquid is one very common application. Most centrifugal pumps capable of handling abrasives have limited head capability (only several hundred feet of head). Reciprocating piston pumps are considered better in abrasive applications than plunger pumps. With piston pumps in abrasive applications, elastomeric pistons are used and cylinders are lined with tungsten carbide or other hardened material.

Important applications are found for both piston and plunger pumps in oil production and pipeline services. They are usually built in a horizontal configuration for smaller sizes, and in vertical configurations for the larger sizes to reduce the loads on the bearings, packing, and crossheads. Vertical designs save floor space, but are relatively more expensive and more difficult to maintain than horizontal designs.

Piston pumps are limited to slower speeds than plunger pumps (piston pumps usually run at 100 rpm or slower) for several reasons. This is partly an effort to reduce the high unbalanced forces from the two throws at 90°. Also, piston pumps are usually designed for longer life than plunger pumps, so the slower speed makes for a more conservative design. The slower speed of a piston pump makes it usually larger in size, and thus more expensive from a capital standpoint, than a plunger selection for a particular rating. However, its slower speed usually makes the piston pump a lower maintenance expense alternative compared to a plunger pump.

Note that the allowable or rated speed of a reciprocating pump is not based solely upon whether the pump is a double-acting piston pump or a single-acting piston or a plunger. The size of the valves limits the speed of the pump for any given piston or plunger size. Sometimes the size of the inlet opening is also a factor in limiting the speed of the pump. Furthermore, the viscosity of the liquid also affects the speed of the pump, as reciprocating pumps must be slowed down for more viscous liquids. The flow of liquid through any check valve in a reciprocating pump is never steady, but is constantly changing from zero to a maximum. Therefore, considerable attention must be given to the selection of a pump that does the job required without cavitation on the inlet stroke. (See Chapter 2, Section VI, for a discussion of NPSH and cavitation.)

12. Plunger Pump

A plunger pump (Figure 1.23) is similar to a piston pump, except the reciprocating member is a plunger rather than a piston. The plunger (Figure 1.24) is single-acting (i.e., only has liquid discharging during the forward stroke), and the plunger is sealed with packing in the cylinder walls.

Generally speaking, plunger pumps are used for higher pressure applications than piston pumps. They are capable of the highest pressures obtainable with a positive displacement pump, with some very special applications achieving pressures greater than 50,000 psi. Plunger pumps generally run at higher speeds than piston pumps. They are therefore usually a lower capital cost alternative to a piston pump if both

Introduction to Pumps 41

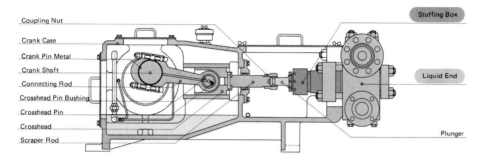

Figure 1.23 Plunger pump. (Courtesy of NIKKISO Pumps America, Inc., Plumsteadville, PA.)

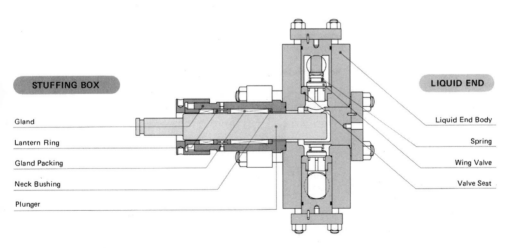

Figure 1.24 Plunger pump liquid end and stuffing box. (Courtesy of NIKKISO Pumps America, Inc., Plumsteadville, PA.)

are being considered. However, the plunger pump would have higher maintenance expense and lower abrasion resistance than a piston alternative.

For all types of reciprocating pumps, rather large pressure pulsations are produced. These pulsations become more smoothed out the higher the number of reciprocating members. Even so, quite often it is necessary to fit pulsation dampener devices downstream of the pump discharge. A pulsation dampener is a vessel that is separated in the middle with a bladder or membrane, and which has air or nitrogen or a neutral gas on the upper half of the bladder. The membrane flexes, and the air compresses and dampens out the pulsations produced as the reciprocating pump strokes.

13. Diaphragm Pump

Diaphragm pumps (Figure 1.25) are similar to piston and plunger pumps, except that the reciprocating motion of the pump causes a diaphragm to flex back and forth, which in turn causes the liquid to flow into and out of the liquid end of the pump. As with all reciprocating pumps, diaphragm pumps require check valves at the inlet and outlet ports (shown as ball checks in Figure 1.25). The diaphragm is usually made of an elastomeric material to allow it to flex. The diaphragm can be mechanically attached to the reciprocating member, or it can be separated and actuated by a reservoir of hydraulic fluid, as shown in Figure 1.25. Contour plates in the pump shown on Figure 1.25 control the travel limits of the diaphragm.

One very common application for diaphragm pumps of the type described above is for metering applications. Metering pumps, or dosing pumps as they are called in Europe, have relatively low flow rates, usually measured in gallons or liters per hour rather than per minute. These pump types are highly accurate in measuring flow (usually having an accuracy of better than ±1%), and the diaphragm makes the pump leak-free and compatible with a variety of liquids. Figure 1.26 shows a hydraulically actuated diaphram metering pump with stroke adjustment capability to vary the flow rate.

Introduction to Pumps 43

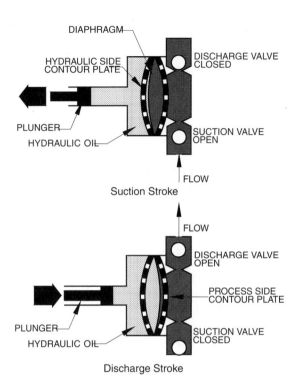

Figure 1.25 Diaphragm pump. (Courtesy of NIKKISO Pumps America, Inc., Plumsteadville, PA.)

Another style of diaphragm pump is the solenoid metering pump, used in light-duty metering applications. This style of diaphragm pump uses an electrical signal to magnetically move the plunger/diaphragm assembly.

Much larger versions of hydraulically actuated diaphragm pumps are used in process services, where their high pressure capability and sealless pumping make them an interesting alternative for special services. These pumps, with metal diaphragms and remote heads, can pump liquids to 900°F.

Figure 1.26 Hydraulically actuated diaphragm metering pump with stroke adjustment. (Courtesy of NIKKISO Pumps America, Inc., Plumsteadville, PA.)

Another type of diaphragm pump is the air-operated, double-diaphragm pump (Figure 1.27). In this pump design, compressed air enters the air chamber behind one of the diaphragms (shown in Figure 1.28), flexing the diaphragm and thus forcing the air or liquid on the other side of the diaphragm out the discharge check valve. Simultaneously, the second diaphragm (Figure 1.28) is pulled inward by a rod connecting the two diaphragms, creating a suction stroke with that diaphragm, with liquid coming in through the inlet check valve on that side of the pump. Then a shuttle valve causes the air distribution to shift, sending air to the other chambers and reversing the stroke of the two diaphragms.

Introduction to Pumps

Figure 1.27 Air-operated, double-diaphragm pump. (Courtesy of Lutz Pumps, Norcross, GA.)

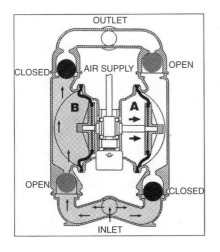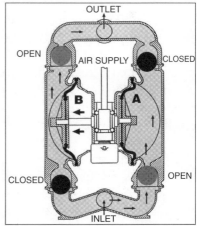

Figure 1.28 Air-operated, double-diaphragm pump. (Courtesy of Wilden Pump & Engineering Co., Coulton, CA.)

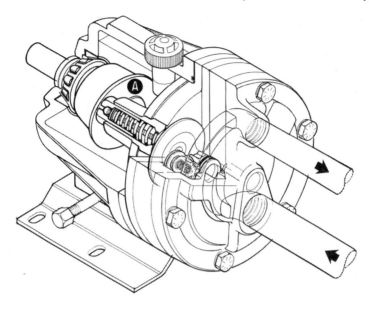

Figure 1.29 Wobble plate pump. (Courtesy of Wanner Engineering, Inc., Minneapolis, MN.)

Any type of diaphragm pump has the distinct advantage, compared with piston or plunger pumps, of being sealless; that is, not requiring any packing assembly or mechanical seal. Air-operated diaphragm pumps offer the additional advantage of being able to accommodate large solids, abrasives, and corrosives. They are self-priming and can run dry. Their versatility makes them a good choice for pumping wastewater, acids, and foods. The shortcomings of air-operated diaphragm pumps are that they require air to operate (this may actually be a benefit if the pump is in an area where compressed air is available but electricity is not), they have limitations on flow and pressure, they produce fairly large pressure pulsations, and they are quite energy inefficient. Some designs also have problems with the air valves stalling or freezing up, and some air valves require periodic lubrication.

Another type of diaphragm pump, shown in Figure 1.29, is known as a *wobble plate* pump. This pump type has the reciprocating action of several pistons or diaphragms caused

Introduction to Pumps

by a rotating plate (labeled A) mounted eccentrically on the shaft. Advantages of this pump type include quite high pressure capability, sealless pumping, self-priming, and the capability of running dry. Disadvantages include relatively low flows and many moving parts.

14. Miniature Positive Displacement Pumps

There are several special classes of miniature positive displacement pumps used for very low flow rates that are required for many specialty Original Equipment Manufacturer (OEM) machines such as lasers and kidney dialysis machines. OEMs require low-cost and small-sized pumps to remain competitive in their marketplace. Several types of small pumps have evolved to support these needs. The two major types of miniature P.D. pumps are described below.

a. *Gear Pumps*

Miniature gear pumps (Figure 1.30) are quite common, especially when moderate pressures (typically less than 150 psi) are required. One of the most common in this category consists of stainless steel construction, composite nonmetallic

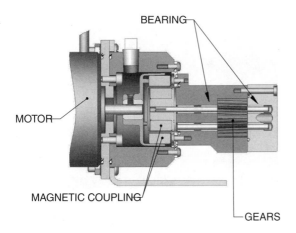

Figure 1.30 Miniature gear pump. (Courtesy of Diener Precision Pumps, Lodi, CA.)

gears/bushings, combined with magnetic drive (sealless construction, discussed more thoroughly in Chapter 5, Section V). These pumps are typically rated from 5 ml/min up to 10 l/min. One of their most important features is pulseless flow. They are ideal for smooth flow of a wide variety of chemicals and coolants. Some of the more common applications include commercial ink jet printers, kidney dialysis machines, critical heating and cooling processes, laboratory instrumentation, dispensing equipment, etc.

These pumps are typically driven by small AC or DC motors. Brushless DC motors are rapidly becoming more popular, due to the safety associated with low voltage and reliability. These pumps are suitable for high-speed operation to keep the package size small, but life will be shorter.

b. Piston Pumps

An alternative for equipment manufacturers are miniature piston pumps (Figure 1.31). They are preferred for high repeatability (<1%) and for applications where linearity with speed is important. Miniature piston pumps tend to serve these needs nicely if the process can live with flow pulsations.

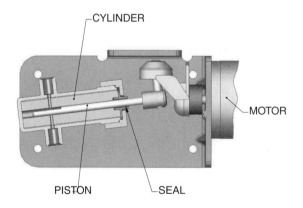

Figure 1.31 Miniature piston pump. (Courtesy of Diener Precision Pumps, Lodi, CA.)

Miniature piston pumps are generally constructed of nonmetallic materials such as ceramics, PEEK, Teflon®, and PVDF. One of the most common technologies, shown in Figure 1.30, adds a rotating motion to the reciprocating motion in such a way that no inlet and outlet check valves are needed. Flow rate can be changed by changing the piston stroke and/or speed. They are frequently driven by a stepper motor for very high positional accuracy, which results in accurate flow or dispense capability.

This type of pump is used for pH balance in applications such as dialysis equipment, laboratory instruments, and pesticide metering.

2

Hydraulics, Selection, and Curves

I. OVERVIEW

This chapter covers the basics of pump hydraulics, particularly as they apply to centrifugal pumps. It shows the reader, step-by-step, how to determine and analyze the criteria to completely select a centrifugal pump. Example problems illustrate what information is needed to size a pump, how to go about picking a pump from manufacturers' catalogs, and how to determine the required driver size.

NPSH (net positive suction head) and *cavitation* are described in great detail, along with a number of examples. NPSH is one of the least-understood principles of pump hydraulics and is the cause for a great many pump problems. It is also often mistakenly blamed for other unrelated pump problems that nevertheless have similar symptoms, such as air in a pump system or misalignment.

Specific speed is an index normally used only by pump designers. However, a better understanding of pump specific speed enables engineers and users to see why certain pumps are used in certain applications, why centrifugal pump curve shapes vary so dramatically, and why pump efficiencies vary so widely from one type of pump to another.

An understanding of how the *affinity laws* work allows a user or engineer to take a pump at one speed and figure out how performance changes when it is run at another speed, or how to take a pump with a particular impeller diameter and ascertain the effects of changing the impeller diameter.

A proper understanding of *system head curves* is very important in determining where on the pump's performance curve the pump operates. The pump runs where the system head curve tells it to operate. A better understanding of how to generate a system head curve, and the knowledge of how the system head curve can vary over the life of the pump, will help prevent many field problems, give a longer life to the pump, and save energy.

This chapter shows why and how multiple pumps perform together in a system, arranged either in parallel or in series.

Finally, there is some discussion of the consequences of using excessive "fudge factors" in sizing centrifugal pumps.

The initial decision that must be made in applying a pump is the decision regarding the type of pump to use. First, a decision must be made as to whether the pump should be of the centrifugal or positive displacement (P.D.) type. A P.D. pump is chosen if one or more of the key application criteria introduced in Chapter 1, Section VI.B, is present. Table 1.1 and Chapter 1, Section VI.C, can then be used to determine the P.D. pump alternatives to consider.

If none of these application criteria are present, or if they are present but are determined to not be a compelling reason to use a positive displacement pump, then a centrifugal pump is chosen.

The next decision is to choose the type of configuration of centrifugal pump to use (e.g., end suction, inline, split case, vertical turbine, submersible, etc.). Unfortunately, making this decision is not a simple process, and requires a full understanding of the application criteria and system constraints, as well as the centrifugal pump configuration options that are available. A great deal of this decision-making process improves with experience, and one only gets better at it by studying various types of pumping equipment and learning more about each type's benefits and shortcomings. Chapters 4, 5, and 7 should prove helpful in this regard.

Hydraulics, Selection, and Curves

Sometimes, special design criteria or standards apply, such as ANSI (American National Standards Institute), API (American Petroleum Institute), nuclear standards (ASME Section III), or sanitary (FDA) requirements. Sometimes, the pump configuration is dictated by the unique layout of the pumping system (e.g., a borehole well, which requires a vertical turbine or submersible pump) or special requirements of the liquid to be pumped (e.g., a highly abrasive slurry). Any one of these requirements places limits on the configurations of pumps that are available and on the number of pump manufacturers that have a pump to offer.

For a great many applications, a "traditional" or "historical" configuration is used (i.e., "the way we've always done it"). With the continuing evolution of pump technology, the traditional approach is not necessarily always the best option.

Sometimes, an economic analysis must be carried out to estimate total lifetime costs to determine the type of pump configuration to use. The cost items to be examined include first costs for the equipment, control system, and the structure in which the equipment is housed; energy costs; and maintenance costs. Chapter 6 includes a more in-depth discussion of pump life-cycle costs.

As an example of the above, consider an application involving emptying a sump of liquid which is oily, corrosive, and abrasive. The choices of pumps might include an end suction horizontal pump mounted above the pit, a submersible pump mounted in the pit, or a vertical column sump or vertical turbine pump partially submerged in the liquid. Each of these is a unique pump configuration, with advantages and disadvantages for the application, differing costs for the equipment, differing expected maintenance concerns, differing operating costs, and differing costs for construction of the structure. This is only one pump application problem, and virtually every application is unique with regard to the pumped liquid's properties, the hydraulic conditions, and the relative importance the user places on first cost, energy costs, and maintenance expenses. The material contained in Chapter 4, where each of the major pump types is discussed as to its strengths and shortcomings, assists in this initial process of deciding the configuration of the pump type to use. Chapter 6 has several examples of an economic analysis that can be made in comparing several configuration alternatives.

II. PUMP CAPACITY

Chapter 1 introduced the characteristic head–capacity curve of a centrifugal pump (see Figure 1.6). The two parameters that must be determined to size a pump are the *capacity* and the *total head*. Capacity is usually expressed in gallons per minute (gpm), or, for larger pumps in cubic feet per second (cfs) using USCS units; and in cubic meters per second, liters per second, or cubic meters per hour using SI units. *Total head*, sometimes simply referred to as *head*, is abbreviated TH or H, and is measured in feet (USCS units) or meters (SI units).

The required capacity of the pump is normally dictated by the requirements of the system in which the pump is located. A process system is designed for a particular throughput. A vessel must be filled (or emptied) in a certain amount of time. An air conditioning system requires a particular flow of chilled water to do the job it is designed to perform.

Regardless of the pump system being designed, it is usually possible to arrive at a design flow rate for the pump. Sometimes there will be a planned *duty cycle* for the pump that will require it to operate at only a fraction of the full design capacity during certain periods of time, and at higher or lower capacities at other times. Examples of this include a process plant with a variable capacity, or a chilled water system designed to meet a variable air conditioning load. The operating duty cycle of the pump should be estimated, using the best available process and operations estimates, as this helps in selecting the best type of pump and control system.

III. HEAD

As introduced in Chapter 1, Section V, a centrifugal pump develops head by raising the velocity of the liquid in the impeller, and then converting some of this velocity into pressure in the volute or diffuser casing by a diffusion process. The amount of head developed in the impeller is approximately:

$$H = \frac{V^2}{2g} \qquad (2.1)$$

where:
V = velocity at the tip of the impeller, in ft/sec
g = acceleration of gravity (32.2 ft/sec²)

The velocity at the impeller tip can also be expressed as:

$$V = \frac{rpm \times D}{229} \qquad (2.2)$$

where:
rpm = pump speed, in revolutions per minute
D = impeller diameter, in inches

Substituting Equation 2.2 above for V in Equation 2.1, results in the following expression:

$$H = \frac{(rpm \times D)^2}{3.375 \times 10^6} \qquad (2.3)$$

Equation 2.3 shows that the head developed by a centrifugal pump is only a function of rpm and impeller diameter, and is not a function of specific gravity of the liquid being pumped. A pump moving a liquid up a static distance of 100 ft always has a required head of 100 ft (ignoring friction for the moment), regardless of the specific gravity of the liquid. If the liquid were water (s.g. = 1.0), a pressure gauge located at the pump discharge would show a pressure, using Equation 1.1, of 100 × 1.0/2.31 = 43.3 psig. If the liquid being pumped up the 100 feet height were oil, with a specific gravity of 0.8, the gauge would read 100 × 0.8/2.31 = 34.6 psig. The point of this is that the pump discharge pressure expressed in psi (or equivalent metric units such as kPa) varies with the specific gravity of the liquid, while the head expressed in feet (or meters) of liquid remains constant for liquids of different density. This is why one should always convert pressure terms into units of feet (or meters) of head when dealing with centrifugal pumps. A pump's head–capacity curve does not require adjustment when the specific gravity of the liquid changes. On the other hand, as will be demonstrated shortly, the horsepower curve does vary with varying specific gravity. Further, as illustrated in Chapter 3, Section II, the head–capacity curve is also affected by the viscosity of the pumped liquid.

To determine the required size of a centrifugal pump for a particular application, all the components of the system head for the system in which the pump is to operate must be added up to determine the pump *total head* (TH). There are four separate components of total system head, and they refer back to the discussion in Chapter 1 on the reasons for raising the pressure of a liquid. The four components of total system head are:

1. Static head
2. Friction head
3. Pressure head
4. Velocity head

Each of these four components of head must be considered for the system in which the pump is to operate, and the sum of these is the total head (TH) of the pump. Note that the last of these components, velocity head, may or may not need to be included in the calculation of the components of system head for sizing a pump, depending on the point of reference for the calculation (where the pressures for pressure head and the levels for static head are measured). The four components of system head are each discussed separately below.

A. Static Head

Static head is the total elevation change that the liquid must undergo. In most cases, static head is normally measured from the surface of the liquid in the supply vessel to the surface of the liquid in the vessel where the liquid is being delivered. The total static head is measured from supply vessel surface to delivery vessel surface, regardless of whether the pump is

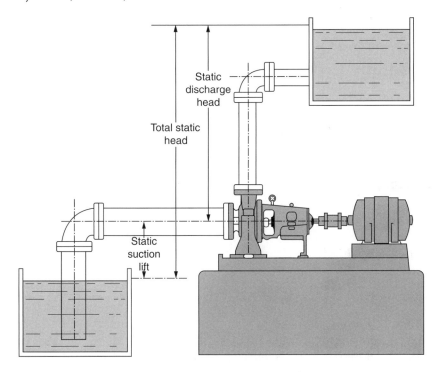

Figure 2.1 Static suction lift, static discharge head, and total static head. (Courtesy of Goulds Pumps, Inc., ITT Industries, Seneca Falls, NY.)

located above the liquid level in the suction vessel (which is referred to as a "suction lift"), or below the liquid level in the suction vessel ("suction head"). Figure 2.1 shows an example of a pump on a suction lift, and defines static suction lift, static discharge head, and total static head. Note that for a pump in a closed loop system, the total static head is zero.

If the pressure head requirements of the system (see III.C below) are given by gauge pressure readings at some point in the system suction and discharge piping rather than at the supply and delivery vessels, then the value of static head is the difference in elevation of the pressure gauges, rather than the difference in elevation between the liquid surface in the supply and delivery vessels.

B. Friction Head

Friction head is the head necessary to overcome the friction losses in the piping, valves, and fittings for the system in which the pump operates. Friction loss in a piping system varies as the square of the liquid's velocity (assuming fully turbulent flow). The smaller the size of the pipe, valves, and fittings for a given flow rate, the greater the friction head loss. In designing a piping system, if smaller sizes of pipes, valves, and fittings are chosen, the cost of the piping system is reduced. However, the trade-off is that this results in higher total pump head due to the increased friction head loss. This, in turn, usually increases pump and driver capital cost, and also increases lifetime energy costs. Another point is that choosing smaller suction lines might cause the pump to cavitate due to the increased suction line friction losses, as is discussed in more detail in Section VI to follow.

In theory, friction losses that occur as liquid flows through a piping system must be calculated by means of complicated formulae, taking into account such factors as liquid density and viscosity, and pipe inside diameter and material. Luckily, these formulae have been reduced to tables and charts that, although somewhat tedious and repetitive, are nevertheless not too complex. Table 2.1 shows a typical pipe friction table for water at 60°F flowing through schedule 40 steel pipe. If the pumped liquid is other than water, or if the pipe schedule or material is other than schedule 40 steel pipe, a different table or an adjustment to the table must be used. Friction data for other pipe materials and inside diameters are often found in engineering data tables, or are sometimes available from pipe manufacturers. Some tables are available for liquids other than water, or computer software data tables can be used. Refer to Chapter 3, Section III, for further discussion of the benefits of using computer software for determining friction head.

To determine the friction head for a particular section of piping, enter Table 2.1 at the planned design capacity and choose a line size. Usually, velocity (labeled "V" in Table 2.1) is used as the criterion for choosing at least a preliminary

Hydraulics, Selection, and Curves

line size, with the trade-off between piping system cost, pump capital cost, and lifetime energy costs being considered. Common velocity guidelines are 4 to 6 ft/sec for suction piping and 6 to 10 ft/sec for discharge piping. The recommended range can vary widely, depending on the particular application. Important factors to consider when establishing the recommended velocity range for a particular application include the abrasiveness of the liquid, pipe material, and settling properties of any suspended solids.

Note that some engineers mistakenly believe that the suction and discharge connection sizes of the pump they are using for a particular application must be the same sizes as those chosen for, respectively, the suction and discharge piping. This is quite often not the case. The pump is designed by the manufacturer to be as compact as possible for a given hydraulic duty point, which sometimes results in velocities in the suction and discharge connections of the pump that are higher than the velocities which are the optimal sizes to be used in the suction and discharge lines (taking into account considerations of avoiding cavitation by improper suction pipe sizing, and avoiding increased pumping costs due to undersizing the discharge piping). It is quite often the case, therefore, that the suction pipe in a pump system is one or more sizes larger than the suction connection on the pump, and that the discharge pipe is one or more sizes larger than the discharge connection on the pump. This means, of course, that appropriately sized piping reducers/expanders must be used at the pump inlet and outlet.

With the design capacity and the chosen preliminary pipe size, the friction tables give the head loss in feet per 100 linear feet of pipe (labeled "h_f" in Table 2.1). Note the two different uses of the term "foot" here, with Table 2.1 showing *feet of head loss* per 100 *linear feet* of pipe. This means that the value found in Table 2.1 must be multiplied by the actual pipe length divided by 100 to get the total friction head loss in a given length of pipe. This is expressed by the formula:

$$H_f \text{ (pipe)} = h_f \times \frac{L}{100} \tag{2.4}$$

Table 2.1 Pipe Friction: Water/Schedule 40 Steel Pipe

U.S. Gallons per Minute	⅛ in. (0.269 in. I.D.)			¼ in. (0.364 in. I.D.)			⅜ in (0.493 in. I.D.)			½ in. (0.622 in. I.D.)			U.S. Gallons per Minute
	V (Ft./Sec.)	$\frac{V^2}{2g}$	h_f (Ft./100 ft.)	V	$\frac{V^2}{2g}$	h_f	V	$\frac{V^2}{2g}$	h_f	V	$\frac{V^2}{2g}$	h_f	
0.2	1.13	0.020	2.72										0.2
0.4	2.26	0.079	16.2										0.4
0.6	3.39	0.178	33.8										0.6
0.8	4.52	0.317	57.4										0.8
1.0	5.65	0.495	87.0	1.23	0.024	3.7							1.0
1.5	8.48	1.12	188	1.85	0.053	7.6	1.01	0.016	1.74				1.5
2.0	11.3	1.98	324	2.47	0.095	12.7	1.34	0.028	2.89				2.0
2.5				3.08	0.148	19.1	1.68	0.044	4.30	1.06	0.017	1.86	2.5
3.0				3.70	0.213	27.2	2.02	0.063	6.15				3.0
3.5				4.62	0.332	40.1	2.52	0.099	8.93	1.58	0.039	2.85	3.5
4.0				6.17	0.591	69.0	3.36	0.176	15.0	2.11	0.069	4.78	4.0
4.5				7.71	0.923	105	4.20	0.274	22.6	2.64	0.108	7.16	4.5
5				9.25	1.33	148	5.04	0.395	31.8	3.17	0.156	10.0	5
6				10.79	1.81	200	5.88	0.538	42.6	3.70	0.212	13.3	6
7				12.33	2.36	259	6.72	0.702	54.9	4.22	0.277	17.1	7
8				13.87	2.99	326	7.56	0.889	68.4	4.75	0.351	21.3	8
9				15.42	3.69	398	8.40	1.10	83.5	5.28	0.433	25.8	9
10							10.1	1.58	118	6.34	0.624	36.5	10
12							11.8	2.15	158	7.39	0.849	48.7	12
14							13.4	2.81	205	8.45	1.11	62.7	14
16							15.1	3.56	258	9.50	1.40	78.3	16
18							16.8	4.39	316	10.6	1.73	95.9	18
20										12.7	2.49	136	20
25										14.8	3.40	183	25

Hydraulics, Selection, and Curves

U.S. Gallons per Minute	¾ in. (0.824 in. I.D.)			1 in. (1.049 in. I.D.)			1¼ in. (1.3880 in. I.D.)			1½ in. (1.610 in. I.D.)			U.S. Gallons per Minute
	V	$\frac{V^2}{2g}$	h_f	V	$\frac{V^2}{2g}$	h_f	V	$\frac{V^2}{2g}$	h_f	V	$\frac{V^2}{2g}$	h_f	
4	2.41	0.090	4.21	1.48	0.034	1.29							4
5	3.01	0.141	6.32	1.86	0.053	1.93							5
6	3.61	0.203	8.87	2.23	0.077	2.68							6
7	4.21	0.276	11.8	2.60	0.105	3.56							7
8	4.81	0.360	15.0	2.97	0.137	4.54	1.29	0.026	0.70				8
9	5.42	0.456	18.8	3.34	0.173	5.65	1.50	0.035	0.93				9
10	6.02	0.563	23.0	3.71	0.214	6.86	1.72	0.046	1.18	1.26	0.025	0.56	10
12	7.22	0.810	32.6	4.45	0.308	9.62	1.93	0.058	1.46	1.42	0.031	0.69	12
14	8.42	1.10	43.5	5.20	0.420	12.8	2.15	0.071	1.77	1.58	0.039	0.83	14
16	9.63	1.44	56.3	5.94	0.548	16.5	2.57	0.103	2.48	1.89	0.056	1.16	16
18	10.8	1.82	70.3	6.68	0.694	20.6	3.00	0.140	3.28	2.21	0.076	1.53	18
20	12.0	2.25	86.1	7.42	0.857	25.1	3.43	0.183	4.20	2.52	0.099	1.96	20
25	15.1	3.54	134	9.29	1.34	37.4	3.86	0.232	5.22	2.84	0.125	2.42	25
30	18.1	5.06	187	11.1	1.93	54.6	4.29	0.286	6.34	3.15	0.154	2.94	30
35				13.0	2.62	73.3	5.37	0.448	9.66	3.94	0.241	4.50	35
40				14.8	3.43	95.0	6.44	0.644	13.6	4.73	0.347	6.26	40
45				16.7	4.33	119	7.52	0.879	18.5	5.52	0.473	8.38	45
50				18.6	5.35	146	8.58	1.14	23.5	6.30	0.618	10.8	50
60				22.3	7.71	209	9.66	1.45	29.5	7.10	0.783	13.5	60
70				26.0	10.5	283	10.7	1.79	36.0	7.88	0.965	16.4	70
80							12.9	2.57	51.0	9.46	1.39	23.2	80
90							15.0	3.50	68.8	11.0	1.89	31.3	90
100							17.2	4.58	89.2	12.6	2.47	40.5	100
120							19.3	5.79	112	14.2	3.13	51.0	120
140							21.5	7.15	138	15.8	3.86	62.2	140
							25.7	10.3	197	18.9	5.56	88.3	
										22.1	7.56	119	

Table 2.1 Pipe Friction: Water/Schedule 40 Steel Pipe (continued)

U.S. Gallons per Minute	2 in. (2.067 in. I.D.) V	$\frac{V^2}{2g}$	h_f	2½ in. (2.469 in. I.D.) V	$\frac{V^2}{2g}$	h_f	3 in. (3.068 in. I.D.) V	$\frac{V^2}{2g}$	h_f	3½ in. (3.548 in. I.D.) V	$\frac{V^2}{2g}$	h_f	U.S. Gallons per Minute
30	2.87	0.128	1.82	2.01	0.063	0.75							30
35	3.35	0.174	2.42	2.35	0.085	1.00							35
40	3.82	0.227	3.10	2.68	0.112	1.28							40
50	4.78	0.355	4.67	3.35	0.174	1.94	2.17	0.073					50
60	5.74	0.511	6.59	4.02	0.251	2.72	2.60	0.105	0.66				60
80	7.65	0.909	11.4	5.36	0.447	4.66	3.47	0.187	0.92				80
100	9.56	1.42	17.4	6.70	0.698	7.11	4.34	0.293	1.57				100
120	11.5	2.05	24.7	8.04	1.00	10.0	5.21	0.421	2.39				120
140	13.4	2.78	33.2	9.38	1.37	13.5	6.08	0.574	3.37				140
160	15.3	3.64	43.0	10.7	1.79	17.4	6.94	0.749	4.51	1.95	0.059	0.45	160
180	17.2	4.60	54.1	12.1	2.26	21.9	7.81	0.948	5.81	2.60	0.105	0.77	180
200	19.1	5.68	66.3	13.4	2.79	26.7	8.68	1.17	7.28	3.25	0.164	1.17	200
220	21.0	6.88	80.0	14.7	3.38	32.2	9.55	1.42	8.90	3.89	0.236	1.64	220
240	22.9	8.18	95.0	16.1	4.02	38.1	10.4	1.69	10.7	4.54	0.321	2.18	240
260	24.9	9.60	111	17.4	4.72	44.5	11.3	1.98	12.6	5.19	0.419	2.80	260
280	26.8	11.1	128	18.4	5.47	51.3	12.2	2.29	14.7	5.84	0.530	3.50	280
300	28.7	12.8	146	20.1	6.28	58.5	13.0	2.63	16.9	6.49	0.655	4.27	300
350				23.5	8.55	79.2	15.2	3.57	19.2	7.14	0.792	5.12	350
400				26.8	11.2	103	17.4	4.68	26.3	7.79	0.943	6.04	400
500				33.5	17.4	160	21.7	7.32	33.9	8.44	1.11	7.04	500
600							26.0	10.5	52.5	9.09	1.28	8.11	600
700							30.4	14.3	74.8	9.74	1.47	9.26	700
800							34.7	18.7	101	11.3	2.00	12.4	800
1000									131	13.0	2.62	16.2	1000
										16.2	4.09	25.0	
										19.5	5.89	35.6	
										22.7	8.02	48.0	
										26.0	10.5	62.3	
										32.5	16.4	96.4	

Hydraulics, Selection, and Curves

U.S. Gallons per Minute	4 in. (4.026 in. I.D.) V	$\frac{V^2}{2g}$	h_f	5 in. (5.047 in. I.D.) V	$\frac{V^2}{2g}$	h_f	6 in. (6.065 in. I.D.) V	$\frac{V^2}{2g}$	h_f	8 in. (7.981 in. I.D.) V	$\frac{V^2}{2g}$	h_f	U.S. Gallons per Minute
140	3.53	0.193	1.16	2.25	0.078	0.38							140
160	4.03	0.253	1.49	2.57	0.102	0.49							160
180	4.54	0.320	1.86	2.89	0.129	0.61							180
200	5.04	0.395	2.27	3.21	0.160	0.74							200
240	6.05	0.569	3.21	3.85	0.230	1.03	2.22	0.077	0.30				240
280	7.06	0.774	4.30	4.49	0.313	1.38	2.66	0.110	0.42				280
320	8.06	1.01	5.51	5.13	0.409	1.78	3.11	0.150	0.56				320
360	9.07	1.28	6.92	5.77	0.518	2.22	3.55	0.196	0.72				360
400	10.1	1.58	8.47	6.41	0.639	2.72	4.00	0.240	0.90				400
450	11.3	2.00	10.5	7.23	0.811	3.42	4.44	0.307	1.09				450
500	12.6	2.47	13.0	8.02	0.999	4.16	5.00	0.388	1.37				500
600	15.1	3.55	18.6	9.62	1.44	5.88	5.55	0.479	1.66				600
700	17.6	4.84	25.0	11.2	1.96	7.93	6.66	0.690	2.34				700
800	20.2	6.32	32.4	12.8	2.56	10.2	7.77	0.939	3.13				800
900	22.7	8.00	40.8	14.4	3.24	12.9	8.88	1.23	4.03				900
1000	25.2	9.87	50.2	16.0	4.00	15.8	9.99	1.55	5.05	2.57	0.102	0.28	1000
1200	30.2	14.2	72.0	19.2	5.76	22.5	11.1	1.92	6.17	2.89	0.129	0.35	1200
1400	35.3	19.3	97.6	22.5	7.83	30.4	13.3	2.76	8.76	3.21	0.160	0.42	1400
1600				25.7	10.2	39.5	15.5	3.76	11.8	3.85	0.230	0.60	1600
1800				28.8	12.9	49.7	17.8	4.91	15.4	4.49	0.313	0.80	1800
2000				32.1	16.0	61.0	20.0	6.21	19.4	5.13	0.409	1.02	2000
2400							22.2	7.67	23.8	5.77	0.518	1.27	2400
2800							26.6	11.0	34.2	6.41	0.639	1.56	2800
3200							31.1	15.0	46.1	7.70	0.920	2.20	3200
3600							35.5	19.6	59.9	8.98	1.25	2.95	3600
4000										10.3	1.64	3.82	4000
										11.5	2.07	4.79	
										12.8	2.56	5.86	
										15.4	3.68	8.31	
										18.0	5.01	11.2	
										20.5	6.55	14.5	
										23.1	8.28	18.4	
										25.7	10.2	22.6	

Table 2.1 Pipe Friction: Water/Schedule 40 Steel Pipe (continued)

U.S. Gallons per Minute	10 in. (10.020 in. I.D.)			12 in. (11.938 in. I.D.)			14 in. (13.124 in. I.D.)			16 in. (15.000 in. I.D.)			U.S. Gallons per Minute
	V	$\frac{V^2}{2g}$	h_f	V	$\frac{V^2}{2g}$	h_f	V	$\frac{V^2}{2g}$	h_f	V	$\frac{V^2}{2g}$	h_f	
800	3.25	0.165	0.328										800
900	3.66	0.208	0.410										900
1000	4.07	0.257	0.500	2.58	0.103	0.173							1000
1200	4.88	0.370	0.703	2.87	0.128	0.210							1200
1400	5.70	0.504	0.940	3.44	0.184	0.296							1400
1600	6.51	0.659	1.21	4.01	0.250	0.395	2.37	0.087	0.131				1600
1800	7.32	0.834	1.52	4.59	0.327	0.509	2.85	0.126	0.185				1800
2000	8.14	1.03	1.86	5.16	0.414	0.636	3.32	0.171	0.247				2000
2500	10.2	1.62	2.86	5.73	0.511	0.776	3.79	0.224	0.317	2.90	0.131	0.163	2500
3000	12.2	2.32	4.06	7.17	0.799	1.19	4.27	0.283	0.395	3.27	0.166	0.203	3000
3500	14.2	3.13	5.46	8.60	1.15	1.68	4.74	0.349	0.483	3.63	0.205	0.248	3500
4000	16.3	4.12	7.07	10.0	1.55	2.25	5.93	0.546	0.738	4.54	0.320	0.377	4000
4500	18.3	5.21	8.88	11.5	2.04	2.92	7.11	0.786	1.04	5.45	0.461	0.535	4500
5000	20.3	6.43	10.9	12.9	2.59	3.65	8.30	1.07	1.40	6.35	0.627	0.718	5000
6000	24.4	9.26	15.6	14.3	3.19	4.47	9.48	1.40	1.81	7.26	0.820	0.921	6000
7000	28.5	12.6	21.1	17.2	4.60	6.39	10.7	1.77	2.27	8.17	1.04	1.15	7000
8000	32.5	16.5	27.5	20.1	6.26	8.63	11.9	2.18	2.78	9.08	1.28	1.41	8000
9000	36.6	20.8	34.6	22.9	8.17	11.2	14.2	3.14	3.95	10.9	1.84	2.01	9000
10,000				25.8	10.3	14.1	16.6	4.28	5.32	12.7	2.51	2.69	10,000
12,000				28.7	12.8	17.4	19.0	5.59	6.90	14.5	3.28	3.49	12,000
14,000				34.4	18.3	24.8	21.3	7.08	8.7	16.3	4.15	4.38	14,000
16,000				40.1	25.0	33.5	23.7	8.74	10.7	18.2	5.12	5.38	16,000
18,000							28.5	12.6	15.2	21.8	7.38	7.69	18,000
20,000							33.2	17.1	20.7	25.4	10.0	10.4	20,000
							37.9	22.4	26.8	29.0	13.1	13.5	
							42.7	28.3	33.9	32.7	16.6	17.2	
										36.3	20.5	21.2	

Hydraulics, Selection, and Curves

U.S. Gallons per Minute	18 in. (16.876 in. I.D.)			20 in. (18.812 in. I.D.)			24 in. (22.624 in. I.D.)			U.S. Gallons per Minute
	V	$\dfrac{V^2}{2g}$	h_f	V	$\dfrac{V^2}{2g}$	h_f	V	$\dfrac{V^2}{2g}$	h_f	
2000	2.87	0.128	0.139							2000
3000	4.30	0.288	0.297							3000
4000	5.74	0.512	0.511							4000
5000	7.17	0.799	0.781							5000
6000	8.61	1.15	1.11							6000
8000	11.5	2.05	1.93	9.23	1.32	1.11				8000
10,000	14.3	3.20	2.97	11.5	2.07	1.70				10,000
12,000	17.2	4.60	4.21	13.8	2.98	2.44				12,000
14,000	20.1	6.27	5.69	16.2	4.06	3.29				14,000
16,000	22.9	8.19	7.41	18.5	5.30	4.26				16,000
18,000	25.8	10.4	9.33	20.8	6.71	5.35				18,000
20,000	28.7	12.8	11.5	23.1	8.28	6.56				20,000
22,000	31.6	15.5	13.9	25.4	10.0	7.91				22,000
24,000	34.4	18.4	16.5	27.7	11.9	9.39	19.2	5.70	3.67	24,000
26,000	37.3	21.6	19.2	30.0	14.0	11.0	20.7	6.69	4.29	26,000
28,000	40.2	25.1	22.2	32.3	16.2	12.7	22.3	7.76	4.96	28,000
30,000	43.0	28.8	25.5	34.6	18.6	14.6	23.9	8.91	5.68	30,000
34,000				39.2	23.9	18.7	27.1	11.4	7.22	34,000
38,000				43.9	29.9	23.2	30.3	14.3	9.00	38,000
42,000							33.5	17.5	11.0	42,000
46,000							36.7	20.9	13.2	46,000
50,000							39.9	24.7	15.5	50,000

Note: The 24 in. column also shows values for 2000–22,000 GPM: V = 3.19, 3.99, 4.79, 6.38, 7.98, 9.58, 11.2, 12.8, 14.4, 16.0, 17.6, 19.2 with $V^2/2g$ = 0.158, 0.247, 0.356, 0.633, 0.989, 1.42, 1.94, 2.53, 3.21, 3.96, 4.79 and h_f = 0.120, 0.181, 0.257, 0.441, 0.671, 0.959, 1.29, 1.67, 2.10, 2.58, 3.10.

From *Engineering Data Book*, 2nd edition, courtesy of Hydraulic Institute, Parsippany, NJ; www.pumps.org and ww.pumplearning.org.

where:
L = pipe length, in feet

The friction loss in valves and fittings is given by the formula:

$$H_f \text{ (valve/fitting)} = K \times \frac{V^2}{2g} \qquad (2.5)$$

The value of $V^2/2g$ for different flow rates and different valve/fitting diameters is found in the pipe friction table (Table 2.1). The value of K, the *resistance coefficient* for the particular valve or fitting, is determined using one of the charts in Figure 2.2, which has a different K chart for each type of valve or fitting. To find K for a particular size of valve or fitting, take the nominal size of the valve or fitting in inches, read up to the heavy line on the chart, and then read on the left scale to get the K factor. For example, from Figure 2.2, K for a 6-inch flanged gate valve is .09. Multiplying the value of K determined from Figure 2.2 times $V^2/2g$ gives the friction loss through the particular valve or fitting expressed in feet.

Note that the K values shown in the Figure 2.2 charts are generic only. If particular valves are already chosen, the valve manufacturer may have more precise resistance coefficients.

The K factors for valves in Figure 2.2 are based on fully open positioning of the valves. For control valves or other valves that may be positioned other than fully open, a value of K must be established for the valve in the partially closed position. Valve manufacturers can help in this effort.

If the system pressure head requirements (see III.C below) are given as gauge readings at some point in the system suction and discharge piping, rather than at the supply and delivery vessels, then the friction head consists only of friction losses in the portion of suction and discharge piping located between the gauges.

C. Pressure Head

Pressure head is the head required to overcome a pressure or vacuum in the system upstream or downstream of the pump.

Hydraulics, Selection, and Curves

It is normally measured at the liquid surface in the supply and delivery vessels. If the pressure in the supply vessel from which the pump is pumping and the pressure in the delivery vessel are identical (e.g., if both are atmospheric tanks), then there is no required pressure head adjustment to TH. Likewise, there is no pressure adjustment to TH for a closed loop system.

If the supply vessel is under a vacuum or under a pressure different than that of the delivery vessel, a pressure head adjustment to TH is required. The pressure or vacuum must be converted to feet. Pressure in pounds per square inch converts to feet by Equation 1.1, rewritten as Equation 2.6:

$$\text{psi} = \frac{\text{feet} \times \text{s.g.}}{2.31} \tag{2.6}$$

where:
psi = pounds per square inch
s.g. = specific gravity

Vacuum, usually expressed in inches of mercury (in. Hg) using USCS units, is converted to feet of head by the formula:

$$\text{Vac. (feet)} = \frac{\text{Vac. (in. Hg)} \times 1.133}{\text{s.g.}} \tag{2.7}$$

Vacuum can also be expressed in inches of water, centimeters of mercury, or other terms, and the conversion to feet or meters of head can be found in the conversion data found in Appendix B at the end of this book.

If the suction vessel is under vacuum, the amount of vacuum (equivalent to gauge pressure, converted to feet) must be added to the delivery vessel gauge pressure (also converted to feet) to get the total pressure adjustment to TH. Alternatively, the vacuum in the suction vessel can be expressed in absolute terms by subtracting the amount of vacuum from the barometric pressure (both terms expressed in inches of mercury). This value is then converted to feet using Equation 2.7, and the result is subtracted from the absolute pressure in the delivery vessel (in feet).

Pump Characteristics and Applications

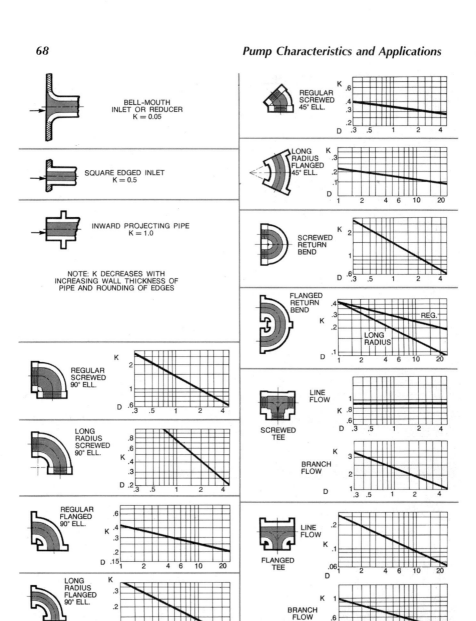

Figure 2.2 Resistance coefficients (K) for valves and fittings. (From *Engineering Data Book*, 2nd edition. Courtesy of Hydraulic Institute, Parsippany, NJ; www.pumps.org and www.pumplearning.org.)

Hydraulics, Selection, and Curves

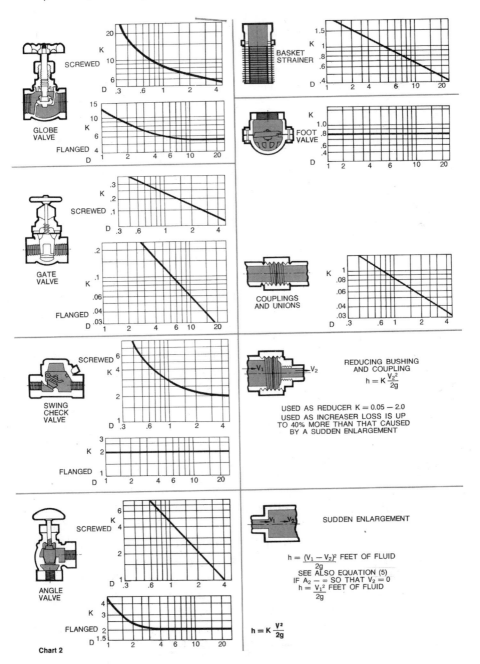

Figure 2.2 (continued)

If the suction vessel is under positive pressure (but different from the pressure of the delivery vessel), then the suction vessel pressure (converted to feet) should be subtracted from the delivery vessel pressure (converted to feet) to get the pressure adjustment to TH. Either gauge pressures or absolute pressures can be used in this case, as long as the pressures in the supply and delivery vessels are expressed consistently as one or the other. If the above seems confusing, the example in Section IV to follow should help clarify this.

If the pressure requirements are given as gauge readings at some point in the system suction and discharge piping, rather than at the liquid surface in the supply and delivery vessels, then the value of static head (see Section III.A above) is the elevation difference between the two gauges. The velocity head (see Section III.D below) is taken at the points of the gauge connections, and the friction head (see Section III.B above) is only the friction losses between the two gauges.

D. Velocity Head

Velocity head is the energy of a liquid as a result of its motion at some velocity V. The formula for velocity head is:

$$H_v = \frac{V^2}{2g} \tag{2.8}$$

This value is found in Table 2.1, expressed in feet of head. The value of velocity head is different at the suction and discharge of the pump, because the size of the suction piping is usually larger than the size of the discharge piping.

Note that the normal procedure in sizing centrifugal pumps measures the required pressure head at the liquid surface in the supply and delivery vessel, as well as establishing static head values from these levels. In this situation, because velocity is zero at the liquid surface in the supply and delivery vessels, velocity head at these points is also zero. Velocity head is only included in the calculation of required pump total head when the pressure head requirements are given as gauge readings at some point in the system suction and discharge piping.

To determine the velocity head component of TH in those situations where it is appropriate, it is necessary to calculate the *change* in velocity head from suction to discharge. Examining Table 2.1, it is seen that in the typical pump configuration, where the suction connection is one size larger than the discharge connection, the change in velocity head across the pump is normally quite small, usually on the order of no more than 1 or 2 ft. With the TH of many pumps being several hundred feet or more, many pump selectors choose to totally ignore the effect of velocity head because the change of velocity head is often less than 1% of TH. However, considering the change of velocity head to be negligible is not always a valid assumption, and a quick check of the velocity head change using Table 2.1 is advisable before dismissing this term. The one situation where velocity head cannot be ignored is when sizing a very low head pump. For example, if the TH of the pump being sized is only, say, 15 ft, then a velocity head change of 1.5 ft represents 10% of the total pump head. This is a significant amount and might affect the pump size or impeller diameter that is required. As another example, consider a pump system with a suction size of 3 in. and a discharge size of 1.5 in., and having a capacity of 120 gpm and a TH of 50 ft. Table 2.1 shows the change of velocity head across this pump to be about 5.1 ft, or slightly over 10% of TH.

IV. PERFORMANCE CURVE

Once the pump configuration and rating (capacity and head) have been determined, as described in the three preceding sections, the next step in the selection process is to decide which pump speeds should be considered. It is quite often the case that two or more operating speeds may be commercially available for a particular pump rating and configuration. Each of these speeds results in a different sized pump, each having different first costs, operating costs, and maintenance costs. Section XIII at the end of this chapter covers the criteria that should be considered to determine which operating speed to consider for a particular application. An example containing some of these criteria is given in Chapter 6.

Table 2.2 A.C. Electric Motor Speeds — 60 Hz

N (# poles)	rpm
2	3600
4	1800
6	1200
8	900
10	720
12	600

The available motor speeds for standard alternating current (A.C.) electric motors are based on the following formula, for 60-cycle current:

$$\text{rpm} = 7200/N \text{ (at frequency = 60 Hz)} \quad (2.9)$$

where:
N = number of poles

Electric motors have an even number of *poles*, starting with two. The commercially available, constant-speed A.C. electric motors with 60 Hz electrical supply are as shown in Table 2.2.

Designations for slower speed motors than shown in Table 2.2 would follow Equation 2.9. The actual operating speeds of motors are slightly less than the values shown in Table 2.2, due to electrical slippage between motor rotor and stator. Thus, the operating speed of a two-pole motor is 3450 to 3550 rpm, the speed for a four-pole motor is 1750 to 1780 rpm, etc.

For 50-cycle current, which is common in Europe and some other parts of the world, Equation 2.9 is revised to:

$$\text{rpm} = 6000/N \text{ (at frequency = 50 Hz)} \quad (2.10)$$

where:
N = number of poles

Accordingly, with 50-cycle current supply, commercially available A.C. electric motor speeds are 3000 rpm, 1500 rpm, etc.

The manufacturer determines which speeds will be offered for each pump type and size, based on a number of

Hydraulics, Selection, and Curves

design and application considerations (Section XIII). In general, the larger the pump impeller (and pump capacity), the slower the pump runs. Also, certain types of applications such as abrasive slurries or paper stock require slower pump speed than clean services.

Once the pump speeds to be considered have been determined, a centrifugal pump selection can be made. However, if a variable-speed pumping system is being considered for a system requiring a range of flow, additional information in the form of a system head curve must be developed before a determination can be made as to the required pump speeds for the various flow requirements. See further discussion on this subject in Chapter 6. For the moment, it is assumed that a constant-speed pump is being selected.

With a constant pump speed, the head–capacity relationship for centrifugal pumps is depicted in Figure 1.6. Most centrifugal pumps, however, have the capability to operate over an extended range of head and flow, by trimming, or cutting the impeller diameter from its maximum size down to some predetermined minimum size. Thus, for a given pump speed, a centrifugal pump produces an envelope of head–capacity performance, as illustrated in Figure 2.3. The upper and lower boundaries of the envelope in Figure 2.3 are dictated by the maximum and minimum impeller diameters that the manufacturer offers for a particular pump size. The right and left boundaries of the envelope are the maximum and minimum flows for each impeller diameter, established by the manufacturer for the particular pump. This is discussed more thoroughly in later sections.

The upper boundary of performance shown in Figure 2.3 is based on the maximum impeller diameter that will physically fit inside the pump casing. The minimum impeller diameter offered by the manufacturer is based on several criteria. Often, simple economics dictate the minimum impeller diameter that the manufacturer offers. A point is reached at which further trimming of impeller diameter makes for an uncompetitive offering, because a competing pump manufacturer is likely able to offer a more competitive pump with a smaller casing. Additionally, the efficiency of the pump usually is

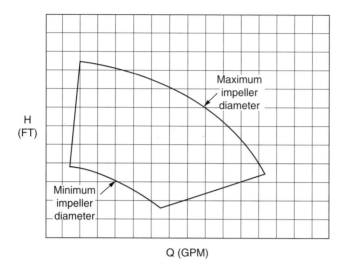

Figure 2.3 Head–capacity envelope for a constant-speed centrifugal pump.

lower if the impeller used is appreciably less than the maximum diameter.

Most pump manufacturers publish performance envelopes for an entire line of pumps at a given speed, as illustrated in Figure 2.4. Once the pump capacity, head, configuration, and speed have been chosen, the preliminary selection of a pump size can be made from a family of envelope curves such as shown in Figure 2.4. Once the pump sizes to be considered are chosen from these envelope curves, the specific pump curves for those pump sizes can be examined in more detail to ascertain information such as the required impeller diameter, efficiency, horsepower, etc.

The convention in the United States for designating the size of a centrifugal pump is as follows:

Suction size × Discharge size × Maximum impeller diameter

All terms are expressed in inches. So, a centrifugal pump with an 8-in. suction flange, a 6-in. discharge flange, and a maximum impeller diameter of 15 in. would be given a size designation

Hydraulics, Selection, and Curves 75

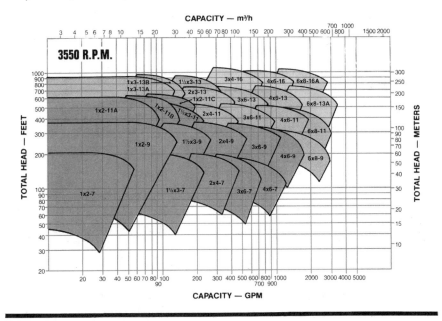

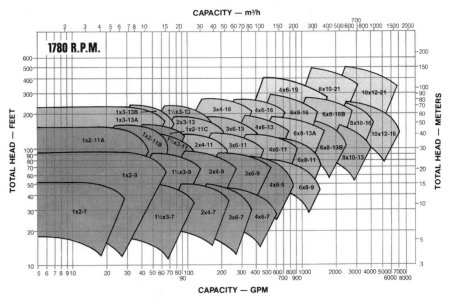

Figure 2.4 Typical family of envelope performance curves for a line of end suction centrifugal pumps, shown at 1800 and 3600 rpm. (Courtesy of Goulds Pumps, Inc., ITT Industries, Seneca Falls, NY.)

of 8 × 6 × 15 (which is read "eight by six by fifteen"). Note that the maximum impeller number is usually nominal. The actual maximum impeller for a pump designated 8 × 6 × 15 may be 15⅛ in. Also note that some manufacturers reverse the size notation, listing the discharge size first. So, the example pump mentioned above would be given the size 6 × 8 × 15 by some companies. Both notations are used in the industry, and because the larger of the first two numbers is always the suction size, this should not cause confusion. Also note that some U.S. manufacturers use a dash instead of a times sign before the final number in the size designation.

The practice in Europe is to express the dimensions for the size designation in millimeters. The size designation is most commonly either expressed in the format first shown above as the standard in the United States (suction size × discharge size × maximum impeller), or is simply shown as discharge — maximum impeller diameter.

The following example serves to illustrate the important concepts which have been introduced thus far in this chapter.

EXAMPLE 2.1: TH calculation/pump selection.

PROBLEM: The piping for the system shown in Figure 2.5 must be sized, the pump TH must be computed, and a centrifugal pump selected.

GIVEN: Capacity = 700 gpm
Liquid = Water at 60°F (s.g. = 1.0)
Pipes = Schedule 40 steel pipe
Atm. press. = 14.7 psia = 29.9 in. Hg
Pump configuration = End Suction
Speeds to consider = 1800/3600 rpm (Refer to Section XIII of this chapter and also Chapter 6, Section II, for a discussion of speeds to consider for a given application.)

SOLUTION: Using Table 2.1, preliminary sizes of the suction and discharge piping are chosen, based on a design velocity of 4 to 6 ft/sec for the suction

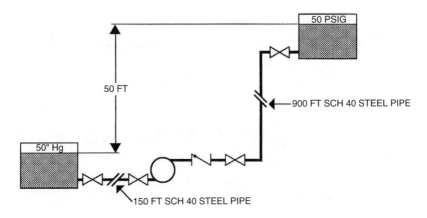

Figure 2.5 System for example problem illustrating selection of line sizes, pump TH, and pump size (Example 2.1); and $NPSH_a$ calculation (Example 2.5).

piping and 6 to 8 ft/sec for the discharge piping. Suction piping size is chosen as 8 in. (Velocity = 4.49 ft/sec per Table 2.1), and discharge piping size is chosen as 6 in. (Velocity = 7.77 ft/sec per Table 2.1).

Note that pressure and static head requirements are given at the liquid surfaces in the suction and delivery vessels. These are the usual reference points for pump TH calculation.

STATIC HEAD: The pump in Figure 2.5 is operating on a suction head. Static head, the total change in elevation from suction vessel surface to delivery vessel surface, equals 50.0 ft.

FRICTION HEAD: Using the selected line sizes, the line lengths shown on Figure 2.5, and data from Table 2.1 and Figure 2.2, the pipe friction head is computed as follows:

Suction:

Line loss
$h_f = 0.80$ for 8 in. pipe (Table 2.1)
$H_f = h_f \times L/100$, from Equation 2.4
$H_f = 0.80 \times 150/100 = 1.20$ ft

Valve loss
Qty. 2 - 8 in. Flanged Gate Valves
$K = 0.07$ (Figure 2.2)
$V^2/2g = 0.31$ (Table 2.1)
$H_f = K \times V^2/2g$ (Equation 2.5) $\times$ Qty
$H_f = 0.07 \times 0.31 \times 2 = 0.04$ ft

Suction friction head = 1.20 + 0.04 = 1.2 ft

Discharge:

Line loss
$h_f = 3.13$ for 6 in. pipe (Table 2.1)
$H_f = h_f \times L/100$, from Equation 2.4
$H_f = 3.13 \times 900/100 = 28.17$ ft

Valve loss
Qty. 2 - 6 in. Flanged Gate Valves
$K = 0.09$ (Figure 2.2)
$V^2/2g = 0.94$ (Table 2.1)
$H_f = K \times V^2/2g$ (Equation 2.5) $\times$ Qty.
$H_f = 0.09 \times 0.94 \times 2 = 0.17$ ft

Valve loss
Qty. 1 - 6 in. Flanged Check Valve
$K = 2.00$ (Figure 2.2)
$V^2/2g = 0.94$ (Table 2.1)
$H_f = K \times V^2/2g$ (Equation 2.5) $\times$ Qty.
$H_f = 2.00 \times 0.94 \times 1 = 1.88$ ft

Fitting loss
Qty. 2 - Regular Flanged 90° Elbows
$K = 0.28$ (Figure 2.2)

$V^2/2g = 0.94$ (Table 2.1)
$H_f = K \times V^2/2g$ (Equation 2.5) $\times$ Qty.
$H_f = 0.28 \times 0.94 \times 2 = 0.53$ ft

Sudden enlargement loss (at delivery tank)
$h = V_1^2/2g$
 (from Figure 2.2 for sudden enlargement)
$h = 0.94$ ft (Table 2.1)

Disch. friction head = 28.17 + 0.17 + 1.88 + 0.53 + 0.94

Disch. friction head = 31.7 ft

Total friction head = 1.2 ft + 31.7 ft = 32.9 ft

PRESSURE HEAD:

Supply vessel pressure = 5 in. Hg
Vac. (ft) = in. Hg. $\times$ 1.133/s.g., from Equation 2.7
Vac. (ft) = 5 $\times$ 1.133/1.0 = 5.7 ft

Delivery vessel pressure = 50 psig
Head (ft) = psi $\times$ 2.31/s.g., from Equation 2.6
Head (ft) = 50 $\times$ 2.31/1.0 = 115.5 ft

Total pressure head = 115.5 + 5.7 = 121.2 ft

Alternately, expressing both pressures in absolute terms, using the given barometric pressure of 29.9 in. Hg, the same result is achieved.

Supply vessel = [(29.9 − 5) $\times$ 1.133]/1 = 28.2 ft

Delivery vessel = 64.7 psia = (64.7 $\times$ 2.31)/1 = 149.4 ft

Total pressure head = 149.4 − 28.2 = 121.2 ft

VELOCITY HEAD: Because the pressure and static head requirements are referenced to the liquid surface in the suction and delivery vessels, the velocity head at the liquid surfaces is zero.

TH = Static + Friction + Pressure + Velocity
TH = 50.0 + 32.9 + 121.2 + 0.0 = 204.1 ft

Using Figure 2.4, the preliminary pump selections are:

1800 rpm: 4 × 6–16
            4 × 6–19
            6 × 8–16
3600 rpm: 3 × 6–9
            4 × 6–9

The individual curves for the preliminary selections should then be examined to evaluate each of the possible choices with regard to horsepower, efficiency, and $NPSH_r$. See discussions on these subjects in Sections V, VI, and XIII to follow. Further analysis should then be done of first cost, power costs to operate, and expected maintenance costs of the alternatives before a final selection is made. Refer to Chapter 6 for an example of this further analysis.

Also, refer to the discussion in Chapter 3, Section III, and the demonstration CD that accompanies this book to learn how computer software can be used to solve the above example problem.

V. HORSEPOWER AND EFFICIENCY

Horsepower refers to the amount of energy that must be supplied to operate a pump. An understanding of how to calculate horsepower and how to read and interpret the horsepower data shown on the pump performance curve is necessary to choose the correct size of driver for the pump. There are several commonly designated expressions for horsepower. *Water horsepower* (WHP) refers to the output of the pump handling a liquid of a given specific gravity, with a given flow and head. The formula for WHP, in USCS units, is:

Hydraulics, Selection, and Curves

$$\text{WHP} = \frac{Q \times H \times \text{s.g.}}{3960} \quad (2.11)$$

where:
Q   = flow rate in gpm
H   = TH in feet
s.g. = specific gravity

The constant 3960 is used when the units are as described in Equation 2.11. The constant is obtained by dividing 33,000 (the number of ft-lb/min in one horsepower) by 8.34 (the number of pounds per gallon of water). If Q is given in cubic feet per second, Equation 2.11 becomes:

$$\text{WHP} = \frac{Q \times H \times \text{s.g.}}{8.82} \quad (2.12)$$

Using SI units, the power WHP in watts is given by:

$$\text{WHP} = 9797 \times Q \times H \times \text{s.g.} \quad (2.13)$$

where:
Q = Flow rate, in cubic meters per second
H = TH, in meters

If Q is given in liters per second, Equation 2.13 becomes:

$$\text{WHP} = 9.797 \times Q \times H \times \text{s.g.} \quad (2.14)$$

Brake horsepower (BHP) is the actual amount of power that must be supplied to the pump to obtain a particular flow and head. It is the input power to the pump, or the required output power from the driver. The formula for BHP, using the same units as Equation 2.11, is:

$$\text{BHP} = \frac{Q \times H \times \text{s.g.}}{3960 \times \eta} \quad (2.15)$$

where:
η = pump efficiency

Other equations for BHP can be written using other USCS or SI units, by taking Equations 2.12, 2.13, or 2.14, and adding pump efficiency, η, in the denominator, as is done in Equation 2.15 above.

BHP is indicated on the pump performance curve as a function of pump capacity, and is used to select an appropriate size of motor (or other driver type) for the pump. Note that the BHP is a function of specific gravity. If the pumped liquid's specific gravity is other than 1.0, the BHP curve should be adjusted accordingly, either by the manufacturer or by the engineer making the motor selection.

Still another horsepower term which is used in studies and discussions of pumping systems is *wire-to-water horsepower*. This term describes the required power input into the driver, and is found by dividing BHP by the motor efficiency. In the case of a pump using a variable-speed device or other auxiliary driving equipment such as a gear box, BHP is divided by the combined efficiency of all of the driver components to obtain the wire-to-water horsepower.

BHP is greater than WHP because of the fact that a pump is not a perfectly efficient machine. There are actually four factors that cause a centrifugal pump to be less than perfectly efficient, as described below.

A. Hydraulic Losses

This term is a summary of internal losses in the impeller and volute or diffuser due to friction in the walls of the liquid passageways and the continual change of direction of the liquid as it moves through the pump.

B. Volumetric Losses

This term refers to the leakage of a usually small amount of liquid from the discharge side of a centrifugal pump to the suction side (the equivalent of slip in a positive displacement pump). The liquid leaks past the wear rings in a closed impeller pump and past the front of the vanes in an open impeller pump. (Refer to Chapter 4, Section II.A.) Volumetric losses increase as internal clearances are opened up due to wear and erosion in the pump. This causes the pump to run less efficiently and increases BHP.

Hydraulics, Selection, and Curves

C. Mechanical Losses

This term refers to the frictional losses that occur in the moving parts of pumps which are in contact (bearings and packing or seals).

D. Disk Friction Losses

If the pump impeller is thought of as a rotating disk, rotating in very close proximity to a fixed disk (the casing), there is a frictional resistance to this rotation known as disk friction.

The pump efficiency is expressed as a decimal number less than one, for example .75 for 75% efficiency. The relative importance of the above four losses varies from one pump type to another. Actual efficiencies for various types of centrifugal pumps can vary widely, over a range from less than 30% to over 90%, for reasons that are explained in more detail in Section XIII.

Comparing Equations 2.11 for WHP and Equation 2.15 for BHP, the only difference between the two is the pump efficiency term. Therefore, the pump efficiency is equal to the ratio of the two:

$$\eta = \frac{WHP}{BHP} = \frac{Q \times H \times s.g.}{3960 \times BHP} \qquad (2.16)$$

The pump manufacturer uses Equation 2.16 to determine the pump efficiency at the time the factory pump performance test is done, as described below.

When a new pump is being designed by a pump manufacturer, there is usually a predetermined objective for the pump's flow and head at the best efficiency point, as well as an expected maximum efficiency. However, it is not until the performance test is run on the prototype that the performance that the manufacturer lists in the catalog for that pump is finally determined. In the early phases of the pump hydraulic design, the designer does calculations to determine the parameters of the design of the impeller and volute or diffuser. These include selection of vane inlet angle, radius of curvature, number of vanes, exit angle, etc.

Note that impeller and volute hydraulic design is beyond the scope of this book. Readers interested in learning more about pump hydraulic design are referred to Refs. [1] through [4] at the end of this book.

With the design parameters selected, the designer can then complete the layout of the impeller and volute or diffuser. This allows creation of pattern and machine drawings for these components. With the completion of the design of the other mechanical components of the pump such as the stuffing box and bearing assembly, a prototype pump can be built and made ready for the performance test.

The manufacturer's pump performance test is usually conducted with the pump mounted on the floor above a large pit or sump filled with water. The pump takes suction from the sump, the flow passes through instrumentation that can measure flow Q and head H, and then the flow is returned back to the sump. (Refer to Chapter 3, Section VI, for a more detailed discussion on how head and flow are measured in the pump test.) The test loop has a throttling valve to allow for variation of the flow and head so that the pump can be run over its full performance range.

Finally, the manufacturer's laboratory test facility has the capability to measure BHP, the power required by the pump. This is done in the laboratory in one of several ways. One common method measures the torque between the pump and motor, and converts this to horsepower by the formula:

$$\text{BHP} = \frac{\text{rpm} \times \text{T}}{5250} \qquad (2.17)$$

where:
T = torque, in ft-lb

Dynamometers are also used to measure torque. A more common approach uses electrical instrumentation to measure the input power drawn by the motor, the wire-to-water horsepower previously discussed. This is then multiplied by the motor efficiency, a known number because the manufacturer is using a calibrated laboratory motor, to produce the motor output horsepower, which is the pump BHP. This is also the

Hydraulics, Selection, and Curves

approach that would be used to measure BHP in a field test of a pump, as described in Chapter 3, Section VI.D.

The pump is turned on in the test loop and the throttle valve is set at an arbitrary position. Then, using the laboratory instrumentation, the values of Q and H are measured, as is BHP using one of the above-described methods. Then, using Equation 2.16 (with s.g. = 1.0 because the test loop contains water), the value of pump efficiency η is determined for that particular point on the pump curve. The data obtained from this test point (Q, H, BHP, η) are recorded, and then the throttle valve is repositioned and a new set of data points is taken. This procedure is repeated over the full range of performance of the pump. Usually, a minimum of five to seven points on the pump curve are measured. The data can then be plotted to create the head–capacity, BHP, and efficiency curves for the pump, using a full-diameter impeller.

Figure 2.6 illustrates typical performance curves generated by the performance test just described. In a typical centrifugal pump, the head–capacity (H–Q) curve typically rises

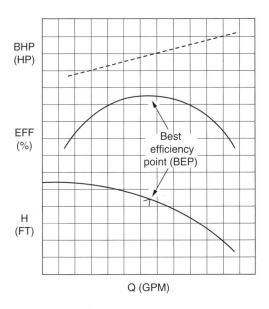

Figure 2.6 H-Q, BHP, and efficiency curves for a pump with a given speed and impeller diameter.

toward shutoff, with the pump developing lower flows at higher heads, and vice versa. The horsepower curve typically is rising as flow increases, although Section VII to follow illustrates that this is not always the case. Finally, the pump efficiency curve shows that the efficiency varies with flow, rising to a peak value known as the best efficiency point (BEP).

It is only after the performance test described above has been completed that the manufacturer finally knows whether the performance objectives for the pump have been reached (i.e., what is the best efficiency flow, head, horsepower and pump efficiency, and how these values vary over the full range of performance of the pump).

The next step for the manufacturer is to perform additional performance tests with reduced impeller diameters. The pump is disassembled, the impeller is machined to a smaller diameter, and then the test described above is repeated. Trim increments may be as little as $1/16$ in. and as much as several inches, depending on the size of the impeller.

After several complete performance tests at different impeller diameters have been conducted by the procedure just described, the manufacturer is able to finally generate the family of curves for the full performance envelope of the pump, which is then published in the manufacturer's catalog. This cataloged family of curves for a pump over its range of offered impeller diameters is illustrated in Figure 2.7.

As exemplified in Figure 2.7, some pump manufacturers display the information on BHP and efficiency on their cataloged curves in a different format from the way Figure 2.6 displays it for a single impeller diameter. In Figure 2.7, the BHP and efficiency data are plotted using *iso curves* (lines of constant BHP and constant efficiency). Other manufacturers simply show a separate efficiency and BHP curve for each impeller trim offered. Either method of displaying BHP and efficiency as a function of pump flow and impeller diameter allows the pump selector to determine the required impeller diameter and motor size for a particular application. Iso-horsepower lines, when they are used, normally show only the commercially available electric motor sizes.

Hydraulics, Selection, and Curves

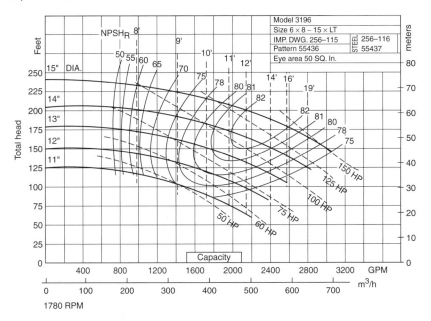

Figure 2.7 Typical manufacturer's published performance curve family for a centrifugal pump operating at a fixed speed and with a range of impeller diameters. (Courtesy of Goulds Pumps, Inc., ITT Industries, Seneca Falls, NY.)

As an example, using the Figure 2.7 pump curve, if the pump design rating is 1800 gpm and 175 ft, the Figure 2.7 curve shows that the required impeller diameter would be just under 14 in. The motor size can then be chosen using the BHP curves in Figure 2.7. If the pump flow is never expected to exceed the design flow rate of 1800 gpm, a motor size of 100 HP can be chosen. Note however that this assumes a specific gravity of 1.0. Remember that if a liquid other than water is being pumped, the BHP curve must be adjusted up or down by the specific gravity of the liquid to be pumped.

As Section IX illustrates, in many cases the pump system allows a pump to operate over a wide range on its H–Q curve. Often, particularly in industrial applications, a motor size is chosen so that the pump can operate over the full range of performance at a given diameter. For the example above, this would lead to a selected motor horsepower of 125 HP. This selection of a motor size to allow operation at any point on the pump curve for a given impeller diameter is known as *nonoverloading motor* selection, and is considered a good selection criterion by most industrial users. A less conservative approach that is acceptable in many lighter-duty applications selects a motor size that is adequate for the design point, and relies on controls or system limitations to keep the pump flow from going beyond that which would overload the motor.

Another approach allows the motor to make use of its *service factor.* The motor service factor is a design margin used in the design of motors, essentially putting more copper in the motor windings to allow the motor to generate more horsepower than the motor is rated for without causing the motor to run excessively hot. Typical service factors for industrial motors are 1.10, 1.15, or 1.2. A 100-HP motor with a 1.15 service factor is actually capable of delivering 115 HP without running so hot that the motor insulation would be harmed or the motor would fail because of excessive heat.

Most conservative industrial users of pumps select motor sizes so that the motor does not make use of the service factor at all (i.e., the motor is chosen to be "nonoverloading" over the entire pump performance range, without making use of the service factor.) This is especially recommended if the pump is to run continuously. This simply means that the motor service factor lets the motor run cooler than it otherwise would. Many lighter-duty fractional horsepower motors have quite high service factors (e.g., 1.6), and it is quite common with residential pumps and other intermittent service or light-duty commercial and industrial applications for the pump to make use of the motor service factor at some points of normal operation on the pump curve.

It is recommended, when sizing and selecting centrifugal pumps, to choose a pump such that the design duty point (head and capacity) is to the left of the BEP on the pump curve. The reason for this is that the vast majority of pumps are oversized, due to the conservatism used by the pump selector in arriving at estimates for total head in the system. Because the actual resulting system head is typically less than that predicted by the engineer at the time of the pump selection process, the pump will tend to move to the right on its performance curve, to a point where the head requirement is less. If the original selection were made to the right of the BEP, the lower than predicted pump head would tend to move the operating point still further to the right on the performance curve. So, it is preferred to make the initial selection to the left of the BEP, so that, if the actual head is less, the pump will move closer to its BEP as it moves to the right on the curve. This is not a hard and fast rule, but the engineer should bear in mind that if a pump is selected at a point well to the right of the BEP on the performance curve, and if the actual pump head is less than predicted, this will allow the pump to move even further to the right, which could lead to problems with overloading the motor or cavitation in the pump.

The one set of curves on Figure 2.7 not yet discussed are the $NPSH_r$ curves. These are discussed in Section VI below.

VI. NPSH AND CAVITATION

A. Cavitation and NPSH Defined

As stated in the overview of this chapter, *NPSH* or (net positive suction head) is probably the most misunderstood aspect of pump hydraulics. It is very important to understand this concept because NPSH problems are among the most common causes of pump failures and are often mistakenly blamed for failures that are completely unrelated.

NPSH must be examined when using centrifugal pumps to predict the possibility of *cavitation*, a phenomenon that has both hydraulic and sometimes destructive mechanical effects on pumps. Cavitation, illustrated in Figure 2.8, is a phenomenon that occurs when vapor bubbles form and move along the vane of an impeller. (What causes the vapor bubbles to form in the first place is discussed shortly.) As these vapor bubbles move along the impeller vane, the pressure around

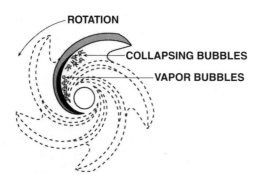

Figure 2.8 Cavitation occurs when vapor bubbles form and then subsequently collapse as they move along the flow path on an impeller.

the bubbles begins to increase. (Figure 1.5 in Chapter 1 shows that the local pressure increases as the flow moves along the path of the impeller vane.) When a point is reached where the pressure on the outside of the bubble is greater than the pressure inside the bubble, the bubble collapses. It does not explode; it implodes. This collapsing bubble is not alone, but is surrounded by hundreds of other bubbles collapsing at approximately the same point on each impeller vane.

The phenomenon of the formation and subsequent collapse of these vapor bubbles, known as *cavitation,* has several effects on a centrifugal pump. First, the collapsing bubbles make a distinctive noise that has been described as a rattling sound, or a sound like the pump is pumping gravel. This can be a nuisance in an extreme situation where a cavitating pump is operating where people are working. This physical symptom is usually the area of least concern with cavitation, however. Of far greater concern is the effect of cavitation on the hydraulic performance and the mechanical integrity of the pump.

The hydraulic effect of a cavitating pump is that the pump performance drops off of its expected performance curve, referred to as *break away,* as illustrated by Figure 2.9, producing a lower than expected head and flow.

An even more serious effect of cavitation is the mechanical damage that can occur due to excessive vibration in the

Hydraulics, Selection, and Curves

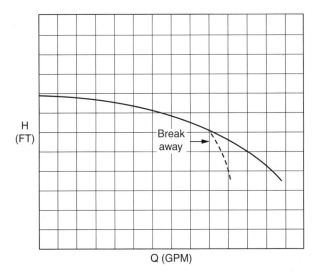

Figure 2.9 Effect of cavitation on the performance of a centrifugal pump.

pump. This vibration is due to the uneven loading of the impeller as the mixture of vapor and liquid passes through it, and to the local shock wave that occurs as each bubble collapses.

The shock waves can physically damage the impeller, causing the removal of material from the surface of the impeller. The amount of material removed varies, depending on the extent of the cavitation and the impeller material. If the impeller is made of ferrous-based material such as ductile iron, material is removed from the impeller due to a combination of corrosion of the ferrous material from the water being pumped and the erosive effect of the cavitation shock waves. If the impeller material is more corrosion resistant but softer, ordinary bronze, for example, the damage that cavitation causes is similar to a peening operation, in which a piece of relatively soft bronze is repeatedly struck with a small ball peen hammer. Materials such as 316 stainless steel, with superior corrosion resistance and ability to work harden under the peening action, have a better ability to resist the metal loss associated with cavitation.

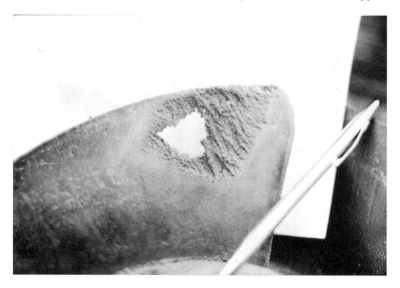

Figure 2.10 Material loss from impeller vane due to cavitation.

In any case, the removal of material, if it occurs at all, proceeds as long as the pump is cavitating. Pits can be formed gradually on the impeller vanes and, in the extreme, the removal of material can actually cause a hole to be eaten clear through an impeller vane, as Figure 2.10 illustrates. This removal of material from the impeller has the obvious effect of upsetting the dynamic balance of the rotating component. The result is similar to what happens if an automobile tire is not properly dynamically balanced, or if it loses one of the balance weights.

It is very important to remember that excessive vibration from cavitation can occur even without the material loss from the impeller described above. This is true because the vibration from cavitation is caused by the uneven loading of the impeller and the local shock wave, as mentioned previously, as well as by the removal of material.

Often, the excessive vibration caused by cavitation subsequently causes a failure of the pump's seal and/or bearings. This is the most likely failure mode of a cavitating pump and the reason why NPSH and cavitation must be properly understood by the system designer and pump user.

What causes the formation of the vapor bubbles in the first place, without which the cavitation would not have a chance to occur? To a person who has never studied thermodynamics, the most obvious way to create vapor bubbles — that is, to make a liquid boil — is by raising the temperature of the liquid. However, this is not what occurs in a cavitating pump because, in the higher flow range where cavitation is likely to occur, the temperature of a liquid as it moves through a centrifugal pump remains very nearly constant.

Another way to make a liquid boil, without increasing its temperature, is if the pressure of the liquid is allowed to decrease. This thermodynamic property of liquids is known as *vapor pressure*. The vapor pressure characteristics of water are illustrated in Table 2.3. Note that vapor pressure can be expressed in psi, or converted to feet of liquid using Equation 2.6. Both values are shown in Table 2.3.

Every liquid has a characteristic vapor pressure that varies with temperature, as Table 2.3 shows for water. Many handbooks carry this data for various liquids. Figure 2.11 shows the vapor pressure for a number of liquids as a function of temperature. As described in Chapter 3, Section III, some computer software generated data tables are available that contain vapor pressure data as well as other properties for many liquids.

For any liquid, as temperature goes up, vapor pressure increases. One way to interpret the vapor pressure data for a liquid is that it shows the temperature at which the liquid boils when it is at a certain pressure. For example, from Table 2.3, we see that at 14.7 psia (atmospheric pressure at sea level), water boils at 212°F (100°C). If the water is subjected to a pressure of 90 psia, the liquid does not boil until it reaches a temperature of 320°F (160°C). This is the principle upon which a pressure cooker is based. With the pressure cooker operating at a pressure above atmospheric pressure, the liquid boils at a much higher temperature than it would in an open pot on the stove, so the food in the pressure cooker cooks faster.

If a liquid is at a certain temperature in a pressurized container, and the pressure in the container is allowed to drop

Table 2.3 Properties of Water at Various Temperatures

Temp. (°F)	Temp. (°C)	Specific Gravity 60°F Reference	Wt. in. lb/cu. ft.	Vapor Pressure (Psi Abs)	Vapor Pressure* Feet Abs. (At Temp.)
32	0	1.002	62.42	0.0885	0.204
40	4.4	1.001	62.42	0.1217	0.281
45	7.2	1.001	62.40	0.1475	0.340
50	10.0	1.001	62.38	0.1781	0.411
55	12.8	1.000	62.36	0.2141	0.494
60	15.6	1.000	62.34	0.2563	0.591
65	18.3	.999	62.31	0.3056	0.706
70	21.1	.999	62.27	0.3631	0.839
75	23.9	.998	62.24	0.4298	0.994
80	26.7	.998	62.19	0.5069	1.172
85	29.4	.997	62.16	0.5959	1.379
90	32.2	.996	62.11	0.6982	1.617
95	35.0	.995	62.06	0.8153	1.890
100	37.8	.994	62.00	0.9492	2.203
110	43.3	.992	61.84	1.275	2.965
120	48.9	.990	61.73	1.692	3.943
130	54.4	.987	61.54	2.223	5.196
140	60.0	.985	61.39	2.889	6.766
150	65.6	.982	61.20	3.718	8.735
160	71.1	.979	61.01	4.741	11.172
170	76.7	.975	60.79	5.992	14.178
180	82.2	.972	60.57	7.510	17.825
190	87.8	.968	60.35	9.339	22.257
200	93.3	.964	60.13	11.526	27.584
212	100.0	.959	59.81	14.696	35.353
220	104.4	.956	59.63	17.186	41.343
240	115.6	.948	59.10	24.97	60.77
260	126.7	.939	58.51	35.43	87.05
280	137.8	.929	58.00	49.20	122.18
300	148.9	.919	57.31	67.01	168.22
320	160.0	.909	56.66	89.66	227.55
340	171.1	.898	55.96	118.01	303.17
360	182.2	.886	55.22	153.04	398.49
380	193.3	.874	54.47	195.77	516.75
400	204.4	.860	53.65	247.31	663.42
420	215.6	.847	52.80	308.83	841.17

Table 2.3 Properties of Water at Various Temperatures (continued)

Temp. (°F)	Temp. (°C)	Specific Gravity 60°F Reference	Wt. in. lb/cu. ft.	Vapor Pressure (Psi Abs)	Vapor Pressure* Feet Abs. (At Temp.)
440	226.7	.833	51.92	381.59	1056.8
460	237.8	.818	51.02	466.9	1317.8
480	248.9	.802	50.00	566.1	1628.4
500	260.0	.786	49.02	690.8	1998.2
520	271.1	.766	47.85	812.4	2446.7
540	282.2	.747	46.51	962.5	2972.5
560	293.3	.727	45.3	1133.1	3595.7
580	304.4	.704	43.9	1325.8	4345.
600	315.6	.679	42.3	1542.9	5242.
620	326.7	.650	40.5	1786.6	6341.
640	337.8	.618	38.5	2059.7	7689.
660	348.9	.577	36.0	2365.4	9458.
680	360.0	.526	32.8	2708.1	11878.
700	371.1	.435	27.1	3093.7	16407.
705.4	374.1	.319	19.9	3206.2	23187.

[a] Vapor pressure in feet of water (Abs.) converted from PSIA using sp. gr. at temperature.

Adapted from *Steam Tables - Thermodynamic Properties of Water, Including Vapor, Liquid, and Solid Phase*, Keenan, 1969. This material is used by permission of John Wiley & Sons, Inc., New York.

below the vapor pressure of the liquid at that particular temperature, the liquid boils. As an example (using Table 2.3), if water at 300°F (148.9°C) is in a vessel which is maintained at a pressure of 100 psia, the water is in a liquid state, i.e., is not boiling. However, if the pressure in the vessel is allowed to drop, when it goes below 67 psia (the vapor pressure at 300°F), the liquid begins to boil.

The above example is exactly analogous to what can occur in a pump system, causing the creation of vapor bubbles and setting up the conditions for the pump to cavitate. In a pump system, as the liquid leaves the supply vessel and approaches the suction of the pump, the local pressure at every point in the suction line varies, due to changes in elevation and

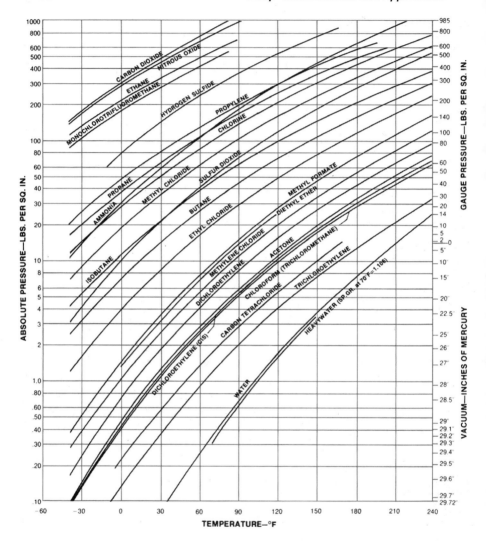

Figure 2.11 Vapor pressure of various fluids as a function of temperature. (Courtesy of Gas Processors Suppliers Association, Tulsa, OK.)

friction in the suction pipe, valves, filters, and fittings. If this combination of changes in local pressure allows the pressure of the liquid to drop below the vapor pressure at the pumping

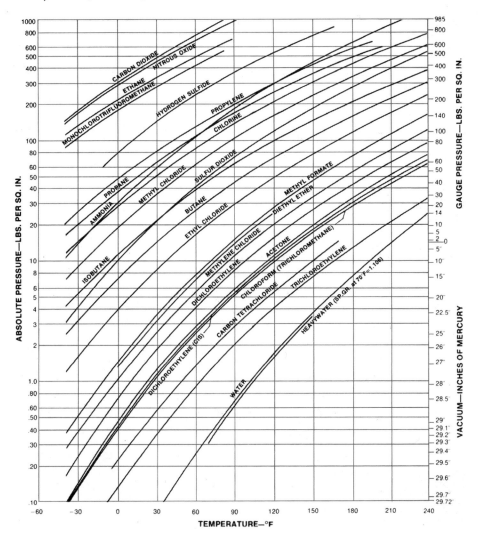

Figure 2.11 (continued)

temperature, vapor bubbles form and the conditions are present for cavitation to commence.

In analyzing a pump operating in a system to determine if cavitation is likely, there are two aspects of NPSH to consider: $NPSH_a$ and $NPSH_r$.

1. $NPSH_a$

Net positive suction head available ($NPSH_a$) is the suction head present at the pump suction over and above the vapor pressure of the liquid. $NPSH_a$ is a function of the suction system and is independent of the type of pump in the system. It should be calculated by the engineer or pump user, and supplied to the pump manufacturer as part of the application criteria or pump specification. The formula for calculating $NPSH_a$ is:

$$NPSH_a = P \pm H - H_f - H_{vp} \qquad (2.18)$$

where:
P = absolute pressure on the surface of the liquid in the suction vessel, expressed in feet of liquid
H = static distance from the surface of the liquid in the supply vessel to the centerline of the pump impeller, in feet; the term is positive if the pump has a static suction head, and negative if the pump has a static suction lift
H_f = friction loss in the suction line, including all piping, valves, fittings, filters, etc., expressed in feet of liquid; this term varies with flow, so $NPSH_a$ must be calculated based on a particular flow rate
H_{vp} = vapor pressure of the liquid at the pumping temperature, expressed in feet of liquid

In a new pump application, $NPSH_a$ (and the static term H in the above formula) must be given to the manufacturer with reference to some known datum point such as the elevation of the pump mounting base. This is because the location of the pump impeller centerline elevation is generally not known when the $NPSH_a$ calculations are made. It is important that the datum point of reference be mentioned in the specification, as well as the calculated value of $NPSH_a$.

Engineers and pump users are sometimes confused as to how to calculate $NPSH_a$ for a closed system, where there is no vented suction tank that establishes a reference point for measuring the static suction head and the pressure at the surface of the liquid, the terms H and P in Equation 2.18. In general, it is not good practice to have a completely closed system with no place for the liquid to expand in the event of temperature fluctuations in the system. Even minor temperature fluctuations can cause the liquid to expand in the system. If there is no place for this liquid to go, it can easily cause system pipes to burst. Good design practice in a completely closed system calls for the introduction of an expansion tank with a bladder in it with air on one side, to allow some expansion and contraction of the liquid due to temperature fluctuations. This not only provides a safety net to keep the pipes from over-pressuring due to the expansion of the liquid, but also provides the reference point needed to begin the $NPSH_a$ calculation. The usual practice is to have the level of the liquid in the expansion tank be the reference point for the calculation of the P and H terms and the beginning point for the determination of H_f in the $NPSH_a$ calculation. An alternative is to have the tank be a vented tank located at the high point in the system. The actual location of the expansion tank is not critical, although usual practice locates it near the suction of the pump. In some cases, if the liquid being pumped is fairly close to its boiling point, it may be necessary to actually pressurize the air side of the expansion tank to increase the overall pressure in the system and provide the liquid with more margin above its boiling point.

2. $NPSH_r$

Net positive suction head required ($NPSH_r$) is the suction head required at the impeller centerline over and above the vapor pressure of the liquid. $NPSH_r$ is strictly a function of the pump inlet design, and is independent of the suction piping system. The pump requires a pressure at the suction flange greater than the vapor pressure of the liquid because merely getting the liquid to the pump suction flange in a liquid state is not sufficient. The liquid experiences pressure losses when it first enters the pump, before it gets to the point on the impeller vane where pressure begins to increase. These losses are caused by frictional effects as the liquid passes through the pump suction nozzle, moves across the impeller inlet, and changes direction to begin to flow along the impeller vanes.

$NPSH_r$ is established by the manufacturer using a special test, and the value of $NPSH_r$ is shown on the pump curve as a function of pump capacity.

For a pump to operate free of cavitation, $NPSH_a$ must be greater than $NPSH_r$. In determining the acceptability of a particular pump operating in a particular system with regard to NPSH, the $NPSH_a$ for the system must be calculated by the system designer or pump user, and then the $NPSH_r$ for the pump to be used must be examined at the same flow rate by looking at this information on the pump curve. This comparison should be made at all possible operation points of the pump, with the worst case usually being at the maximum expected flow, also called the *runout* flow.

Some manufacturers use iso curves to show $NPSH_r$ (e.g., Figure 2.7), similar to the treatment of BHP and pump efficiency. Regardless of how it is shown on the curve, $NPSH_r$ increases at higher flow rates, due to the increased amount of friction loss inside the pump inlet before the liquid reaches the impeller. For some pumps, $NPSH_r$ is also higher with flow unchanged but impeller diameter reduced, as illustrated on Figure 2.17. This is an important point to check, particularly if a reduction in impeller diameter is being contemplated for an installed pump.

To be conservative, the value of $NPSH_a$ should normally be calculated at its minimum. This means that it should be calculated based on the lowest liquid level in the supply vessel (minimum static head or maximum static lift), with the highest friction losses in the suction system (usually at the highest planned capacity), and at the highest expected liquid temperature. At the same time, however, one should not be so conservative in $NPSH_a$ calculations as to unnecessarily restrict the type of pump that can be used. As an example, if a system is being considered that has large changes in static head, and where the low level would only incur infrequently and for very short periods of time, it may not be the best advice to use this low liquid level to calculate the value of $NPSH_a$ given to the pump manufacturer. This is especially the case when doing so might force a different pump than would otherwise be preferred. This could result in a pump selection that might

Hydraulics, Selection, and Curves

cost more, use more energy, or require more maintenance. This points out the importance of good communication between the pump user and supplier when the pump selection is made.

B. Calculating $NPSH_a$: Examples

Several examples illustrate how to calculate $NPSH_a$, how to determine if a pump might experience cavitation, and what to do about it if it does. Also refer to the discussion in Chapter 3, Section III, and the demonstration CD that accompanies this book to learn how computer software can be used to perform these calculations.

EXAMPLE 2.2: Calculating $NPSH_a$ and comparing with published $NPSH_r$.

PROBLEM: Calculate $NPSH_a$ for this system and verify the adequacy of the selected pump.

GIVEN: The system shown in Figure 2.1
Pump shown in Figure 2.7
Suction lift = 12 ft
Design capacity (Q) = 2000 gpm
Design pump total head = 175 ft
Liquid = water at 80°F (s.g. = 1.0)
H_f = 3 ft (Usually, this is calculated using friction tables as illustrated in Example 2.1 in Section IV of this chapter.)
P = 14.2 psia (atmospheric pressure at the pump site)

SOLUTION: $NPSH_a = P \pm H - H_f - H_{vp}$ (Equation 2.18)
P = 14.2 psia (given)
= (14.2 × 2.31)/1.0 ft
= 32.8 ft
H = −12 ft (given; negative because it is a suction lift)
H_f = 3 ft (given)
H_{vp} = 1.2 ft (Table 2.3 at 80°F)
$NPSH_a$ = 32.8 − 12 − 3 − 1.2 = 16.6 ft

From Figure 2.7, at 2000 gpm, $NPSH_r = 11.2$ ft, so $NPSH_a > NPSH_r$ and the chosen pump is acceptable from an NPSH point of view (i.e., does not cavitate).

EXAMPLE 2.3: Calculating $NPSH_a$ and comparing with published $NPSH_r$.

PROBLEM: Calculate $NPSH_a$ for this system and verify the adequacy of the selected pump.

GIVEN: The problem presented in Example 2.2 above, except using water at 160°F

SOLUTION: $NPSH_a = P \pm H - H_f - H_{vp}$ (Equation 2.18)
s.g. = 0.98 (Table 2.3 at 160°F)
P = 14.2 psia (given)
  = (14.2 × 2.31)/.98 ft
  = 33.5 ft
H = –12 ft (given; negative because it is a suction lift)
H_f = 3 ft (given)
H_{vp} = 11.2 ft (Table 2.3 at 160°F)
$NPSH_a$ = 33.5 – 12 – 3 – 11.2 = 7.3 ft

From Figure 2.7, at 2000 gpm, $NPSH_r = 11.2$ ft, so $NPSH_a < NPSH_r$ and the chosen pump is unacceptable from an NPSH point of view (i.e., the pump would cavitate). In fact, most pumps in the flow range of 2000 gpm require more than 7.3 ft of NPSH, so it would be difficult to locate a standard commercially available pump for this service as described.

C. Remedies for Cavitation

To make the proposed pump in Example 2.3 above acceptable from an NPSH point of view, there are a number of system modifications that might be considered to increase the calculated $NPSH_a$. These alternatives might include system changes that affect any of the four terms of Equation 2.18, or some combination of them. The most obvious system change would be to change the value of H to a smaller negative number by raising the minimum liquid level in the suction supply vessel, or by moving the pump to a lower elevation.

A second system change might involve reducing the value of H_f by making the suction piping shorter (moving the pump closer to the supply vessel). Alternative system modifications could involve making the suction pipe size larger, eliminating fittings or valves in the suction line, or, if there is a filter in the suction line, changing the filter design to a type that does not permit as high a differential pressure buildup between cleaning operations.

A third system change might involve reducing the temperature of the process liquid (because it is the higher temperature that causes the pump in Example 2.3 to be unacceptable from an NPSH point of view).

Finally, the first term in Equation 2.18 could be increased by pressurizing the tank with a blanket of air, nitrogen, or other compatible gas above the liquid surface. The gas blanket could be of fairly low pressure (only several pounds of pressure). However, a blanket gas as a substitute for a static head tank in a pump system should be used with caution. If some of the blanket gas is permitted to dissolve in the liquid (a function of the gas solubility), the dissolved gas can be liberated in the low pressure area at the entrance of the pump, and this can reduce the effective $NPSH_a$. The amount of additional margin between $NPSH_a$ and $NPSH_r$ to account for the release of this dissolved gas can be substantial (Ref. [6]). Note that to the author's knowledge, none of the software programs discussed in Chapter 3 for modeling piping systems take into account the possible effects of dissolved gas on the $NPSH_a$ calculation.

If no combination of the changes to the system suggested in the preceding paragraphs can be done, or if they are not adequate to raise the value of $NPSH_a$ to a level where it exceeds $NPSH_r$, there may be other options to consider. Recall from the definitions of $NPSH_a$ and $NPSH_r$, which are given prior to the examples above, that both $NPSH_a$ for a system and $NPSH_r$ for a pump vary with flow, with $NPSH_a$ decreasing and $NPSH_r$ increasing at higher flows. This relationship is

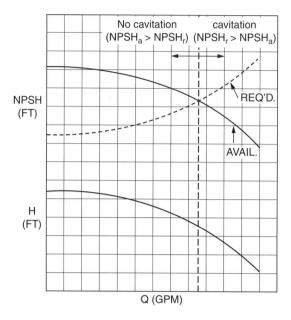

Figure 2.12 Variation of $NPSH_a$ and $NPSH_r$ with flow through the system.

shown in Figure 2.12. At flows higher than the intersection point of the two curves in Figure 2.12, the $NPSH_r$ would exceed the $NPSH_a$ and so the pump in that system would cavitate. This suggests one rather rudimentary way to solve the problem of a pump in the field that is cavitating, or the problem found in Example 2.3 above where a proposed pump is unacceptable from an NPSH point of view. If a lower flow rate is acceptable from a process point of view, it may be possible to merely throttle the pump back to a lower flow rate by closing down on a valve in the pump discharge, until the flow is reduced to the point where $NPSH_r$ is reduced and $NPSH_a$ is increased enough that the pump no longer cavitates.

If none of the above proposed solutions are adequate to solve the problem, then the only way to eliminate the cavitation problem described here is with a different pump. There may be a commercially available pump that has a lower

Hydraulics, Selection, and Curves

$NPSH_r$. If a pump is available to make the same head and flow but at a slower speed, the slower speed pump would most likely have a lower $NPSH_r$ (although it would also be a physically larger pump, and would likely be more expensive). Another possibility that may be available without going to a slower speed is a double suction pump. (Refer to Chapter 4, Section II.B.)

There are also specially designed low $NPSH_r$, single suction pumps available from certain manufacturers in limited styles and sizes of process pumps. These pumps achieve the lower value of $NPSH_r$ by having a special impeller design with an increased inlet (eye) area. While this solution may be the most economical choice, or perhaps the only alternative in certain circumstances, care should be taken when using these special impeller designs. The larger inlet area can restrict the flow range over which the pump can operate successfully to a smaller range than a comparable pump with a "standard"-sized impeller inlet. The geometry of the impeller inlet is described by a dimensionless variable called *suction specific speed*. This subject is discussed again in Section VII and in Chapter 4, Section II.C.

It may be necessary to divide the flow between two pumps operating in parallel to solve the problem raised by this example. This would be especially worth considering if there is a range of flow requirement. Or, it may be possible to add a second pump in series upstream of the first one, acting as a booster pump to increase the $NPSH_a$ at the inlet of the second pump. Applications involving multiple pumps in parallel or series are discussed further in Sections X and XI to follow.

If all possible affordable alternatives have been explored for modifying the system and for changing the pump, and still no solution has been found, in some cases there is no choice but to live with the cavitation. For example, if the expected flow, temperature, friction losses, and liquid levels are such that it is estimated that the pump will cavitate during only 5 to 10% of the time it operates, the designer or user may simply choose to live with this. Also, it may be possible to minimize the damage from the cavitation by selecting an impeller material that is more resistant to cavitation damage, such as a high chrome stainless steel.

Also, in considering the possibility of designing a system with a pump which may be operating in a cavitating mode for certain periods of time, recall that part of the cavitation damage occurs as a result of the shock wave caused when the vapor bubbles collapse. It turns out that the amplitude of the shock wave produced in a cavitating pump is directly proportional to the value of the pump $NPSH_r$. Thus, a pump that requires only 6 ft of NPSH would cause much less damage when cavitating than would a cavitating pump of the same metallurgy that requires 25 ft of NPSH.

D. More $NPSH_a$ Examples

Several more examples help to further solidify the reader's understanding of NPSH.

Example 2.4: Calculation of $NPSH_a$ and suction pressure.

PROBLEM: Calculate $NPSH_a$ for this system.

GIVEN: The system shown in Figure 2.13
Suction head = 7 ft
Liquid = water at 300°F
H_f = 2 ft
P = 75 psia
Atmospheric pressure = 14.7 psia

SOLUTION: $NPSH_a = P \pm H - H_f - H_{vp}$ (Equation 2.18)
s.g. = 0.92 (Table 2.3 at 300°F)
P = 75 psia (given)
  = (75 × 2.31)/0.92 ft
  = 188.3 ft
H = 7 ft (given; positive because it is a suction head)
H_f = 2 ft (given)
H_{vp} = 168.2 ft (Table 2.3 at 300°F)
$NPSH_a$ = 188.3 + 7 − 2 − 168.2 = 25.1 ft

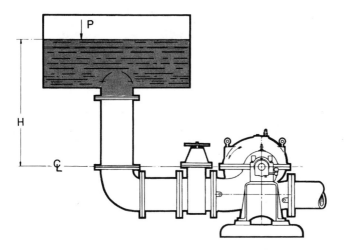

Figure 2.13 System for example problem on calculating $NPSH_a$ and suction pressure (Example 2.4). (Courtesy of Goulds Pumps, Inc., ITT Industries, Seneca Falls, NY.)

If the pressure in the supply vessel in the proceeding problem were changed to 67 psia instead of 75 psia, the first term in the $NPSH_a$ calculation would be $67 \times 2.31/0.92 = 168.2$ ft. This would exactly cancel out the fourth term, H_{vp}, because 67 psia is also the vapor pressure of water at the pumping temperature of 300°F. In that case,

$$NPSH_a = 7 - 2 = 5 \text{ ft}$$

In the case of the example above, the pressure in the supply vessel could get no lower than 67 psia because if it did, the vessel contents would no longer be in a liquid state. The situation just considered, where the pressure in the supply vessel was changed to 67 psia, illustrates an application involving pumping a *saturated liquid*. This is a common application in industry, for example, a condensate pump in a power plant. When a saturated liquid is to be pumped, the first and fourth terms of the $NPSH_a$ formula (P and H_{vp}) cancel out each other, so the $NPSH_a$ is only equal to:

$$NPSH_a = H - H_f \text{ (for saturated liquids)} \quad (2.19)$$

Because H_f is always subtracted, the only term that adds positively to $NPSH_a$ is H, which is why the condensate pump must be located at the lowest possible elevation in the plant to get the value of H as large as possible.

Some people mistakenly believe that as long as a pump has a reasonable amount of suction pressure (regardless of temperature), then cavitation is not a concern. This is not true, as an example illustrates. Suppose we wish to calculate what suction pressure would be measured by a gauge located at the centerline of the suction of the pump in Example 2.4 above, but operating with a supply vessel pressure of 67 psia. Ignoring velocity head, which is a very small term, the expression for suction pressure is:

$$\text{Suction pressure} = P + H - H_f \quad (2.20)$$

For P = 67 psia (168.2 ft), H = 7 ft, and H_f = 2 ft

$$\text{Suction pressure} = 168.2 + 7 - 2 = 173.2 \text{ ft}$$

Converting to psi, 173.2 × 0.92/2.31 = 69.0 psia = 54.3 psig

Although suction pressure is shown above to be well over 50 psig, $NPSH_a$ was calculated in Example 2.4 as only 5 ft, so cavitation may be a problem with this system. The important point here is that one should not confuse the existence of seemingly adequate suction pressure with the question of whether or not a pump cavitates in a given service. Cavitation does not depend on suction pressure, but rather on the relative size of $NPSH_a$ for a system and $NPSH_r$ for a pump operating in that system.

EXAMPLE 2.5: Calculation of $NPSH_a$ with suction source under vacuum.

PROBLEM: Calculate $NPSH_a$ for this system.

GIVEN: The system shown in Figure 2.5
Suction head = 10 ft
Liquid = water at 140°F
H_f = 1.4 ft

Vacuum in supply tank = 5 in. Hg
Atm. press. = 14.5 psia = 29.5 in. Hg

SOLUTION: $NPSH_a = P \pm H - H_f - H_{vp}$ (Equation 2.18)
s.g. = 0.985 (Table 2.3 at 140°F)
$P = (29.5 - 5) \times 1.133/0.985 = 28.2$ ft
(Equation 2.7)
H = 10 ft (given; positive because it is a suction head)
H_f = 1.4 ft (given)
H_{vp} = 6.8 ft (Table 2.3 at 140°F)
$NPSH_a = 28.2 + 10 - 1.4 - 6.8 = 30.0$ ft

E. Safe Margin $NPSH_a$ vs. $NPSH_r$

An often-asked question is: "What is a safe margin to maintain between $NPSH_a$ and $NPSH_r$?" Unfortunately, like so many questions related to pumps, the answer must begin with "That depends...." Section VI.C above covers the fact that for pumps with very low $NPSH_r$ values and with impellers constructed of cavitation resistant materials, it is possible to operate under some amount of cavitation for an extended period of time without causing damage to the pump impeller (although the other symptoms of cavitation — noise, vibration, and the drop off of the H–Q curve from its expected path — would still be present). This situation is the exception rather than the rule, however. For the majority of applications, it is good practice to have a safe margin between the available and required NPSH.

Fortunately, most pump applications have a safe margin, with an $NPSH_a$ of at least 10 ft above that required by most commercially available pumps. In most circumstances, it is usually only process applications with liquids close to their boiling point, or high flow applications (above 15,000 gpm) that present NPSH problems.

The way that the $NPSH_r$ curve is developed by the manufacturer involves a special test, done in a different laboratory test setup from the pump performance test described in Section V. The NPSH test is usually done in a closed loop piping system where the manufacturer has the capability not only to measure head and flow, but also to vary the $NPSH_a$. Varying the $NPSH_a$ is usually accomplished by means of pulling a vacuum at the top of the supply tank (reducing the value of P in the $NPSH_a$ equation) while maintaining a constant flow and continuously monitoring head. When the $NPSH_a$ is reduced to the point that the pump begins to cavitate, the pump curve begins to drop off, as described earlier and shown in Figure 2.9. Because flow is being held constant during the test, this means that the pump TH begins to drop.

According to the convention established in the Hydraulic Institute (H.I.) Standards (Ref. [2]), when the head drops by 3%, the pump is presumed to be in full cavitation; 3% was chosen by the writers of the H.I. Test Code because when cavitation first begins, the head fluctuates up and down and is difficult to measure. This figure (3%) was chosen by the H.I. as an amount of head reduction that would unmistakably mean that the pump is cavitating. When the head drops by 3%, the pump is defined to be in full cavitation and the $NPSH_a$ in the laboratory setup at that point is set equal to $NPSH_r$. This establishes one point on the $NPSH_r$ curve for that pump.

The pump control valve is then repositioned and the test is repeated at a different flow rate. After several test points are taken, the $NPSH_r$ curve can be drawn for the full impeller diameter. This NPSH test is repeated for each impeller trim when the manufacturer is developing the catalog curve.

If a pump's $NPSH_r$ curve is developed by the test described above, and if the pump is operated in a system where $NPSH_a$ exactly equals $NPSH_r$, the pump would actually be cavitating and the performance curve would drop off, with head dropping by 3% from what it should be. This emphasizes the need for a margin between $NPSH_a$ and $NPSH_r$.

Note that for some pumps, the $NPSH_r$ curve with a trimmed impeller rises more steeply at higher flows than it does with a full diameter impeller (see Chapter 3, Figure 3.2). This higher $NPSH_r$ often occurs with impeller diameter reductions greater than about 10%. Users should be aware of this if, for example, they choose to trim an impeller in the field to lower the head produced by the pump at a given flow. If this pump has a characteristic $NPSH_r$ curve that is steeper with trimmed impellers, the $NPSH_r$ may actually be higher at the same flow rate with the trimmed impeller than it was with a full diameter impeller.

The effect of impeller trim on $NPSH_r$ can be predicted by studying the manufacturer's family of curves for the pump in question. If the $NPSH_r$ curves are presented as iso curves, they are completely vertical if there is no effect on $NPSH_r$ from trimming the impeller (as in the case of Figure 2.7). They curve to the left for trimmed impeller diameters if the pump is one that has higher $NPSH_r$ at trimmed impeller diameters (as in the case of Figure 3.2). Unfortunately, not all pump manufacturers consistently show on their curves the increased $NPSH_r$ with reduced impeller diameter.

Another consideration when deciding the margin that should be maintained between $NPSH_a$ and $NPSH_r$ is that testing and experience have shown that damage due to cavitation when pumping cold water is more severe than it is when pumping hot water and certain other liquids. Figure 2.14 shows a NPSH Reduction Chart, which allows the determination of a correction factor to reduce cataloged $NPSH_r$ (or increase calculated $NPSH_a$) under certain conditions. If the liquid is one of the liquids shown in this chart, entering the temperature or vapor pressure permits determination of a correction factor that effectively allows a closer margin between $NPSH_a$ and $NPSH_r$. However, users of this chart are cautioned that there are a number of limitations that apply to the use of the chart.

No NPSH reduction should exceed the lesser of 50% of the $NPSH_r$ with cold water, or 10 ft. Reductions are not applicable or may be different if there is entrained air or other noncondensable gas in the liquid, or if the system has transient changes in temperature or pressure. Vapor pressures of hydrocarbon mixtures should be determined by the bubble point method at pumping temperature, rather than using the Reid vapor pressure or the vapor pressure of the lightest fraction. Finally, use of the chart for liquids other than those

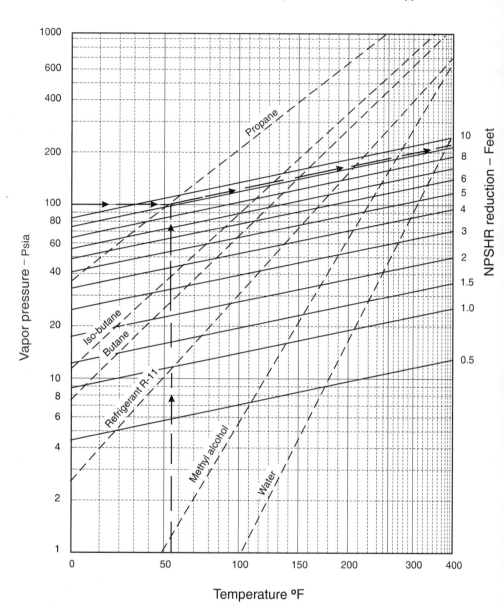

Figure 2.14 NPSH reduction chart. (Courtesy of the Hydraulic Institute, Parsippany, NJ; www.pumps.org and www.pumplearning.org.)

Hydraulics, Selection, and Curves

specifically shown on the chart must be considered experimental only. (See Ref. [1].)

Because of all the restrictions and hedges given for the use of this chart, many users and engineers have simply decided to not take advantage of the additional correction to NPSH that the chart allows. The corrections should not be made in any event without consulting with the pump manufacturer, who may have more NPSH test data for that particular pump. Not using the correction factors allowed by the NPSH reduction chart is a more conservative approach and should add an additional margin of safety between $NPSH_a$ and $NPSH_r$ for some liquids.

Mention has already been made of the fact that if there are dissolved gases in the liquid, then depending on the solubility of the gases, there may be a substantial additional margin required between $NPSH_a$ and $NPSH_r$ to account for the release of the dissolved gases in the low pressure area at the impeller inlet (Ref. [6]).

Another point to consider is how oversized is the pump's bearing system (bearing load capability and shaft diameter) for the application. A very conservative bearing system for a given application could tolerate greater loads on the impeller from cavitation without deflecting the shaft at the mechanical seal area than a less conservative bearing system design. As an example, a typical ANSI process pump line (see Chapter 4, Section XIV.B) might have 20 to 25 sizes of pumps, but only use three designs of bearing frames for the entire line. Thus, each bearing frame is used for a range of approximately eight pump sizes. When a bearing frame is being used on the smallest of the eight sizes that it can accommodate, its design is much more conservative than if it is used for the largest of the eight sizes. So, when it is used for the smallest of the eight sizes, it should be able to tolerate relatively greater impeller loads due to cavitation without causing vibration or deflection of the shaft at the mechanical seal.

In summary, when considering the margin that should be maintained between $NPSH_a$ and $NPSH_r$ for a particular application, the questions to ask include:

- How conservatively was $NPSH_a$ calculated for the system, and for what percentage of the pump's duty cycle is this low value of $NPSH_a$ actually present?
- What is the level of NPSH and thus the relative amplitude of the cavitation shock waves?
- What is the pump impeller material, and how resistant is it to cavitation damage?
- Is the liquid pumped one of those that allows a reduction of $NPSH_r$ by the chart in Figure 2.14?
- How oversized is the bearing system for the application?
- Does the pump system make use of a gas blanket that may become dissolved in the liquid and subsequently liberated in the low pressure area of the impeller inlet?

Depending on the answers to these questions, the recommended minimum margin between calculated $NPSH_a$ and $NPSH_r$ can range from 0 to 35%. A conservative rule of thumb is that $NPSH_a$ should exceed $NPSH_r$ by a minimum of 5 ft, or be equal to 1.35 times the $NPSH_r$, whichever is the greater value. For example, for an $NPSH_r$ of 10 ft, $NPSH_a$ should be a minimum of 15 ft. As pointed out at the beginning of this discussion, good engineering practice is to have a safe margin at all times if possible, and to add more if it can be easily and economically done. As to the acceptable margin for any particular application, the material in this chapter should give the engineer and pump user the tools to help make informed decisions in this regard.

F. NPSH for Reciprocating Pumps

Before leaving the subject of NPSH, a final point should be made regarding an additional factor to be calculated when a reciprocating positive displacement pump is being considered. This type of pump is also subject to cavitation damage and thus a calculation of $NPSH_a$ must be made and supplied to the pump manufacturer. With this type of pump, there is an additional term in the formula for $NPSH_a$, called *acceleration head*. Acceleration head is caused by the rapid deceleration

Hydraulics, Selection, and Curves

and acceleration of the liquid in the suction line as the suction check valves open and close in the pump. The calculated value of acceleration head must be subtracted from the calculated $NPSH_a$ when a reciprocating pump is being considered. Acceleration head H_a (in feet) is computed for a given pump and system combination as follows (Ref. [1]):

$$H_a = \frac{L \times V \times \text{rpm} \times C}{g \times K} \quad (2.21)$$

where:
L = length of suction pipe, ft
V = velocity in suction pipe, ft/sec
rpm = pump rpm
C = factor for number of pistons or plungers

No. Pistons/Plungers	C
2	0.115
3	0.066
4	0.080
5	0.040
6	0.055
7	0.028

g = acceleration of gravity, 32 ft/sec^2
k = factor for liquid type

Liquid	K
Water	1.4
Petroleum	2.5
Liquid with entrained gas	1.0

The acceleration head for reciprocating pump systems can often be as large a factor as all of the other terms in the $NPSH_a$ formula combined, so it should by no means be ignored. If the analysis of a proposed reciprocating pump and system indicates that NPSH may be a problem, possible solutions might include reducing the length or increasing the diameter of the suction line, slowing down the pump, or installing a pulsation dampener in the suction line. Pulsation dampeners were described in Chapter 1, Section VI.C.12 to reduce discharge pulsations caused by reciprocating pumps; but in this case, their function is to help absorb the acceleration and deceleration of the liquid in the suction line.

VII. SPECIFIC SPEED AND SUCTION SPECIFIC SPEED

Specific speed (N_s) is a design index primarily used by pump designers. It is a dimensionless index used to describe the geometry of pump impellers and to classify them as to their type. An understanding of how to calculate and interpret the specific speed for a particular pump provides greater insight into the reasons why pump impellers are shaped so differently, why different pumps have differently shaped performance curves, and why there is such wide variation in the value of efficiency at the BEP for different pumps. Furthermore, Chapter 6, Section II, shows how it is possible to use specific speed to select pumps for maximum efficiency.

The formula for pump specific speed N_s, in USCS units, is:

$$N_s = \frac{N \times \sqrt{Q}}{H^{3/4}} \quad (2.22)$$

where:
N = pump speed, rpm
Q = capacity at BEP, full diameter, gpm
H = pump head per stage at BEP, full diameter, ft

In the U.S. pump industry, Q in Equation 2.22 is taken as the full pump capacity, for either a *single suction* or a *double suction impeller* (Chapter 4, Section II.B). This is not the case for *suction specific speed* (discussed later in this section), where the value of Q in Equation 2.23 for a double suction impeller is one half the total pump capacity. European practice uses one half of total pump capacity for both terms with a double suction impeller.

Table 2.4 Specific Speed N_s for Selected Pumps

Q	H	N	N_s
120	300	3550	540
350	300	3550	1250
1000	100	1780	1780
70	20	3550	3140
9000	40	1180	7040
50000	20	590	13960

An analysis of the units in Equation 2.22 reveals that the term N_s is not truly dimensionless, although it would become so with the addition of g, the acceleration of gravity, into the equation's denominator. The convention in the centrifugal pump industry is to omit the g term but still treat N_s as dimensionless. In SI units, the specific speed is designated N_{sm} and is usually based on capacity expressed in cubic meters per hour and head in meters. Therefore, $N_s = 0.8609 N_{sm}$. If capacity is expressed in cubic meters per second and head in meters, then $N_s = 51.65\ N_{sm}$.

The specific speed of a particular pump can be calculated from the pump curve; picking N, Q, and H off the curve at full diameter, best efficiency point; and applying Equation 2.22. Once N_s for a particular pump has been calculated, its value will not change, even if the pump is run at a different speed. Obviously, if the pump is run at a different speed, the pump's total head and capacity do change but the specific speed does not change, because it is defined by the equation above. In fact, it is the fact that the specific speed will not change that is the basis for the derivation of the pump affinity laws (Section VIII), which allow the pump performance to be predicted for changes in pump speed or impeller diameter.

Table 2.4 shows the calculated value of specific speed based on Equation 2.22 for some arbitrarily selected pump BEP conditions. The data is arranged in order of increasing N_s, the last column in the table. The data is not, however,

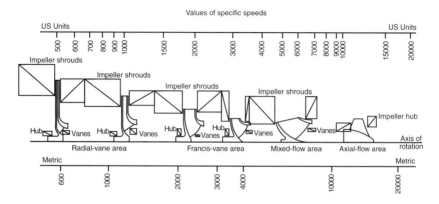

Figure 2.15 Impeller profile vs. N_s. (Courtesy of the Hydraulic Institute, Parsippany, NJ; www.pumps.org and www.pumplearning.org.)

arranged in order of increasing Q or decreasing H. This is because the formula for N_s depends on all three variables, rather than on any one variable. For example, the fourth entry in Table 2.4 has the smallest Q value of all the entries. Yet because the value of H is so small for that pump, the value of N_s is higher than the others above it in the table.

Figure 2.15 illustrates typical impeller profiles for the range of specific speeds found for centrifugal pumps. At the low end of the range, impellers develop head by moving the liquid radially from the shaft centerline. These low specific speed pumps are called *radial flow* pumps. Radial flow pumps have the characteristic of relatively low flow and high head. At the opposite end of Figure 2.15, impellers develop head through axial forces, and so these high specific speed pumps are referred to as *axial flow* pumps, or *propeller* pumps. Axial flow pumps have the characteristic of relatively high flow and low head. As N_s increases, the ratio of the impeller outlet diameter to inlet diameter decreases. This ratio becomes 1.0 for a true axial flow impeller.

Pumps that are neither pure radial flow nor pure axial flow are called *mixed flow*, and represent some combination of radial flow and axial flow. Actually, because the total specific

Hydraulics, Selection, and Curves

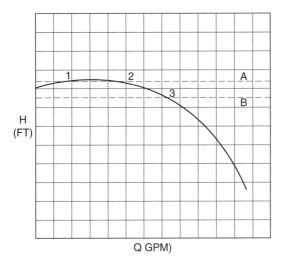

Figure 2.16 Drooping H–Q curve.

speed range for centrifugal pumps is a continuous spectrum from several hundred to about 20,000, pumps with N_s ranging from several hundred to about 2000 are often referred to as radial flow; pumps with N_s greater than about 8000 are called axial flow; and pumps with N_s between these two ranges are called mixed flow.

One of the characteristics of specific speed is its effect on the shape and slope of the pump head–capacity and BHP curves. Radial flow pumps have the flattest H–Q curves, with the head at zero flow (called the *shutoff* head) often no more than about 120% of the head at BEP. The lower the specific speed, the flatter the pump H–Q curve. Note that the slope of the H–Q curve is also affected by certain design parameters in the impeller design such as the number of vanes and the vane angles.

Low specific speed pumps sometimes exhibit a drooping characteristic at shutoff, as illustrated in Figure 2.16. This may lead to unstable operation in certain systems, although it may be perfectly acceptable for use in other circumstances. The dashed lines shown as A and B in Figure 2.16, and the conditions that might lead to instability with a drooping H–Q

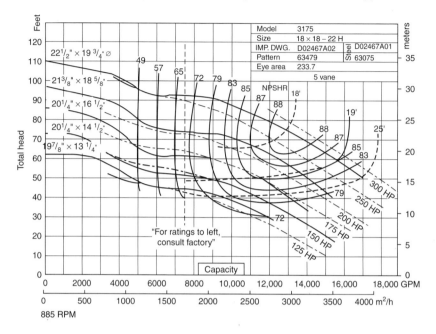

Figure 2.17 Dip in H–Q curve. (Courtesy of Goulds Pumps, Inc., ITT Industries, Seneca Falls, NY.)

curve such as shown in this figure, are discussed in Section IX on system head curves.

Mixed flow pumps have steeper H–Q curves than radial flow pumps. The head at shutoff for mixed flow pumps is around 160% of the head at BEP. Axial flow pumps have the steepest H–Q curves of all, with shutoff head being in the range of 300% of head at BEP. Another characteristic of some mixed flow pumps is a dip in the H–Q curve (Figure 2.17). This may or may not be a problem, depending on the shape of the system head curve (see Section IX). Some manufacturers simply do not show the dip on their published curves, but stop the H–Q curve short of going back to zero flow, with a notation that the pump should not be run in the unstable region.

The BHP curve shape is also affected by specific speed. Radial flow pumps exhibit increasing horsepower with increasing pump flow, with the maximum BHP occurring at

Hydraulics, Selection, and Curves   121

the maximum flow at which the pump can operate in the system (called *runout*). Because most process and transfer pumps are in the radial flow specific speed range, this is the shape of the horsepower curve with which the majority of pump engineers and users are most familiar. Mixed flow pumps have a flatter horsepower curve, and axial flow pumps have their horsepower curve shaped just the opposite of radial flow pumps, with the highest horsepower at the lowest flow. The BHP at shutoff for an axial flow pump is in the range of twice the BHP at BEP. Axial flow pumps are generally not run at low flows, in part because of this higher horsepower at lower flows. Furthermore, if it has a BHP curve that rises toward shutoff, the pump is started with the pump discharge valve open rather than closed or nearly closed, which is the usual valve position when the pump is started. In most cases, the motor for axial flow pumps is not sized to handle the higher horsepower at lower flows. Attempting to start the pump with the valve closed would cause the motor to overload.

Suction specific speed is a nondimensional index used to describe the geometry of the suction side of an impeller, or its $NPSH_r$ characteristics. The term "suction specific speed" is designated S, and its formula, very similar to the formula for specific speed, is:

$$S = \frac{N \times \sqrt{Q}}{NPSH_r^{3/4}} \qquad (2.23)$$

As with specific speed, the terms in the equation above are taken at BEP, full diameter. The value of Q in Equation 2.23 for a double suction impeller (see Chapter 4, Section II.B) is taken as one half the total pump capacity.

Typical values of S for most standard designed impellers are in the range of 7000 to 9000. While there are special impeller designs available from some manufacturers having higher S values (Section VI.C and Chapter 4, Section II.C), various sources recommend that, except in special circumstances, pumps be chose with S values below a maximum value. The maximum recommended value varies with the source, but a common recommendation is a maximum value for S of 8500 (Ref. [2]). Using higher values of S as an impeller design criteria will result in lower $NPSH_r$, but could introduce problems of recirculation (discussed in Chapter 8, Section III.B) if the pump operates at flows less than the BEP flow.

VIII. AFFINITY LAWS

The pump affinity laws are rules that govern the performance of a centrifugal pump when the speed or impeller diameter is changed. The basis for the derivation of the affinity laws is that a pump's specific speed, once calculated, does not change. If the performance of a pump at one speed and impeller diameter are known, it is possible to predict the performance of the same pump if the pump's speed or impeller diameter is changed.

There are two sets of affinity laws. With the impeller diameter, D, held constant, the first set of laws is as follows:

$$\frac{Q_1}{Q_2} = \frac{N_1}{N_2} \tag{2.24}$$

$$\frac{H_1}{H_2} = \left(\frac{N_1}{N_2}\right)^2 \tag{2.25}$$

$$\frac{BHP_1}{BHP_2} = \left(\frac{N_1}{N_2}\right)^3 \tag{2.26}$$

With the speed, N, held constant, the second set of laws is:

$$\frac{Q_1}{Q_2} = \frac{D_1}{D_2} \tag{2.27}$$

$$\frac{H_1}{H_2} = \left(\frac{D_1}{D_2}\right)^2 \qquad (2.28)$$

$$\frac{BHP_1}{BHP_2} = \left(\frac{D_1}{D_2}\right)^3 \qquad (2.29)$$

where:
Q = capacity
H = total head
BHP = brake horsepower
N = pump speed
D = impeller outside diameter

The manufacturer or user makes use of the second set of affinity laws above to calculate the exact impeller trim on a pump to make a particular rating, if the performance at specific impeller diameters bracketing the rating is known, or if the performance at only one impeller diameter is known and performance at another diameter is to be determined. Realistically, for most applications, the required impeller trim can be "eyeballed" by interpolating between the diameters shown on the family of H–Q curves for the pump.

Note that the use of the affinity laws to calculate impeller trims generally produces results that dictate more of an impeller trim than is actually required to produce the desired head and flow reduction, with the amount of error being as high as 15 to 20% in some cases. Lower specific speed pumps generally tolerate a greater amount of impeller trim and still follow the affinity laws than do higher specific speed pumps. The reasons for the deviation from the expected results are complex, related to the fact that the impeller shrouds are not completely parallel with each other in most pumps, that the vane exit angle is altered as the impeller is trimmed, and other design-related reasons that are beyond the scope of this book.

Perhaps a more significant and practical use of the affinity laws is the use of the first set of laws, Equations 2.24, 2.25, and 2.26, to determine performance for a pump which operates at more than one speed. The increasing popularity of variable-speed pumping systems (see Chapters 6 and 7) requires the capability of generating new H–Q and BHP curves for a pump running at a different speed than the published or tested curve for that pump.

The following practical example serves to illustrate the value of the affinity laws.

Example 2.6: Use of affinity laws to determine required speed of pump.

PROBLEM: Determine what speed the pump in Figure 2.7 should be run with full diameter impeller to make a rating of 3000 gpm and 225 ft. The solution to this problem cannot be determined by inspection. All that is known from studying Figure 2.7 is that the pump must be run faster than the speed shown on Figure 2.7, which is 1780 rpm. There is only one speed that causes the full diameter pump curve to fall on the desired operating point.

This type of problem may need to be solved if a pump is being transferred to a new service, or in a variable-speed pump application, where published performance conditions exist at one speed and it is desired to operate the pump at a higher speed to make a different rating point.

GIVEN: To solve this problem, Equations 2.24 and 2.25 are used. For this problem, N_1 must be calculated, and the known terms are:

$Q_1 = 3000$ gpm
$H_1 = 225$ ft
$N_2 = 1780$ rpm

(Note that subscripts 1 and 2 above could have been reversed, with subscript 1 referring to the conditions at 1780 rpm.)

SOLUTION: This problem requires a trial and error, or iterative, solution. To solve for N_1, the steps are:

1. Estimate a trial solution for N_1.
2. Using the given values of Q_1, H_1, and N_2, solve Equation 2.24 for Q_2 and Equation 2.25 for H_2.
3. If the resulting data point (Q_2 and H_2) falls on the curve represented by the N_2 speed (1780 rpm), then the trial solution is correct. If it does not, repeat steps 1 through 3 again with a new trial solution.

Using the above procedure, a trial solution for N_1 = 2200 rpm is proposed. Solving for Q_2 and H_2:

$Q_2 = Q_1 \times (N_2/N_1)$ (from Equation 2.24)
$Q_2 = 3000 \times (1780/2200) = 2427$ gpm
$H_2 = H_1 \times (N_2/N_1)^2$ (from Equation 2.25)
$H_2 = 225 \times (1780/2200)^2 = 147$ ft

The resulting data point of 2427 gpm and 147 ft falls below the Figure 2.7 full diameter curve at 1780 rpm, so a new trial solution must be made and the process repeated. A second trial solution of 2000 rpm for N_1 is attempted as follows:

$Q_2 = Q_1 \times (N_2/N_1)$ (from Equation 2.24)
$Q_2 = 3000 \times (1780/2000) = 2670$ gpm
$H_2 = H_1 \times (N_2/N_1)^2$ (from Equation 2.25)
$H_2 = 225 \times (1780/2000)^2 = 178$ ft

The resulting data point does fall on the Figure 2.7 full diameter curve at 1780 rpm, so the second trial solution for N_1 is the correct one, and the pump must be run at 2000 rpm.

If, for the second iteration, a trial solution had been chosen which was too slow a speed (such as 1900 rpm), the resulting data point Q_2 and H_2 would have been found to fall

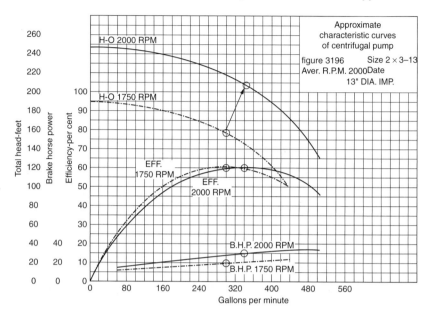

Figure 2.18 Effects of increasing pump speed on centrifugal pump performance. (Courtesy of Goulds Pumps, Inc., ITT Industries, Seneca Falls, NY.)

above the Figure 2.7 full diameter curve at 1780 rpm, and a third iteration would have been required.

Once the required operating speed has been determined as shown in the above example problem, a new set of curves at the higher speed can be developed. Figure 2.18 illustrates how pump H–Q, BHP, and efficiency curves change with increasing speed. The H–Q and BHP curves at the higher speed in Figure 2.18 are developed by taking arbitrary rating points (values of Q, H, and BHP) from the lower speed curves and applying Equations 2.24, 2.25, and 2.26 at the higher speed.

The efficiency curve for the higher speed pump is developed by taking values of Q, H, and BHP from the higher speed pump curve, once it is developed (see preceding paragraph), and solving Equation 2.15 for efficiency η. Note that the value of BEP efficiency stays constant, but the entire efficiency curve shifts to the right at higher speeds, or to the left at lower speeds.

Hydraulics, Selection, and Curves

Also refer to the discussion in Chapter 3, Section III, and the demonstration CD that accompanies this book to learn how computer software can be used to solve the above problem.

Caution should be exercised when operating pumps at higher speeds than their published curves. Ordinarily, the manufacturer should be consulted to verify that running the pump at the higher speed does not exceed the design limits of the pump (i.e., the upper horsepower limit of the shaft and the load limit of the bearings). Also, the required motor size will increase.

Another point to consider when running a pump at higher speeds than published for that particular pump is the fact that $NPSH_r$ increases with increasing speed, a factor not considered in the affinity law equations previously given. There is not complete agreement among manufacturers as to the amount of variation of $NPSH_r$ with speed changes. A conservative approach says that $NPSH_r$ varies with speed like head does, that is, with the square of the speed change. In other words, at constant diameter:

$$\frac{NPSH_1}{NPSH_2} = \left(\frac{N_1}{N_2}\right)^2 \quad (2.30)$$

Therefore, if a speed increase is being considered, the new $NPSH_r$ curve should be calculated and an evaluation made of the potential of cavitation at the higher operating speed.

Some manufacturers say that $NPSH_r$ varies with the speed ratio to the 1.5 power, rather than the 2nd power as shown by Equation 2.30. So, a conservative approach suggests that if one is planning to raise the pump speed, one should assume that Equation 2.30 applies; and if one is planning to lower the pump speed, one should substitute a power of 1.5 in place of 2 in Equation 2.30.

IX. SYSTEM HEAD CURVES

Previous sections of this chapter have discussed the characteristic head–capacity curve that describes the performance of a centrifugal pump with a particular speed and impeller diameter. When pump manufacturers build a pump, they usually select an impeller trim diameter based upon a design operating point (head and capacity) specified by the buyer. This rated head and flow, incidentally, is usually the only point on the pump curve that the manufacturer guarantees, unless other operating points are specified by the buyer and agreed upon by the manufacturer. The rated capacity and head usually appear on the nameplate of the pump. The only trouble is that the pump is a rather illiterate machine and does not know what head and flow it is supposed to deliver. The pump only knows the shape of its head–capacity curve based on its speed and impeller geometry, and that it operates wherever on the H–Q curve the system tells it to operate. This may be an operating point quite different from the rated point shown on the nameplate. Furthermore, the point at which the pump operates on its H–Q curve may change as system conditions change. A better understanding of these principles can be gained from a study of *system head curves.*

The system head curve is a plot of the head requirement of the system as it varies with flow. System head usually consists of a combination of static and frictional components, although the relative value of these two components varies considerably from one system to another. For example, a typical boiler feed system is composed of mostly static head (boiler pressure is equivalent to a static head), which does not vary with flow; while a typical pipe line system is composed of mostly frictional head, which does vary with flow. As another example, a closed loop system is composed entirely of friction head. Other examples fall everywhere between these extremes.

The best way to understand system head curves is with a few illustrations. The first example is a system that does not actually exist in the real world, a system with no friction whatsoever. Figure 2.19 illustrates such a system, where pipes are extremely short and extremely large in diameter, and with no valves, so that friction losses are negligible over the entire flow range of the pump.

The head requirement for this system, regardless of the flow, is that the liquid must be pumped up a static distance

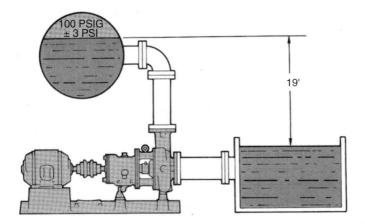

Figure 2.19 An all-static system with no friction, used to demonstrate the development of a system head curve. (Courtesy of Goulds Pumps, Inc., ITT Industries, Seneca Falls, NY.)

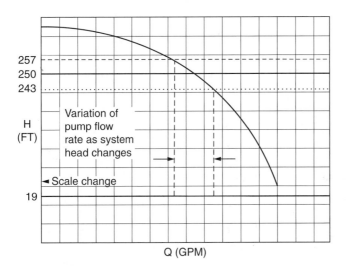

Figure 2.20 System head curve for the all-static system shown in Figure 2.19, with pump H–Q curve superimposed.

of 19 ft, and then pumped into the vessel that is maintained at a pressure of 100 psig ± 3 psi. Figure 2.20 shows the system head curve for this system. The system head curve, plotted on a head vs. capacity scale, is a horizontal line for this all-static system. The lower horizontal line represents the static elevation change. The middle of the three horizontal lines grouped together represents the system head curve when the tank pressure is 100 psig. In that case, for all flows, the system head (with s.g. = 1.0) is:

$$\text{System head} = 19 + (100 \times 2.31)/1.0 = 250 \text{ ft}$$

The other two horizontal lines grouped together in Figure 2.20 represent the new system head curves when the pressure in the discharge vessel is, respectively, 97 and 103 psig. When the pressure is 97 psig in the discharge vessel, the system head is:

$$\text{System head} = 19 + (97 \times 2.31)/1.0 = 243 \text{ ft}$$

When the discharge vessel pressure is 103 psig, the system head is:

$$\text{System head} = 19 + (103 \times 2.31)/1.0 = 257 \text{ ft}$$

Also shown in Figure 2.20 is the H–Q curve for a pump operating in this system. The general rule here is that the pump operates at the point of intersection of the pump H–Q curve and the system head curve. The pump seeks the point on its H–Q curve where it delivers the head required by the system. Figure 2.20 shows that as the pressure in the delivery vessel changes from 97 to 103 psig, the intersection of the system head curve and the pump H–Q curve shifts, causing the pump to deliver a varying flow rate.

Figure 2.21 shows that the combination of an all-static system (no friction) and a low specific speed pump (which produces a very flat H–Q curve as discussed in Section VII above) can cause wide swings in flow with relatively small changes in the system head requirement. This makes this combination of pump and system type difficult to control and to maintain a constant flow without the use of a flow control valve. In general, a centrifugal pump with a steeper H–Q curve is easier to control in a system, particularly a system with a flat or nearly flat system head curve, because variations in system head do not produce wide swings in flow.

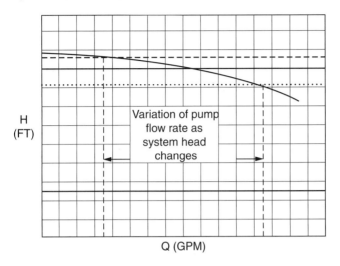

Figure 2.21 The combination of a flat system head curve (no friction or very little friction) with a relatively flat H–Q pump curve produces wide flow swings with relatively small system head variation.

To continue the discussion of system head curves, consider the more real-world system shown in Figure 2.22, which includes both a static component (elevation change of 130 ft) and a friction component (friction losses through system piping and fittings). For this system, the system head requirement as a function of flow is shown in Figure 2.23. The system head curve in this figure is labeled curve X, and the pump H–Q curve is superimposed on the same scale. The system head curve is generated by starting at H = 130 ft (the static component of head, and the total head requirement at flow rate = 0), selecting an arbitrary value of flow Q, and calculating friction losses through the entire piping system at this flow rate. This procedure is repeated for several flow rates until a smooth curve for the system head curve can be drawn as shown in Figure 2.23.

As before, the rule is that the intersection of the system head curve and the pump H–Q curve, labeled point a in Figure 2.23, dictates where the pump shown operates when

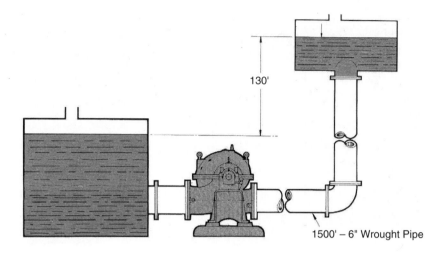

Figure 2.22 A system composed of both static and friction head requirements. (Courtesy of Goulds Pumps, Inc., ITT Industries, Seneca Falls, NY.)

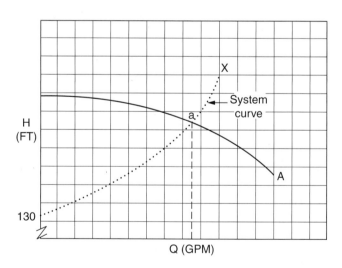

Figure 2.23 System head curve for a system composed of static and friction components, with pump H–Q curve superimposed.

Table 2.5 Calculation of Total System Head at Various Flow Rates Is Simplified, once Friction Head Is Determined for a Single Flow Rate, since the Friction Head Varies as the Square of the Flow Rate

Flow Rate (gpm)	Static Head (ft)	Friction Head (ft)	Total Head (ft)
1000	130	100	230
800	130	$(800/1000)^2 \times 100 = 64$	194
600	130	$(600/1000)^2 \times 100 = 36$	166
1200	130	$(1200/1000)^2 \times 100 = 144$	274

running in a system described by the system head curve labeled X in Figure 2.23.

It is possible to build a complete system head curve for a simple system (one that does not have any branches) after only the head at one flow point and the static head are known. The reason for this is that the friction head loss in a pipe line varies as the square of the flow going through the pipe line. So once the friction loss has been calculated for a single flow rate, it is possible to construct a table to determine the friction head loss at other flow rates using this squared relationship. Adding that friction loss to the static component of head gives the total head for the system at the other flow points, and therefore other points on the system head curve are produced. This is illustrated in Table 2.5. Given that the friction head loss is a second-order curve, it is also possible to use a log scale on the system head curve, which means that a straight line can be drawn from the static head at 0 flow through the one calculated head-flow point to generate a system head curve.

In the all-static example shown in Figures 2.19 and 2.20, the system head curve is seen to shift as the pressure in the delivery vessel changes. Similarly, in the system shown in Figure 2.22 and 2.23, which includes friction as well as static components, the system head curve can again change shape. This, in turn, causes the point of intersection with the pump H–Q curve to change. As before, this causes the pump to operate at a different point on its H–Q curve. Either the static component of a system head curve might shift as shown in

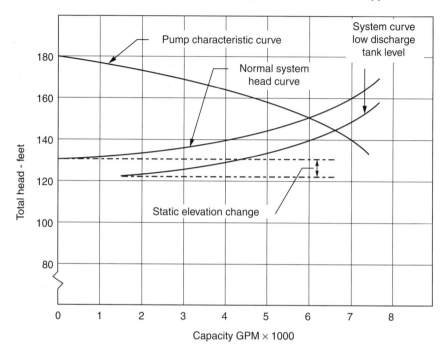

Figure 2.24 Change in static component of system head curve. (Courtesy of Goulds Pumps, Inc., ITT Industries, Seneca Falls, NY.)

Figure 2.24, or the friction component of the system head curve might change, as shown in Figure 2.25. Changes in the static component of the system head curve might occur due to changes in the level of either the suction supply vessel or the delivery vessel, or due to variations in pressure in the supply vessel or in the delivery vessel as illustrated in the all-static example.

Changes in the frictional component of the system head curve might be due to several causes. These might include the buildup of solids on a filter in the system, the throttling of a valve in the system, or the gradual buildup of calcification or other corrosion products on the inside of system piping. Any of these three causes, or some combination of them, might cause the frictional component of the system head curve to change over time, which in turn would move the point of operation on the pump's H–Q curve.

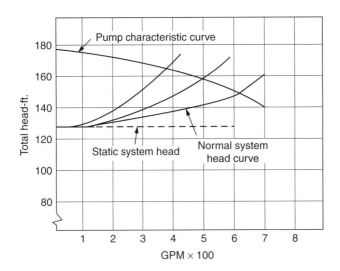

Figure 2.25 Change in friction component of system head curve.

Actually, both the static and the frictional components of a system head curve might change over a period of time within a system, for the reasons discussed above, causing variations in the point of operation of the pump on its curve. A good engineering analysis of a system during the design phase considers the range of these swings of operation of the pump on its head–capacity curve, and considers the following for all points:

- Is the motor sized to handle the horsepower at all points?
- Is there adequate margin of NPSH at all points?
- Is the capacity of the pump at all operating points adequate to meet process requirements?
- Does the pump operate at a point on its H–Q curve that is healthy for the pump on a continuous basis? Consider minimum flow, bearing loads, etc. (Refer to the discussion on minimum flow in Chapter 8, Section III.B.)

Returning for a moment to Figure 2.16, which illustrates a drooping H–Q curve, it should now be clear why a system head curve described by curve A in Figure 2.16 would cause unstable operation for the pump with a drooping H–Q curve. In that case, because there are two points of intersection of system head curve A and the pump H–Q curve, the pump would try to operate both at point 1 and at point 2 on its H–Q curve, causing flow to surge back and forth. On the other hand, if the pump were installed in a system described by curve B, the pump would produce stable flow at point 3 on its H–Q curve. This illustration of possible unstable operation resulting from a drooping H–Q curve in conjunction with a flat system head curve is one reason why boiler feed pumps (which operate in a system that is nearly all static and thus are described by a nearly flat system head curve) are generally specified to have a constantly rising (not drooping) H–Q curve.

Another purpose for developing a system head curve is for analyzing power consumption using variable-speed pumping in a system with a variable flow demand. This analysis is illustrated in Chapter 6, Section IV.

System head curves can become quite complex when they are developed for systems with multiple branches, especially where flow is going through more than one branch at the same time. Consider the system shown in Figure 2.26. The system might be designed and operated in a way that has the valves in the branches positioned such that all of the pump flow goes through only one branch at a time (e.g., branch B and C are blocked off, while branch A is open). This arrangement of a system is common, for example, where flow in a process might go through a process stream, recycle back to the suction supply vessel, or be pumped directly to a storage vessel. In that case, there are actually three separate system head curves, each of which may have a different static head requirement and a different friction curve that varies with flow. Thus, the pump very likely operates at different points on its H–Q curve, depending on the branch through which the flow is directed, unless the flow is equalized by using orifices or control valves to adjust the friction losses such that the total head loss (static plus friction) in all three branches is equal.

If the system shown in Figure 2.26 is designed and operated such that the flow can be directed to all three branches

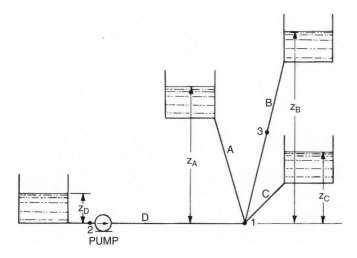

Figure 2.26 System with multiple branches. (From *Pump Handbook*, I.J. Karassik et al., 1986. Reproduced with permission of McGraw-Hill, Inc., New York.)

at the same time, a combined system head curve must be developed to determine the resulting flow produced by the pump and to determine how the flow splits between the branches. The system head curves for the individual branches A, B, C, and D must be developed separately as described earlier, then combined graphically, to get a system head curve for the entire system. Figure 2.27 illustrates how the system curves are added geographically. The separate system head curves for each of the three branches that are in parallel (A, B, and C) are combined by taking arbitrary values of head and adding the flow from each of the three system curves at each head value. This produces the curve labeled A+B+C in Figure 2.27. This part of the system is in series with branch D, and so the curve labeled A+B+C is graphically added to the system curve labeled D by taking arbitrary values of flow and adding the head from each of the two curves. This produces the curve labeled (A+B+C)+D, which is the system head curve for the entire system.

The pump H–Q curve (labeled curve E in Figure 2.27) intersects with the system head curve (A+B+C)+D at point

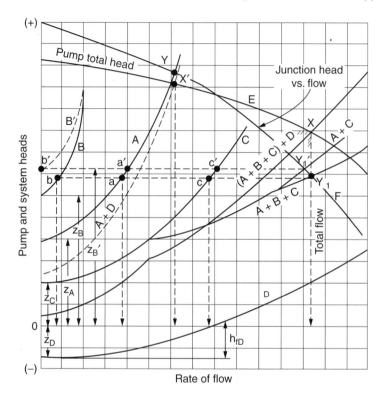

Figure 2.27 Combined system head curves for multiple branch system. (From *Pump Handbook*, I.J. Karassik et al., 1986. Reproduced with permission of McGraw-Hill, Inc., New York.)

X, and this therefore represents the flow delivered by the pump into the system. The curve representing the head at the junction (point 1 on Figure 2.26) as it varies with flow (labeled curve F in Figure 2.27) is produced by observing that the head at the junction point 1 is equal to z_D (the suction head in Figure 2.26, a constant), plus the TH of the pump (which varies with flow), less the friction head in branch D (which also varies with flow).

Finally, the key to solving this problem is the observation that the flow splits at the junction point 1 in such a way that the total head requirements (static plus friction) in each of the three branches A, B, and C are equal to each other and to the head at junction point 1. So, from point X on Figure 2.27 (the flow of the pump), a line is drawn straight down to curve F, intersecting at the point labeled Y_1. This is the head at the junction when all three lines are open and the pump is running. Drawing a line from this point Y_1 straight left until it intersects with the branch system curves labeled A, B, and C (at points a, b, and c, respectively) produces the flow through each of these branches. This shows that the lowest flow rate is produced in branch B, the branch with the highest static head to overcome. If it is desired to direct more flow through branch B, the system and probably the pump would have to be modified to achieve this. Conversely, the highest flow rate is produced in branch C, the branch with the lowest static head.

The above example of a multiple branch system can become very cumbersome very quickly as more branches are added to the system. The graphical solution described above (shown in Figure 2.27) is tedious to carry out and not practical for more than three or four branches. This illustrates one of the benefits of computer software based piping network analysis, which is discussed more thoroughly in Chapter 3, Section III.

Another example of a common system involving multiple branches with flow going through more than one branch simultaneously is a closed loop, chilled water system in an HVAC system. Here, balancing valves in each parallel branch are commonly used to do a final field balance of the system to achieve the required flow through each branch of the network.

X. PARALLEL OPERATION

Parallel operation of pumps is illustrated in Figure 2.28, where one pump, some of the pumps, or all of the pumps can be operated at the same time. The primary purpose of operating pumps in parallel is to allow a wider range of flow than would be possible with a single, fixed-speed pump for systems with widely varying flow demand. Examples of applications for parallel pumping include municipal water supply and

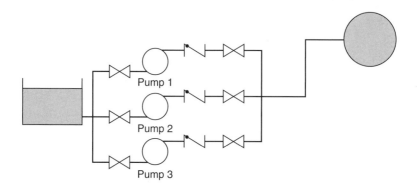

Figure 2.28 Parallel operation.

wastewater pumps, HVAC system chilled water pumps, main process pumps in a variable capacity process plant, and condensate pumps in a steam power plant. Usually, there are no more than three or four pumps operating in parallel.

When parallel pumps are being considered for a system design, the pumps must be carefully matched to each other and to the system to ensure that the pumps are always operating at a healthy point on their H–Q curves, and to ensure that the system is such that true benefits are achieved from the parallel pumping arrangement. (This does not always turn out to be the case, as will be demonstrated shortly.)

To analyze a parallel pumping arrangement, it is necessary to construct the system head curve, as explained in Section IX above. Then a combined pump curve must be developed depicting the head–flow relationship for the pumps while pumping in parallel. Once these two curves are constructed, the same rule for total system flow as discussed in Section IX applies, that is, that the total flow through the system is represented by the intersection of the system head curve with the combined pump curve.

Figure 2.29 shows how the combined pump curve is developed, with a system having either two or three identical pumps operating in parallel. Curve A in Figure 2.29 represents the H–Q curve for any one of the pumps. Curve B is the combined pump curve, which represents two pumps operating

Hydraulics, Selection, and Curves

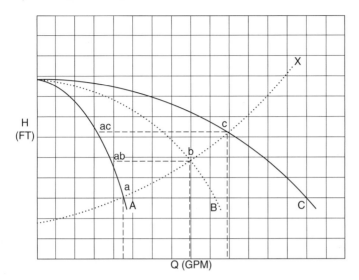

Figure 2.29 H–Q curve for a centrifugal pump, combined curves for two or three identical pumps operating in parallel, and system head curve.

at the same time in parallel. This curve is produced by starting at the shutoff point for curve A and moving down the H scale. At arbitrarily selected values of head, the flow on curve A at that head is doubled, which produces a point on curve B. This is done for about six to eight points, and then a smooth curve is drawn through these points, creating curve B. Curve C is similarly produced to show the combined pump curve when all three identical pumps are operated in parallel.

The system head curve is shown as curve X in Figure 2.29. The point of intersection of pump curve A with system head curve X (labeled point a in Figure 2.29) represents the flow delivered to the system when only one pump is operated. The point of intersection of combined pump curve B with system head curve X (labeled point b) represents the total flow delivered by the pumps into the system when two of them are operated at the same time in parallel. Finally, the point of intersection of curve C with system head curve X (labeled point c) represents the total flow when all three pumps are operated in parallel.

Figure 2.29 shows that the resultant flow when two pumps are operated in parallel in a system is not double the flow that one pump alone produces when operating by itself, a sometimes mistaken impression. If the system curve is completely flat (i.e., if there is no friction losses in the system), then the flow would double when both pumps are operating. But because the system head curve curves up due to friction in the system, two pumps operating in parallel do not deliver double the flow of a single pump operating alone in the system. Putting the third pump on line even further diminishes the increment of flow added to the system, as Figure 2.29 illustrates.

With two identical pumps operating, the point where each pump is operating on its own H–Q curve is obtained by moving left from the intersection point of the combined two-pump curve B with the system head curve X (point b), until the dashed line meets pump curve A. This point (labeled point ab) represents where each pump is operating on its own H–Q curve when these two identical pumps are operated in parallel. Similarly, the dashed line moving left from the intersection of the combined three-pump curve C with system head curve X (point c), intersects the pump curve A at the point where it would operate when three pumps operate in the system (labeled point ac). In general, with parallel pumping, each pump runs out the furthest on its own H–Q curve when that pump operates alone in the system (or when the fewest number of pumps allowed to be operated are running). The pumps run the furthest back on their own H–Q curves when the maximum number of pumps are operated in parallel in the system.

If the system head curve is too steep, a situation that could be caused by an undersized piping system or by some other undersized component in the system that acts as a bottleneck, then it turns out that very little benefit is achieved by operating pumps in parallel in that system. This is illustrated in Figure 2.30, where the system head curve is deliberately shown to be quite steep. As Figure 2.30 illustrates, the amount of incremental flow gained by adding the second and third pump to the system is so small as to render the

Hydraulics, Selection, and Curves

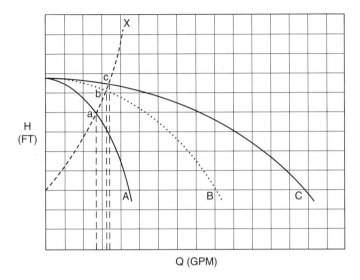

Figure 2.30 When the system head curve is very steep, operating a second pump in parallel with the first produces only a marginal increase in flow.

benefit from parallel pumping negligible. In this case, capital would have been better spent, if possible, in flattening the system head curve, reducing bottlenecks, and increasing piping sizes, rather than in adding additional parallel operating pumps to the system.

When considering a system with pumps in parallel, all combinations of pumps and variations of system head curve should be considered to establish a duty cycle for the pumps, that is, for how long a period of time each of the pumps will operate at particular points on its own H–Q curve. The health of the pumps at these operating points should be considered with regard to NPSH, horsepower, bearing load, and minimum flow. This will also permit selection of the optimal pump to minimize total energy consumed in pumping during the pump's life cycle.

The situation with two nonidentical pumps operating in parallel is shown in Figure 2.31. In this figure, the curves of the two different pumps, marked A and B, are combined in curve C, by adding the flow of the two pumps at arbitrarily

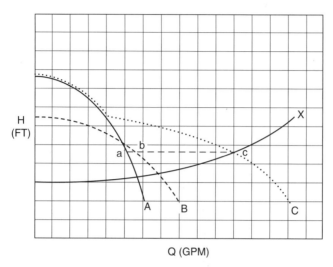

Figure 2.31 Combined pump curve for two nonidentical pumps in parallel, and system head curve.

selected values of head. The combined curve is not a smooth one due to the way it is generated as the graphical sum of curves A and B. The combined curve C actually follows curve A for a while, because the maximum head for curve B is lower than the maximum head for curve A. The same rule applies as before. The intersection point of the combined curve and the system head curve (point c) determines the total flow of the two pumps when they are running in parallel. The horizontal dashed line going left from point c intersects the individual H–Q curves at the point where each of these two pumps operates when they are operated together in parallel, labeled points a and b, respectively.

Figure 2.32 illustrates that operating two nonidentical pumps in parallel can present problems when the two nonidentical pumps are mismatched for the system in which they are working. Figure 2.32 is the same as Figure 2.31, except that the system curve in Figure 2.32, labeled curve Y, is steeper than the one shown in Figure 2.31. An attempt to run these two pumps in parallel in the system represented by system head curve Y would mean that the pump represented

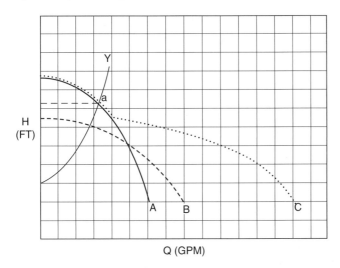

Figure 2.32 A mismatch occurs when attempting to operate these two nonidentical pumps in parallel in the system shown.

by curve A would operate at the intersection point of the combined curve and the system head curve (labeled point a in Figure 2.32), which is the same point it would operate at when pumping alone in the system. The pump represented by curve B, on the other hand, would never be able to develop enough head to overcome the system back pressure if both pumps were running. Thus, attempting to run the two pumps in parallel would cause the pump represented by curve B to be running at full speed but delivering no flow into the system, just as if a valve at that pump's discharge were completely closed. (The pump is said to be *dead headed*.) In addition to delivering no flow to the system, the pump would be operating at shut-off, a very unhealthy point for continuous operation of most pumps.

Even if the system head curve shape in this example were such that the intersection point between the system head curve and the combined pump curve C caused pump B to operate at a very low flow, although not at shut-off, this flow still might be below the recommended minimum continuous flow for that pump. (Refer to the discussion in Chapter 8, Section

III.B, on minimum flow.) This type of mismatch of nonidentical pumps in a system should obviously be avoided.

The question is sometimes asked, while preparing the system head curve shown in the parallel pump arrangement in Figure 2.29, how to account for the friction losses in the branched part of the system that includes the three pumps (see Figure 2.28). If the friction loss through each of these branches (with approximately one third of the total flow rate through passing through each branch) is insignificant compared to the total friction loss in the rest of the system, the splitting of the flow in these three branches can simply be ignored. On the other hand, if the friction loss in each of the branches is a significant part of the total system friction loss, it should not be ignored. In that case, the simplest approach is to make an adjustment to the pump curve itself. An adjusted pump curve is produced by calculating, at an arbitrary value of flow, the head loss through one of the three parallel branches, and then subtracting this head from the head on the pump curve at that flow, to produce a point on the adjusted pump curve. This is done at several flow points, and then the adjusted pump curve can be drawn. This adjusted pump curve is then used to build the two-pump and three-pump combined pump curves in the same manner as before.

If the system is further complicated by the fact that the friction loss in one of the parallel branches is significantly different from that of another branch (in the case, for example, where one of the three pumps is located a good distance away from the other two, but still piped in parallel), then this is handled exactly as described above, except the pump curves are adjusted separately. For each of the branches, an adjusted pump curve is produced, only this time the adjusted pump curves would be different for the two pumps. From there the combined parallel flow pump curves would be generated using the respective adjusted pump curves, treating the pumps as if they were two different-sized pumps.

XI. SERIES OPERATION

Series operation of pumps is illustrated in Figure 2.33. In series operation, the discharge of one pump feeds the suction

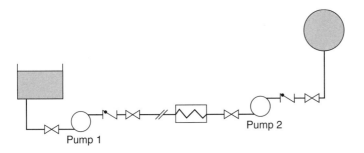

Figure 2.33 Series operation.

of a second pump. When two or more pumps operate in series, the flow through all of the pumps is equal because whatever flows through one pump must flow through the next pump in series (provided there are no side streams).

Operating pumps in series is done for several reasons. One reason for operating pumps in series is to ensure that commercially available equipment can be used in a particular system, while at the same time reducing system costs. For example, if the Figure 2.33 system is a long pipe line with a large amount of friction loss across the entire pipe line, attempting to deliver the flow through the pipe line with a single pump would result in a pump with an extremely high head, and thus extremely high horsepower. The required horsepower might possibly be so high, in fact, that a single pump would not be commercially available to do the job. Some pipe lines are hundreds or even thousands of miles in length, and may have a great many pumps in series as part of the pipe line. In addition to ensuring that the pumps are of a size that are commercially available, breaking up the line with a number of pumps in series reduces the required design pressure for the system piping, valves, instrumentation, etc.

This same reasoning accounts for the use of several pumps in series in a steam power plant (condensate pump, condensate booster pump, and boiler feed pump). In theory, a single pump could pump from the condenser through the feedwater heaters and into the boiler. However, the required discharge pressure of this one pump would be so high that the feedwater heaters and other system components would be outrageously expensive due to their extremely high design pressure.

Considering the above pipe line application again, another very important reason for having multiple pumps in series is that this allows a variation of flow through the pipe line by varying the quantity of pumps that are pumping at one time. (How to calculate the effect of serial pumping on system flow rate is explained shortly.) Also, the use of multiple pumps in series allows the pipe line operator flexibility to deal with a variety of pumped products having a wide range of specific gravity and viscosity.

A final application of two pumps in series is to ensure adequate available NPSH for the second pump. In this case, the first pump of two in series might be a fairly low head pump (and thus a fairly low horsepower pump). But, in calculating $NPSH_a$ for the second pump in series, Equation 2.18 for $NPSH_a$ would have an additional term, namely the TH of the first pump. An example of this is a central power station boiler feed pump, which might operate at a speed of 6000 or 7000 rpm to develop its required head to achieve boiler pressures. Because of this high operating speed, and assuming good impeller design practice, this might result in an $NPSH_r$ value of as much as 100 ft for the boiler feed pump. A high $NPSH_a$ value is provided by adding a low-speed condensate booster pump to the system.

If pumps are used in series in a system, a combined pump curve can be generated, analogous to the combined curve discussed for parallel pumping in Section X (although the combined curve is generated differently, of course). Again, the same general rule applies, namely that the intersection of the combined pump curve with the system head curve determines the total flow delivered to the system, and allows one to determine where on its own H–Q curve each of the pumps is running.

Figure 2.34 shows how to develop a combined pump curve for two or three identical pumps operated in series. Each pump has curve A for its H–Q curve. Curve B, the combined curve for two pumps, is developed by starting at

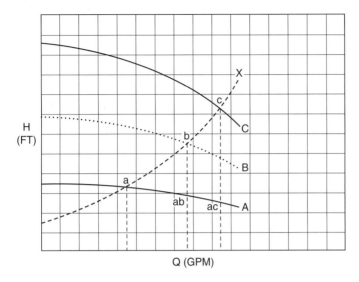

Figure 2.34 H–Q curve for a centrifugal pump, combined curves for two or three identical pumps operating in series, and system head curve.

zero flow, and moving to the right on the flow axis. At arbitrarily selected flow values, the TH at each flow value is doubled, producing a point on curve B. This process is repeated for a number of points, thereby generating curve B. The system curve X is generated as described in Section IX, and the intersection of the system curve X with pump curve B (intersection point labeled b) determines the flow pumped through each pump and through the system when two pumps are running in series. Curve C and intersection point c for three identical pumps in series are similarly produced.

As with parallel pumping, adding pumps in series increases the amount of flow going through the system. Pumps in series behave opposite to the way that pumps in parallel operate, in the sense that the pumps run out the farthest on their H–Q curves when the highest possible number of pumps are operated in series (point ac in Figure 2.34), and they run back the farthest on the curve when only one pump is running (point a in Figure 2.34).

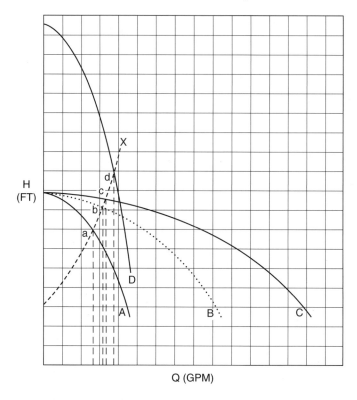

Figure 2.35 Series operation of the pumps shown in Figure 2.30 produces more flow with a steep system head curve than parallel operation.

Return again to Figure 2.30, which shows that when the system head curve is very steep, operating a second pump in parallel with the first produces only a marginal increase in flow. If these same pumps are piped in series, rather than in parallel, they would produce a higher flow through the system. Figure 2.35 shows that only two such pumps piped in series (curve D is the combined pump curve) delivers more flow through the system (intersection point d) than three pumps operating in parallel (intersection point c). Therefore, a general rule is that if the system head curve is relatively steep, series pumping is probably more effective than parallel pumping for increasing the flow range of the pumps. The

Hydraulics, Selection, and Curves

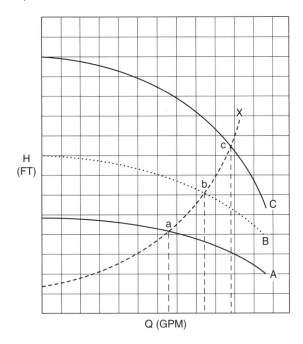

Figure 2.36 Combined pump curve for two nonidentical pumps in series, and system head curve.

converse is also true, namely that for fairly flat system head curves, parallel pumping is probably more effective than series pumping to produce a wide flow range.

Figure 2.36 shows the combined pump curve (curve C) for two nonidentical pumps (curves A and B) operating in series. The combined curve is generated using the same procedure as outlined above for identically sized pumps. At the arbitrarily selected flow rates, the TH values of the two pumps are added together to produce the combined curve data point. It is common in pipe lines to have several differently sized pumps, allowing the operators the widest possible range of flow and/or variation of products pumped.

One final note on series pumping. If a valve in the line downstream of the last pump in a series installation is inadvertently closed completely, all of the pumps in the line move to their shutoff head, which means that the pressure in the system is considerably higher than normal, and possibly higher than the design pressure of the system piping and other components. In consideration of this possibility, a pressure relief system or other pressure limits should be incorporated into the system design to keep the pressure from exceeding design limits.

XII. OVERSIZING PUMPS

A very common approach in sizing centrifugal pumps is to calculate capacity and head, as described in previous sections of this chapter, and then to adjust these terms by some sort of "fudge factor." There are often good reasons to adjust the design parameters of the pump in this way. Experience has shown many engineers that the system they are designing for a process plant today will be too small in a few years, and management will want to get more capacity out of the plant. This is an argument for adding a fudge factor to the capacity in sizing the pump. As for pump TH, many piping systems are likely to have a buildup of corrosion products on the pipe interior walls over time, which, as discussed in Section IX, would require an increased pump TH to maintain the same flow. This argues in favor of adding a safety factor to the calculated TH.

While both of these adjustments might seem reasonable, the amount of fudge factor is often arbitrarily chosen, without regard to the effect on capital and operating costs. Also, sometimes the engineer applies a fudge factor, and then the engineer's supervisor will add an additional fudge factor. Then, the pump vendor may raise the TH a bit more, just to make certain that there is "enough pump to do the job." Another practice sometimes followed that might cause the fudge factor to be excessive is to take the fudge factor as a percentage of calculated total head, rather than as a percentage of only the friction head. Only the friction head would increase over time as a buildup of corrosion products causes the pipe diameter to decrease. The static component of head would be unaffected.

The costs of adding on these fudge factors can be significant, as the following example illustrates.

Hydraulics, Selection, and Curves

Suppose a rating of 2000 gpm and 150 ft has been calculated for a pump handling water (s.g. = 1.0). A fudge factor of 15% is applied to both head and flow to account for expected future conditions, making the rating for which the pump is actually purchased 2300 gpm and 172.5 ft (2000 gpm and 150 ft, each multiplied by 1.15). The pump shown in Figure 2.7 is chosen for the service. Had the original rating of 2000 gpm and 150 ft been used in sizing the pump, an impeller diameter of approximately 13 3/8 in. would satisfy the condition. With this impeller diameter, the efficiency (from Figure 2.7) is 81.5%, and BHP is computed as 93 HP at the design point of 2000 gpm and 150 ft using Equation 2.15. A 100-HP motor should provide nonoverloading service over the entire range of this pump curve.

Instead of the above selection, the 15% fudge factor is applied, and the new rating of 2300 gpm and 172.5 ft is used to select the pump. It turns out that with this new rating, the same pump shown in Figure 2.7 can be used, but with a larger impeller diameter of approximately 14 3/8 in. Because the design capacity of 2000 gpm is all that is needed when the pump is first commissioned, presumably the pump would be throttled back on its curve to a flow rate of 2000 gpm. At that point on the pump curve with a 14 3/8-in. impeller, TH is about 185 ft, efficiency is about 81.9%, and BHP is computed at 114 HP. A 125-HP motor looks like it will not provide nonoverloading service at the far right end of the curve with this larger impeller diameter, so either the pump must operate slightly in the service factor in this event, or else a 150-HP motor must be chosen. In either event, the chosen motor must be larger than the 100-HP motor that satisfied the full pump range with the smaller impeller diameter.

The decision to add the 15% adjustment to head and flow in sizing the pump has negative consequences for the owner of this pump for several reasons. The owner's capital cost is higher because the pump with the larger impeller requires a larger motor. (The pump itself probably would cost no more because it is the same pump in either case with only a different impeller diameter.) Of greater significance, however, is the fact that from the moment the pump is put into service, the pump with the larger impeller consumes 114 HP to deliver 2000 gpm, vs. the 93 HP that the pump with the smaller impeller would have consumed to deliver the same 2000 gpm. This additional 21 HP is simply wasted energy.

It is possible to calculate the annual cost of the energy wasted if the larger impeller diameter had been chosen by converting the 21 HP mentioned above to kilowatts (1 HP = 0.746 KW, 21 HP = 15.7 KW), then multiplying the result by the owner's cost of power in $/KW-HR, and by the anticipated number of hours of operation of the pump in a year. Cost of power varies widely from location to location, and depends on the amount of consumption and several other factors. In the United States, most industrial and commercial users pay from $0.07 to $0.12 per KW-HR. Assuming a cost of power of $0.10/KW-HR, and 6000 hours of operation per year (roughly operating 75% of full-time operation), the annual cost to the pump owner of the energy wasted by choosing the larger impeller diameter is:

$$(15.7 \text{ KW}) \times (\$0.10/\text{KW-HR}) \times (6000 \text{ HR/year}) = \$9420/\text{year}$$

The annual cost to the owner for the energy wasted by choosing the larger impeller can be nearly as much as the capital cost of the pump itself! Clearly, it does not make good economic sense to waste such a large amount of energy beginning from the day the pump is commissioned, merely in anticipation of the need for more capacity and head some years in the future.

A third negative consequence of excessively oversizing the pump is that this oversized pump requires more throttling to achieve lower flow rates than would a smaller pump or a pump with a smaller impeller. The lowest flow rate required for this system (at start-up or recycle, for example) would then be a smaller percentage of BEP flow than would be the case if the chosen pump had a smaller impeller diameter. During these periods of low flow, the negative consequences of operating the pump at a flow rate that is too small a percentage of BEP (such as excessive shaft deflection, premature bearing failure, or premature seal failure) would be more severe than would be the case if the smaller diameter impeller were

Hydraulics, Selection, and Curves 155

instead chosen. Refer to related discussions on this topic in the Chapter 8, Section III.B, discussion of minimum flow.

Oversizing a pump can lead to significant negative consequences, as the preceding discussion has summarized. On the other hand, to totally ignore the possibility of needing higher flow and head from a pump at some time in the future is not wise either. An alternative that makes sense in the example shown here would be to buy the pump with the smaller 13⅜ in. diameter impeller, but spend the extra money to buy the larger 125-HP motor for the pump. Then, five years or so later when the higher head and flow are required, all that is necessary is to install the larger impeller. The bigger motor is purchased in the beginning so that it will not have to be changed out at a later date to accommodate the anticipated larger impeller. Sometimes, retrofitting a larger motor can be troublesome if it requires a larger pump bedplate or different electrical requirements. It may be easier to simply put in the larger motor right away. The difference in capital cost for the larger motor size may not be that significant. Also, the 125-HP motor in this example operating at a lower percentage of full load generally should not suffer more than a point of efficiency loss compared to a 100-HP motor.

One last lesson this example teaches is the wisdom of not choosing the maximum impeller diameter for an application, allowing the possibility of putting in a larger impeller at a later date to stretch the capacity and head of the pump if needed. Using good pump selection practices, the impeller chosen for a particular application should normally not exceed 90 to 95% of the maximum size.

XIII. PUMP SPEED SELECTION

The operating speed for each of the centrifugal pumps considered in this chapter has been treated as a given. Practically speaking, for a given application, there may be a range of speeds considered by the user and offered by the manufacturer for pumps that are suitable for the service. This leads to the question of how the pump manufacturer determines what are the appropriate speeds to offer for a particular pump type and size. Or alternatively, given a particular capacity and head requirement, how does one consider which speeds should be considered for a given application?

There are several criteria that should be considered in combination, as follows.

A. Suction Specific Speed

Suction specific speed was introduced in Section VII. Using the recommended value of 8500 for suction specific speed as a criterion, Equation 2.23 can be used to choose a maximum operating speed for a given application. This is done by taking the $NPSH_a$ from the planned installation and subtracting a reasonable margin between $NPSH_a$ and $NPSH_r$ (Section VI.E) to arrive at a value of $NPSH_r$ to use in Equation 2.23. Using the design capacity Q, Equation 2.23 can then be solved for N, the maximum speed. A motor with nominal speed below this calculated maximum would then be chosen. This will not be the only choice for pump speed, but will generally result in the smallest sized pump that will suit the application, and the lowest first cost alternative. Other criteria, as described in the following sections, might dictate the use of a slower pump speed.

B. Shape of Pump Performance Curves

The speed chosen for a particular application (along with the number of stages if the pump is multi-stage) affects the specific speed N_s of the pump chosen. As Section VII demonstrated, this has an effect on the shape of the pump H–Q and BHP curves. It may be desirable, for example, for a particular application to avoid having a high head or high horsepower at shutoff, both of which could result if a pump having the higher value of N, and thus higher N_s, is chosen. Reasons for not wanting the higher specific speed pump might be due to the greater control problems that the steeper pump curve could cause, or concerns with start-up motor overload problems with two pumps in parallel. These considerations can result in the choice of a speed that is less than the maximum one indicated by the criterion of suction specific speed discussed in Section XIII.A above.

Hydraulics, Selection, and Curves

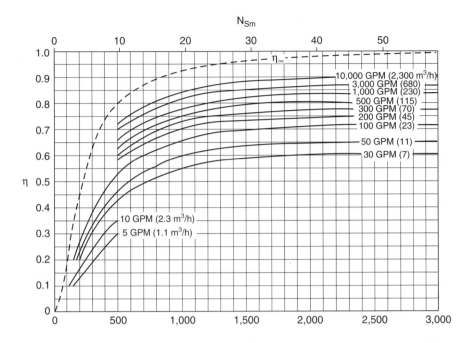

Figure 2.37 Pump efficiency as a function of specific speed and capacity. (From *Pump Handbook*, I.J. Karassik et al., 1986. Reproduced with permission of McGraw-Hill, Inc., New York.)

C. Maximum Attainable Efficiency

Refer to Figure 2.37, which shows the range of efficiencies to be expected for centrifugal pumps as a function of specific speed and capacity. Although these are only theoretical values of the maximum efficiency obtainable, they do represent what is generally obtainable in the centrifugal pump industry with good design and manufacturing practices. The curves shown in Figure 2.37 each represent a different best efficiency point capacity. This figure shows that, in general, the higher the BEP capacity of the pump, the greater the expected best efficiency of that pump. And, for a given pump capacity, the larger the design specific speed of the pump, the greater the expected pump efficiency (up to a point at least). This figure helps explain why the efficiency of very small centrifugal pumps is relatively low, while that of very large centrifugal pumps may be greater than 90%. Note that the Figure 2.37 chart only goes to a maximum specific speed of 3000. Other similar presentations of the data shown in this figure indicate that at specific speeds above about 3000, the efficiency curves droop down from their peak values.

Figure 2.37 suggests that there is an optimum specific speed range to achieve the highest BEP for the pump chosen for a particular application. Chapter 6, Section II, illustrates in several examples how the selection of a higher operating speed (or number of stages) affects specific speed, which in turn can affect the efficiency of the pump chosen for the application. Depending on the specific speeds, this might mean that the higher speed choice is the most efficient, or that the lower speed choice is most efficient, because as mentioned above, the Figure 2.37 curves droop above N_s of 3000.

D. Speeds Offered by Manufacturers

In general, the speed choices to consider are limited to those that are commercially offered by the manufacturers of that pump type. In addition to the criteria described above, other limitations on maximum speed offered by the manufacturer for a given pump type might include maximum available shaft size for the pump, largest bearing system offered for a given pump, casing pressure limits, or other mechanical design limitations. Speeds higher than those published by the manufacturer should only be considered after receiving confirmation from the manufacturer of the acceptability of the speed based on these mechanical design considerations. Consideration of operating speeds other than the synchronous motor speeds offered by the manufacturer (e.g., through the use of variable speed drives, which are discussed in Chapter 6, Section IV) should include a verification that the planned operating speed is sufficiently far away (at least 20%) from the pump natural frequency and the shaft critical speed. Operating a pump at a speed too close to the natural frequency of the pump or the rotating element produces unacceptable vibration levels. It is more likely to be a potential problem with larger equipment, such as vertical turbine pumps (see Chapter 4, Section XI).

E. Prior Experience

The selecting engineer should consider the prior operation and maintenance experience with similar pumps in the speeds being considered at the facility where the pump will run. Prior problems, or lack thereof, with pumps in similar or related applications may signal the acceptability of a higher speed choice for a particular application. In looking for examples of prior experience, consideration should include the liquid being pumped (especially the amount of abrasives or level of corrosiveness), past success with keeping similarly sized pumps at the higher speed properly aligned and balanced, and the facility's experience with seals and bearings with similarly sized pumps at the higher operating speed.

3

Special Hydraulic Considerations

I. OVERVIEW

This chapter discusses some special hydraulic considerations that are beyond standard application problems. It includes a discussion of viscosity, describing terms and units used to measure viscosity. An example shows the reader how to adjust centrifugal pump performance when handling viscous liquids, and how to judge whether or not it makes sense to even try to use a centrifugal pump with a highly viscous liquid.

Section III of this chapter is devoted to a discussion of using computer software to design, analyze, and optimize piping systems and to select pumps.

Section IV describes the importance of proper piping design and installation to avoid cavitation, air entrainment, and other field problems.

Field testing — that is, producing a pump's performance curve by measuring total head, flow, and power for an installed pump — is demonstrated in Section VI. The usefulness of field testing in pump preventive maintenance and troubleshooting is explained. Some practical considerations, such as what is the best way to measure flow and head in the field and what types of devices are available to do this, are included. This includes a brief description of the most common types of field testing equipment, as well as a summary of the merits and shortcomings for each type of device.

II. VISCOSITY

The viscosity of a liquid can be thought of as the resistance of one layer of liquid to movement against another layer, or the internal friction within a liquid that resists a shear force. Viscous liquids can be found in everyday life, with examples including food items like syrup and catsup, personal care products such as shampoo and lotion, and automobile engine oil.

There are several categories of liquid based on their behavior when subjected to shear forces. *Newtonian liquids* are "well behaved," that is, they act like water. Their viscosity remains constant as rate of shear increases.

Some liquids, called *pseudo-plastic* liquids, have a decreasing viscosity with increasing rates of shear. These liquids, examples being grease and paint, are not difficult to pump because they tend to thin out when subjected to the rather high rates of shear inside a centrifugal pump.

Plastic liquids act just like pseudo-plastic ones, except that a certain force must be applied before the liquid begins to move. An example of a plastic liquid is tomato catsup.

Dilatant liquids behave in just the opposite way of pseudo-plastic liquids. That is, the viscosity increases in these liquids with increasing rates of shear. These liquids, examples being clay slurries and candy compounds, cannot generally be pumped with centrifugal pumps.

Viscosity is usually expressed in units of centipoise, centistokes, or SSU (Seconds Saybolt Universal). These terms are related by the following formulae:

$$\text{Centistokes} = \text{centipoise}/\text{s.g.} \qquad (3.1)$$

$$\text{Centistokes} = 0.22\ \text{SSU} - 180/\text{SSU} \qquad (3.2)$$

$$\text{SSU} = \text{Centistokes} \times 4.635\ (\text{for} >70\ \text{centistokes}) \qquad (3.3)$$

There are many other units used to express viscosity of liquids. These have been developed by different industries to describe their particular liquid products. Table 3.1 can be used for converting from other viscosity units into centistokes or SSU.

Special Hydraulic Considerations

Table 3.2 lists the viscosity of some common liquids, in units of SSU and centistokes, as well as the specific gravity of these liquids. Note that the values of viscosity are given at several temperatures, with viscosity decreasing as temperature increases. Note also that the range of viscosities in Table 3.2 is quite wide, ranging from less than 100 to 500,000 SSU.

If a centrifugal pump is being considered for pumping a viscous liquid, the pump performance curve must be adjusted for the effect of the viscosity. In general, pumping a viscous liquid causes some reduction in both head and capacity, and usually a significant reduction in efficiency. Figure 3.1 shows a chart that can be used for correcting pump performance when handling a viscous liquid. This chart is restricted for use only with Newtonian liquids and radial flow pumps.

To make a preliminary pump selection for a viscous application, enter the bottom of the chart in Figure 3.1 with the required viscous pump capacity (in gpm). Then move vertically on the chart to the sloped line representing the required head per stage of the pump, expressed in feet of the viscous liquid. For example, for a two-stage pump with a viscous rating of 500 gpm and 400 ft, enter the chart at 500 gpm, and move up to the 200-ft line.

Next, move horizontally on the chart (either to the right or to the left) to the sloped line representing the viscosity of the liquid (shown in the Figure 3.1 in both centistokes and in SSU). Finally, move vertically to the top of the chart, intersecting the correction factor curves C_η, C_Q, and C_H, which represent the correction factors for efficiency, capacity, and head, respectively. There are four separate head correction factors in Figure 3.1; but for the preliminary pump selection, use the correction factor curve labeled $1.0 \times Q_{NW}$.

The desired capacity and head of the pump when handling the viscous liquid are then divided by the respective correction factors C_Q and C_H obtained above (in decimal form), to determine the equivalent water rating of the pump, allowing a pump size to be selected. Once the pump size and impeller diameter are selected, the BEP of this pump with the chosen impeller diameter is used to re-enter the Figure 3.1

Table 3.1 Viscosity Conversion

Seconds Saybolt Universal (SSU)	Kinematic Viscosity (Centistokes)	Seconds Saybolt Furol (SSF)	Seconds Redwood 1 (Standard)	Seconds Redwood 2 (Admiralty)	Degrees Engler	Degrees Barbey	Seconds Parlin Cup #7	Seconds Parlin Cup #10	Seconds Parlin Cup #15	Seconds Parlin Cup #20	Seconds Ford Cup #3	Seconds Ford Cup #4
31	1.00	—	29	—	1.00	6200	—	—	—	—	—	—
35	2.56	—	32.1	—	1.16	2420	—	—	—	—	—	—
40	4.30	—	36.2	5.10	1.31	1440	—	—	—	—	—	—
50	7.40	—	44.3	5.83	1.58	838	—	—	—	—	—	—
60	10.3	—	52.3	6.77	1.88	618	—	—	—	—	—	—
70	13.1	12.95	60.9	7.60	2.17	483	—	—	—	—	—	—
80	15.7	13.70	69.2	8.44	2.45	404	—	—	—	—	—	—
90	18.2	14.44	77.6	9.30	2.73	348	—	—	—	—	—	—
100	20.6	15.24	85.6	10.12	3.02	307	—	—	—	—	—	—
150	32.1	19.30	128	14.48	4.48	195	—	—	—	—	—	—
200	43.2	23.5	170	18.90	5.92	144	40	—	—	—	—	—
250	54.0	28.0	212	23.45	7.35	114	46	—	—	—	—	—
300	65.0	32.5	254	28.0	8.79	95	52.5	15	6.0	3.0	30	20
400	87.6	41.9	338	37.1	11.70	70.8	66	21	7.2	3.2	42	28
500	110.0	51.6	423	46.2	14.60	56.4	79	25	7.8	3.4	50	34
600	132	61.4	508	55.4	17.50	47.0	92	30	8.5	3.6	58	40

Special Hydraulic Considerations

700	154	71.1	592	64.6	20.45	40.3	106	35	9.0	3.9	67	45
800	176	81.0	677	73.8	23.35	35.2	120	39	9.8	4.1	74	50
900	198	91.0	762	83.0	26.30	31.3	135	41	10.7	4.3	82	57
1000	220	100.7	896	92.1	29.20	28.2	149	43	11.5	4.5	90	62
1500	330	150	1270	138.2	43.80	18.7	—	65	15.2	6.3	132	90
2000	440	200	1690	184.2	58.40	14.1	—	86	19.5	7.5	172	118
2500	550	250	2120	230	73.0	11.3	—	108	24	9	218	147
3000	660	300	2540	276	87.60	9.4	—	129	28.5	11	258	172
4000	880	400	3380	368	117.0	7.05	—	172	37	14	337	230
5000	1100	500	4230	461	146	5.64	—	215	47	18	425	290
6000	1320	600	5080	553	175	4.70	—	258	57	22	520	350
7000	1540	700	5920	645	204.5	4.03	—	300	67	25	600	410
8000	1760	800	6770	737	233.5	3.52	—	344	76	29	680	465
9000	1980	900	7620	829	263	3.13	—	387	86	32	780	520
10000	2200	1000	8460	921	292	2.82	—	430	96	35	850	575
15000	3300	1500	13700	—	438	2.50	—	650	147	53	1280	860
20000	4400	2000	18400	—	584	1.40	—	860	203	70	1715	1150

From *Engineering Data Book*, 2nd edition. Courtesy of Hydraulic Institute, Parsippany, NJ; www.pumps.org and www.pumplearning.org.

Table 3.2 Viscosity of Common Liquids

Liquid	*Sp. Gr. at 60°F	Viscosity		At F
		SSU	Centistokes	
Freon	1.37 to 1.49 @ 70°F		.27–.32	70
Glycerine (100%)	1.26 @ 68°F	2,950	648	68.6
		813	176	100
Glycol:				
Propylene	1.038 @ 68°F	240.6	52	70
Triethylene	1.125 @ 68°F	185.7	40	70
Diethylene	1.12	149.7	32	70
Ethylene	1.125	88.4	17.8	70
Hydrochloric acid (31.5%)	1.05 @ 68°F		1.9	68
Mercury	13.6		.118	70
			.11	100
Phenol (carbolic acid)	0.95 to 1.08	65	11.7	65
Silicate of soda	40 Baume	365	79	100
	42 Baume	637.6	138	100
Sulfuric acid (100%)	1.83	75.7	14.6	68
Fish and Animal Oils:	0.918	220	47.5	130
Bone Oil		65	11.6	212

Special Hydraulic Considerations

Cod oil	0.928	150	32.1	100
		95	19.4	130
Lard	0.96	287	62.1	100
		160	34.3	130
Lard oil	0.912 to 0.925	190 to 220	41 to 47.5	100
		112 to 128	23.4 to 27.1	130
Menhaden oil	0.933	140	29.8	100
		90	18.2	130
Neatsfoot oil	0.917	230	49.7	100
		130	27.5	130
Sperm oil	0.883	110	23.0	100
		78	15.2	130
Whale oil	0.925	163 to 184	35 to 39.6	100
		97 to 112	19.9 to 23.4	130
Mineral Oils:				
Automobile Crankcase Oils				
(Average Midcontinent Paraffin Base):				
SAE 10	**0.880 to 0.935	165 to 240	35.4 to 51.9	100
		90 to 120	18.2 to 25.3	130
SAE 20	**0.880 to 0.935	240 to 400	51.9 to 86.6	100
		120 to 185	25.3 to 39.9	130

* Unless otherwise noted.
** Depends on origin or percent and type of solvent.

Table 3.2 Viscosity of Common Liquids (continued)

Liquid	*Sp. Gr. at 60°F	Viscosity		
		SSU	Centistokes	At F
SAE 30	**0.880 to 0.935	400 to 580 185 to 255	86.6 to 125.5 39.9 to 55.1	100 130
SAE 40	**0.880 to 0.935	580 to 950 255 to 80	125.5 to 205.6 55.1 to 15.6	100 130 210
SAE 50	**0.880 to 0.935	950 to 1,600 80 to 105	205.6 to 352 15.6 to 21.6	100 210
SAE 60	**0.880 to 0.935	1,600 to 2,300 105 to 125	352 to 507 21.6 to 26.2	100 210
SAE 70	**0.880 to 0.935	2,300 to 3,100 125 to 150	507 to 682 26.2 to 31.8	100 210
SAE 10W	**0.880 to 0.935	5,000 to 10,000	1,100 to 2,200	0
SAE 20W	**0.880 to 0.935	10,000 to 40,000	2,200 to 8,800	0
Automobile Transmission Lubricants:				
SAE 80	**0.880 to 0.935	100,000 max	22,000 max	0
SAE 90	**0.880 to 0.935	800 to 1,500 300 to 500	173.2 to 324.7 64.5 to 108.2	100 130

Special Hydraulic Considerations

	Specific gravity	SSU	°F
SAE 140	**0.880 to 0.935	950 to 2,300	130
		120 to 200	210
		205.6 to 507	
		25.1 to 42.9	
SAE 250	**0.880 to 0.935	Over 2,300	130
		Over 200	210
		Over 507	
		Over 42.9	
Crude oils:			
Texas, Oklahoma	0.81 to 0.916	40 to 783	60
		34.2 to 210	100
		4.28 to 169.5	
		2.45 to 45.3	
Wyoming, Montana	0.86 to 0.88	74 to 1,215	60
		46 to 320	100
		14.1 to 263	
		6.16 to 69.3	
California	0.78 to 0.92	40 to 4,840	60
		34 to 700	100
		4.28 to 1,063	
		2.4 to 151.5	
Pennsylvania	0.8 to 0.85	46 to 216	60
		38 to 86	100
		6.16 to 46.7	
		3.64 to 17.2	
Diesel engine lubricating oils (based on average midcontinent paraffin base):			
Federal Specification No. 9110	**0.880 to 0.935	165 to 240	100
		90 to 120	130
		35.4 to 51.9	
		18.2 to 25.3	
Federal Specification No. 9170	**0.880 to 0.935	300 to 410	100
		140 to 180	130
		64.5 to 88.8	
		29.8 to 38.8	

* Unless otherwise noted.
** Depends on origin or percent and type of solvent.

Table 3.2 Viscosity of Common Liquids (continued)

Liquid	*Sp. Gr. at 60°F	Viscosity		At F
		SSU	Centistokes	
Federal Specification No. 9250	**0.880 to 0.935	470 to 590 200 to 255	101.8 to 127.8 43.2 to 55.1	100 130
Federal Specification No. 9370	**0.880 to 0.935	800 to 1,100 320 to 430	173.2 to 238.1 69.3 to 93.1	100 130
Federal Specification No. 9500	**0.880 to 0.935	490 to 600 92 to 105	106.1 to 129.9 18.54 to 21.6	130 210
Diesel Fuel Oils:				
No. 2 D	**0.82 to 0.95	32.6 to 45.5 39	2 to 6 1 to 3.97	100 130
No. 3 D	**0.82 to 0.95	45.5 to 65 39 to 48	6 to 11.75 3.97 to 6.78	100 130
No. 4 D	**0.82 to 0.95	140 max 70 max	29.8 max 13.1 max	100 130
No. 5 D	**0.82 to 0.95	400 max 165 max	86.6 max 35.2 max	122 160
Fuel Oils:				
No. 1	**0.82 to 0.95	34 to 40 32 to 35	2.39 to 4.28 2.69	70 100

Special Hydraulic Considerations

No. 2	**0.82 to 0.95	36 to 50	3.0 to 7.4	70
		33 to 40	2.11 to 4.28	100
No. 3	**0.82 to 0.95	35 to 45	2.69 to 0.584	100
		32.8 to 39	2.06 to 3.97	130
No. 5A	**0.82 to 0.95	50 to 125	7.4 to 26.4	100
		42 to 72	4.91 to 13.73	130
No. 5B	**0.82 to 0.95	125 to	26.4 to	100
		400	86.6	122
		72 to 310	13.63 to 67.1	130
No. 6	**0.82 to 0.95	450 to 3,000	97.4 to 660	122
		175 to 780	37.5 to 172	160
Fuel Oil–Navy Specification	**0.989 max	110 to 225	23 to 48.6	122
		63 to 115	11.08 to 23.9	160
Fuel Oil–Navy II	1.0 max	1,500 max	324.7 max	122
		480 max	104 max	160
Gasoline	0.68 to 0.74		0.46 to 0.88	60
			0.40 to 0.71	100
Gasoline (Natural)	76.5 degrees API		0.41	68
Gas Oil	28 degrees API	73	13.9	70
		50	7.4	100

* Unless otherwise noted.
** Depends on origin or percent and type of solvent.

Table 3.2 Viscosity of Common Liquids (continued)

Liquid	*Sp. Gr. at 60°F	Viscosity		At F
		SSU	Centistokes	
Insulating Oil: Transformer, switches and circuit breakers		115 max 65 max	24.1 max 11.75 max	70 100
Kerosene	0.78 to 0.82	35 32.6	2.69 2	68 100
Machine Lubricating Oil (Average Pennsylvania Paraffin Base):				
Federal Specification No. 8	**0.880 to 0.935	112 to 160 70 to 90	23.4 to 34.3 13.1 to 18.2	100 130
Federal Specification No. 10	**0.880 to 0.935	160 to 235 90 to 120	34.3 to 50.8 18.2 to 25.3	100 130
Federal Specification No. 20	**0.880 to 0.935	235 to 385 120 to 185	50.8 to 83.4 25.3 to 39.9	100 130
Federal Specification No. 30	**0.880 to 0.935	385 to 550 185 to 255	83.4 to 119 39.9 to 55.1	100 130
Mineral Lard Cutting Oil: Federal Specification Grade 1		140 to 190 86 to 110	29.8 to 41 17.22 to 23	100 130

Special Hydraulic Considerations

Federal Specification Grade 2		190 to 220	41 to 47.5	100
		110 to 125	23 to 26.4	130
Petrolatum	0.825	100	20.6	130
		77	14.8	160
Turbine Lubricating Oil:				
Federal Specification (Penn Base)	0.91 Average	400 to 440	86.6 to 95.2	100
		185 to 205	39.9 to 44.3	130
VEGETABLE OILS:				
Castor Oil	0.96 @ 68 F	1,200 to 1,500	259.8 to 324.7	100
		450 to 600	97.4 to 129.9	130
China Wood Oil	0.943	1,425	308.5	69
		580	125.5	100
Cocoanut Oil	0.925	140 to 148	29.8 to 31.6	100
		76 to 80	14.69 to 15.7	130
Corn Oil	0.924	135	28.7	130
		54	8.59	212
Cotton Seed Oil	0.88 to 0.925	176	37.9	100
		100	20.6	130
Linseed Oil, Raw	0.925 to 0.939	143	30.5	100
		93	18.94	130

* Unless otherwise noted.
** Depends on origin or percent and type of solvent.

Table 3.2 Viscosity of Common Liquids (continued)

Liquid	*Sp. Gr. at 60°F	Viscosity		At F
		SSU	Centistokes	
Olive Oil	0.912 to 0.918	200	43.2	100
		115	24.1	130
Palm Oil	0.924	221	47.8	100
		125	26.4	130
Peanut Oil	0.920	195	42	100
		112	23.4	130
Rape Seed Oil	0.919	250	54.1	100
		145	31	130
Rosin Oil	0.980	1,500	324.7	100
		600	129.9	130
Rosin (Wood)	1.09 Avg.	500 to 20,000	108.2 to 4,400	200
		1,000 to 50,000	216.4 to 11,000	190
Sesame Oil	0.923	184	39.6	100
		110	23	130
Soja Bean Oil	0.927 to 0.98	165	35.4	100
		96	19.64	130
Turpentine	0.86 to 0.87	33	2.11	60
		32.6	2.0	100

SUGAR, SYRUPS, MOLASSES, ETC.				
Corn Syrups	1.4 to 1.47	5,000 to 500,000 1,500 to 60,000	1,100 to 110,000 324.7 to 13,200	100 130
Glucose	1.35 to 1.44	35,000 to 100,000 4,000 to 11,000	7,700 to 22,000 880 to 2,420	100 150
Honey (Raw)		340	73.6	100
Molasses "A" (First)	1.40 to 1.46	1,300 to 23,000 700 to 8,000	281.1 to 5,070 151.5 to 1,760	100 130
Molasses "B" (Second)	1.43 to 1.48	6,400 to 60,000 3,000 to 15,000	1,410 to 13,200 660 to 3,300	100 130
Molasses "C" (Blackstrap or final)	1.46 to 1.49	17,000 to 250,000 6,000 to 75,000	2,630 to 5,500 1,320 to 16,500	100 130
Sucrose Solutions (Sugar Syrups):				
60 Brix	1.29	230 92	49.7 18.7	70 100
62 Brix	1.30	310 111	67.1 23.2	70 100
64 Brix	1.31	440 148	95.2 31.6	70 100
66 Brix	1.326	650 195	140.7 42.0	70 100
68 Brix	1.338	1,000 275	216.4 59.5	70 100

* Unless otherwise noted.
** Depends on origin or percent and type of solvent.

Table 3.2 Viscosity of Common Liquids (continued)

Liquid	*Sp. Gr. at 60°F	Viscosity		
		SSU	Centistokes	At F
70 Brix	1.35	1,650	364	70
		400	86.6	100
72 Brix	1.36	2,700	595	70
		640	138.6	100
74 Brix	1.376	5,500	1,210	70
		1,100	238	100
76 Brix	1.39	10,000	2,200	70
		2,000	440	100
TARS:				
Tar-Coke Oven	1.12+	3,000 to 8,000	600 to 1,760	71
		650 to 1,400	140.7 to 308	100
Tar-Gas House	1.16 to 1.30	15,000 to 300,000	3,300 to 66,000	70
		2,000 to 20,000	440 to 4,400	100
Road Tar:				
Grade RT-2	1.07+	200 to 300	43.2 to 64.9	122
		55 to 60	8.77 to 10.22	212
Grade RT-4	1.08+	400 to 700	86.6 to 154	122
		65 to 75	11.63 to 14.28	212
Grade RT-6	1.09+	1,000 to 2,000	216.4 to 440	122
		85 to 125	16.83 to 26.2	212

Special Hydraulic Considerations

Material	Specific Gravity*	Viscosity SSU	Viscosity Centistokes	Temperature °F
Grade RT-8	1.13+	3,000 to 8,000 150 to 225	660 to 1,760 31.8 to 48.3	122 212
Grade RT-10	1.14+	20,000 to 60,000 250 to 400	4,400 to 13,200 53.7 to 86.6	122 212
Grade RT-12	1.15+	114,000 to 456,000 500 to 800	25,000 to 75,000 108.2 to 173.2	122 212
Pine Tar	1.06	2,500 500	559 108.2	100 132
MISCELLANEOUS				
Corn Starch Solutions:				
22 Baume	1.18	150 130	32.1 27.5	70 100
24 Baume	1.20	600 440	129.8 95.2	70 100
25 Baume	1.21	1400 800	303 173.2	70 100
Ink-Printers	1.00 to 1.38	2,500 to 10,000 1,100 to 3,000	550 to 2,200 238.1 to 660	100 130
Tallow	0.918 Avg.	56	9.07	212
Milk	1.02 to 1.05		1.13	68

* Unless otherwise noted.
** Depends on origin or percent and type of solvent.

Table 3.2 Viscosity of Common Liquids (continued)

Liquid	*Sp. Gr. at 60°F	Viscosity		
		SSU	Centistokes	At F
Varnish-Spar	0.9	1425	313	68
		650	143	100
Water-Fresh	1.0		1.13	60
			0.55	130

* Unless otherwise noted.
** Depends on origin or percent and type of solvent.
From *Engineering Data Book*, 2nd edition. Courtesy of Hydraulic Institute, Parsippany, N.J., www.pumps.org and www.pumplearning.org.

Special Hydraulic Considerations

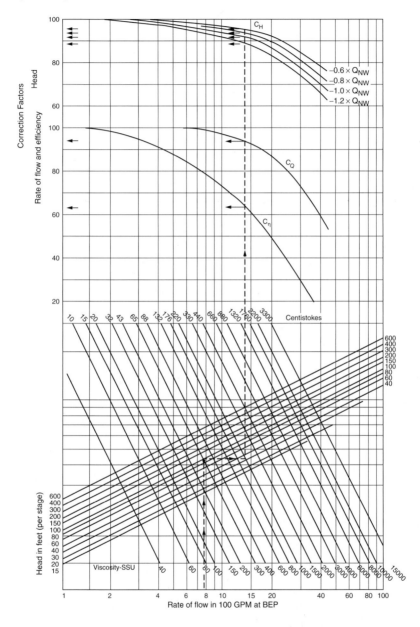

Figure 3.1 Performance correction chart for pumps handling viscous liquids. (Courtesy of Hydraulic Institute, Parsippany, NJ; www.pumps.org and www.pumplearning.org.)

correction chart to obtain revised correction factors. These final correction factors are then used to create a new pump curve for the pump handling the viscous liquid.

The following example of selecting a pump for a viscous application illustrates the use of the Figure 3.1 performance correction chart.

EXAMPLE 3.1: Selecting a pump for a viscous service, calculating effect of viscosity on performance, and developing new pump curves.

PROBLEM: Select a pump to deliver 750 gpm at 100 ft total head of a liquid having a viscosity of 1000 SSU and a specific gravity of 0.90 at the pumping temperature. (From Table 3.2, SAE 40 or SAE 50 automobile crankcase oil at about 100°F has a viscosity of about 1000 SSU.)

SOLUTION: Enter the Figure 3.1 chart with 750 gpm, go up to 100 ft head, over to 1000 SSU, and then up to the preliminary correction factors.

$C_Q = 0.95$
$C_H = 0.92$ (for $1.0 \times Q_{NW}$)
$C_\eta = 0.635$

Note that these preliminary correction factors allow a rough estimate of the effect on performance of the viscous liquid described in this example. In this case, pumping this liquid will reduce capacity by about 5%, head by about 8%, and efficiency by about 36.5%, compared to the pump's performance when handling water.

Using the above preliminary correction factors, the equivalent water rating is:

$Q_W = 750/0.95 = 790$ gpm
$H_W = 100/0.92 = 109$ ft

Special Hydraulic Considerations

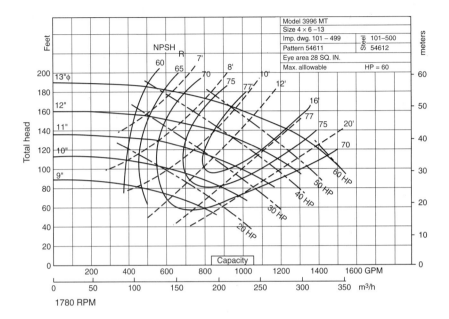

Figure 3.2 Pump chosen for viscosity problem, Example 3.1. (Courtesy of Goulds Pumps, Inc., ITT Industries, Seneca Falls, NY.)

Using this equivalent water rating, the pump whose curve is shown in Figure 3.2 is chosen for the service, with an impeller diameter of 10¾ in. For this pump, the best efficiency point on the water curve at the selected impeller diameter (Q_{NW}) is 850 gpm and 106 ft. Now re-enter the correction chart Figure 3.1 with this BEP data from the water curve. Enter the chart at 850 gpm, go up to 106 ft, over to 1000 SSU, and then up to the revised correction factors.

$C_Q = 0.96$
$C_H = 0.96$ (for $0.6 \times Q_{NW}$)
$C_H = 0.95$ (for $0.8 \times Q_{NW}$)
$C_H = 0.93$ (for $1.0 \times Q_{NW}$)
$C_H = 0.90$ (for $1.2 \times Q_{NW}$)
$C_\eta = 0.65$

Table 3.3 Viscous Pump Performance: Example 3.1

	$0.6 \times Q_{NW}$	$0.8 \times Q_{NW}$	$1.0 \times Q_{NW}$	$1.2 \times Q_{NW}$
Q_W	510.0	680.0	850.0	1020.0
H_W	125.0	117.0	106.0	93.0
η_W	67.5	74.2	77.0	75.0
C_Q	0.96	0.96	0.96	0.96
C_H	0.96	0.95	0.93	0.90
C_η	0.65	0.65	0.65	0.65
Q_V	490.0	653.0	816.0	979.0
H_V	120.0	111.0	99.0	83.7
η_V	43.9	48.2	50.1	49.0

Now, on the water curve (Figure 3.2 with 10¾ in. impeller diameter), pick off head and efficiency at the following flows:

$0.6 \times Q_{NW} = 0.6 \times 850 = 510$ gpm
$0.8 \times Q_{NW} = 0.8 \times 850 = 680$ gpm
$1.0 \times Q_{NW} = 1.0 \times 850 = 850$ gpm
$1.2 \times Q_{NW} = 1.2 \times 850 = 1020$ gpm

Using these data points from the water curve and the revised correction factors from above, Table 3.3 can be constructed to generate four points on the viscous pump curve. Water curve data points are labeled Q_W, H_W, and η_W. Viscous curve data points, labeled Q_V, H_V, and η_V, are obtained by multiplying the water data points by their respective correction factors.

Finally, the viscous brake horsepower is calculated for each of the above four data points, using the formula for horsepower, Equation 2.15; the given specific gravity, 0.9; and the values for Q_V, H_V, and η_V from Table 3.3.

Special Hydraulic Considerations

$$BHP_V = (Q_V \times H_V \times \text{s.g.})/(3960 \times \eta_V)$$
$$= 30.4 \text{ (for } 0.6 \times Q_{NW})$$
$$= 34.2 \text{ (for } 0.8 \times Q_{NW})$$
$$= 36.6 \text{ (for } 1.0 \times Q_{NW})$$
$$= 38.0 \text{ (for } 1.2 \times Q_{NW})$$

With the viscous performance values of Q_V, H_V, and η_V from Table 3.3 and the above values for BHP_V, a complete viscous performance curve can now be drawn.

Note that the preceding discussion and example made no mention of NPSH. Very little information is available to predict the effect of viscosity on $NPSH_r$. For values of viscosity below 2000 SSU, it is safe to use the $NPSH_r$ from a curve based on water. For values above 2000 SSU, experimentation or other experience with similar applications are about the only choices for determining the suitability of the application. The best approach is to have the largest margin possible between $NPSH_a$ and $NPSH_r$.

The correction chart shown in Figure 3.1 can be used to very quickly decide what will be the penalty in efficiency for pumping a viscous liquid with a centrifugal pump. This is a quick way to decide if it makes sense to consider a centrifugal pump for the application, or whether a positive displacement pump should be used for the application. Remember that it may be possible to heat the liquid to reduce its viscosity and thus make a centrifugal pump a more viable option. Also, the computer software discussed in Section III to follow and the demonstration CD that accompanies this book can be used to correct centrifugal pump performance for viscous liquids.

Figure 3.3 shows a correction chart similar to the one shown in Figure 3.1, but for small pumps only (capacities less than 100 gpm). The Figure 3.3 chart is used in the same manner as the example above, except that for smaller pumps there is only one head correction factor C_H.

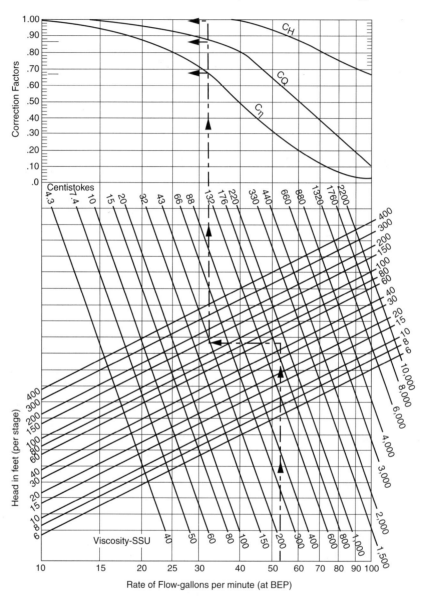

Figure 3.3 Performance correction chart for small pumps (capacities of 100 gpm or less) handling viscous liquids. (Courtesy of the Hydraulic Institute, Parsippany, NJ; www.pumps.org and www.pumplearning.org.)

III. SOFTWARE TO SIZE PUMPS AND SYSTEMS

A. General

Chapter 2 described manual techniques for sizing the components of a pumping system, calculating pump TH, and then selecting a pump for this service. The chapter explained how to develop a system head curve to study how a pump will perform when pumping through various branches of the piping system. Finally, Chapter 2 explained how to account for the use of multiple pumps in the system. A thorough review of Chapter 2 reveals that the sizing of pumps and systems can be tedious and time consuming. This is particularly true if the system contains multiple branches, loops, or if multiple pumps are used in the system. Furthermore, if the pumped liquid is something other than water, tables to determine friction losses through the pipes, valves, and fittings may not be readily available. The same dilemma might exist if the pipe wall thickness and material is different from standard schedule 40 steel.

Because of these limitations, many pump system designers must make their best approximation of pump head in these situations. This usually results in the pump being oversized, and the adjustments to make the pump operate at the desired capacity are done with control valves or orifices.

As the example in Chapter 2, Section XII, showed oversizing a pump usually results in a more expensive pump, and a sizable increase in operating cost. Furthermore, because of the tedious and repetitive calculations that must be performed by the engineer in designing a piping system, few systems ever actually are optimized when they are being designed.

Fortunately, there are solutions to these dilemmas. Several computer software packages are available that significantly reduce the time it takes to design a piping system, while improving the capability to optimize the system design. The software packages vary widely in their ease of use, their capability to model complex systems, their flexibility to handle a range of liquids and pipe materials, their user interface features, and their capability to optimize the system design.

B. Value of Piping Design Software

Piping software provides a simulated view of how the entire piping system operates. An engineer can use this information to both design and optimize the piping system. A complete piping system model can be used at the operating plant to provide everyone with a clear picture of how the system operates. How the piping system model is used depends on the duties and responsibilities of the user:

- Design engineers can use the information to size and optimize individual pipe lines. They are interested in calculating the pressure drop or head loss in the pipe lines, and in optimizing the design.
- Rotating equipment specialists are interested in calculating the design point needed for pump selection. After selecting the pump, they want to see how it operates in the piping system.
- Plant or project engineers are interested in seeing how a piping system operates, as well as in maintaining proper system operation as system changes are made.
- Plant maintenance personnel use the model to identify, isolate, and correct problems in the piping system that cause excessive wear to pumps and control valves.
- Finally, plant operators are interested in seeing how the system operates under a variety of expected operating conditions.

Ultimately, an accurate piping system model is a valuable resource for everyone involved with the system, and can provide a clear picture for the life of the plant.

C. Evaluating Fluid Flow Software

When evaluating fluid piping software, it is important to determine what features are necessary to meet the needs of the people who will use it, and then choose the program that meets the majority of these requirements.

A variety of piping system software is currently available. To be effective, any piping software being considered should incorporate the following features:

Special Hydraulic Considerations 187

- Provide an easy-to-understand schematic showing how the pipe lines, pumps, tanks, and components are connected.
- Perform the hydraulic network analysis needed to calculate the balanced flow rates and pressures for each pipe line in the system.
- Provide the results in a format that is easy to understand, and can also be shared by everyone involved with the design, operation, or maintenance of the system.

The remainder of this section describes the features found in a variety of fluid flow programs. Each section describes the basic features needed to meet minimum requirements, as well as "enhanced" features that provide extra value to the user. All the programs listed in Table 3.4 meet all of the basic requirements. At the time of this writing, none of the programs listed in Table 3.4 contain all the enhanced features. Because software developers are continually adding new features, no effort has been made to identify what enhanced features are available for each program, and that is left to the person evaluating the software.

Because of the differences in features and ease of use of the available programs, interested readers are urged to study the product literature and demonstration disks available from the software developers. One such demo CD for the PIPE-FLO software has been included with this book.

D. Building the System Model

The first step in modeling a fluid piping system is to create the system model. Because building the system model is the most time-consuming process, it is important to select a program that streamlines this task as much as possible. The ability to build the piping system model using a piping schematic drawing is a basic feature of any fluid piping program. This includes the ability to place piping system items such as pumps, components, heat exchangers, tanks, and controls on the piping schematic. The various system items are connected with individual pipe lines.

Table 3.4 Software for Designing or Analyzing Piping Systems and Components

Program	Manufacturer	Website	Phone Number
Design Flow Solutions	ABZ Inc.	www.abzinc.com	(800)747-7401
Fathom	Applied Flow Technology	www.aft.com	(800)589-4943
Flow of Fluids Premium	Crane Valve Co.	www.cranevalve.com	(888)845-4992
FluidFlow Info	Flite Software Ltd.	www.fluidflowinfo.com	44 28 71279227 (United Kingdom)
KYPIPE 2000	KYPIPE LCC Software Center	www.kypipe.com	(859)263-2234
PIPE-FLO Professional	Engineered Software, Inc	www.eng-software.com	(800)786-8545
PIPENET	Sunrise Systems Ltd.	www.sunrise-sys.com	44 1223 441311 (United Kingdom)
SINET	Epcon International	www.epcon.com	(800)367-3585

Special Hydraulic Considerations

Not only does the piping drawing show all of the items in the system, but it also provides the connectivity information needed by the software to set up the equations needed to perform the balanced flow rate and pressure calculations.

Another basic requirement is the ability of the program to automatically trace the loops in the system, set up the equations, and supply the initial guesses needed to perform the calculations.

The user can build the piping schematic by selecting items from menus or button bars, placing the items on the piping schematic, and entering information about each item. As the various items are added to the piping schematic, the user has visual indication as to what is added and what remains to be added.

The data needed for the fluid flow calculations are entered in dialog boxes. If the program relies on manual entry of data, for example an inside pipe diameter, the user must look up the information in a handbook and type in the value. If the program uses pipe data tables to look up pipe properties, the user selects the pipe material, a corresponding pipe schedule or wall thickness from a displayed list, and then chooses an available pipe nominal size. The program then looks up the inside diameter from the pipe table. The program should also be able to look up the fluid properties (density, viscosity, and vapor pressure) at the operating temperature from data tables that are imbedded in the software.

Following are the enhanced features found in some of the programs listed on Table 3.4.

1. Copy Command

The ability to copy and paste items on the piping schematic simplifies the creation of a piping system model. For example, when modeling a discharge manifold, the ability to copy pipe segments and paste them into a manifold saves time in building the model. The copy feature should copy both the object on the piping drawing and the underlying design information. Some programs support a group copy command, capable of copying multiple objects at one time and pasting them to the piping schematic.

2. Customize Symbols

Each object placed on the piping schematic typically has a unique symbol shape. The user can easily identify the pumps, tanks, and controls just by looking at the drawing. If the program supports multiple symbol shapes for each type of item, users can further customize the look of their piping schematic. Some of the programs listed even allow users to create their own symbol shapes.

3. CAD Drawing Features

The ability to zoom and pan around the piping schematic makes it easy to get around large piping system models. The ability to insert new pipelines and piping system objects into the existing piping schematic greatly streamlines the building of the model. The ability to move items around the screen and add new vertices to pipelines lets the user give the piping schematic the look and feel of an established drawing. The ability to place notes on the piping schematic, change the size of the symbols and text, and change the pipe colors or line thickness on the schematic increases the presentation value of the program.

4. Naming Items

Each item in a piping system typically has a name or equipment identifier. Many of the programs listed allow the user to attach a name to the object and to display that name on the piping schematic. Using the user company's naming convention on the piping system model makes it easier for everyone to understand the model.

5. Displaying Results

The ability to display the calculated results on the piping schematic makes it easier to understand the operation of the piping system. Seeing the flow rate printed on the piping schematic next to the pipeline, or the Total Head of a pump displayed next to the pump, provides the user with a clear picture of how the piping system operates.

Special Hydraulic Considerations

6. The Look of the Piping Schematic

If the piping schematic has the look of a generally recognized drawing, it is easier for everyone to recognize the system model. Many of the programs listed allow the user to create a piping schematic that looks like a standard process flow diagram or P&ID (Piping & Instrumentation Drawing). Often, piping drawings are laid out in isometric view. This type of drawing provides an indication of elevation on the various items on the drawing. If the program has an isometric grid feature, it is very easy to develop a professional-looking piping isometric.

E. Calculating the System Operation

The primary function of the piping system software is to calculate the balanced flow rates and pressures for the entire system. The program must also factor in the effect of pumps, tanks, components, and control valves. Reviewing the calculated results, the user can see how everything works together as a system.

Hydraulic network analysis techniques are used to calculate the balanced flow rates and pressures. Without getting into the technical details, the programs perform the calculations in the following ways:

1. Determine how all the pipe lines in the system are connected and then set up the necessary simultaneous equations needed to perform the balancing calculations.
2. Establish an initial guess to balance the flow rates in the system.
3. The initial guess is used to calculate the pressure drops around the various loops in the system.
4. If the loop pressure drops are not balanced, the program improves on the flow rate guess, and then repeats the pressure calculations.
5. Once the flow rates and pressure drop calculations have converged to a solution, the results are displayed.

All programs listed in Table 3.4 calculate the balanced flow rates and pressures as outlined above. A list of enhanced calculation features that provide additional value to the program is included below.

1. Sizing Pipe Lines

If the user is working on a new piping system or adding a pipe to an existing system, it is often helpful to size a single pipe line and determine its pressure drop. This allows the user to optimize the pipe line prior to adding it to the piping system model, by comparing the pumping energy and initial installed cost for alternate pipe sizes. Many of the programs listed have a pipe line-sizing feature built in, or have a separate pipe line-sizing program.

2. Calculating Speed

The speed at which the program calculates the balanced flow rates and pressures is based on the speed of the computer running the program, the efficiency of the calculation method used in the program, and the speed of the programming language used in writing the program. For small piping systems of 20 or 30 pipe lines, the listed programs all have acceptable calculation speeds. When working with larger piping system models, say 100 pipe lines or more, the program calculation speed can become an important factor in software selection.

3. Showing Problem Areas

The ability to highlight results in the piping model that are outside the expected ranges provides the user with the ability to quickly identify and correct problem areas in the system. For example, if the program provides a list of pipe lines with high fluid velocities, the user can see which pipe lines are undersized for the required flow rate.

4. Equipment Selection

Often, the calculated results derived by the program are used for equipment selection. Knowing the TH and $NPSH_a$ is a

Special Hydraulic Considerations

requirement for proper pump selection. The ability to select pumps and control valves from manufacturers' catalogs is a valuable program feature. Once the equipment is selected, the user should be able to easily add the pumps and control valves to the model to provide a picture of how the selected equipment operates in the system.

5. Alternate System Operational Modes

A system is designed to meet specific flow rate and pressure requirements. The original design values are also used to size pipe lines, pumps, and control valves. During normal plant operations, the piping system may run at different conditions: lower flow rates, different tank levels, etc. The ability to see how the system operates under these expected conditions provides an invaluable record of the system design.

F. Communicating the Results

Sharing information with others is a vital part of any piping project. Once the system is modeled, everyone in the project can share the results. This requires the printing of the calculated results in a format that everyone can use.

Typically, hydraulic analysis software calculates the pressures and flow rates in pipe lines and shows how pumps, components, and control valves are operating. The software should provide all the information needed for someone to determine how the system is operating. All the programs listed in Table 3.4 have the ability to send the calculated results to a printer.

Below is a list of enhanced results features that provide additional value to the program.

1. Viewing Results within the Program

The calculated results are the primary output of the program. This information is used to select equipment, optimize pipe lines, balance the system, or identify when the results are outside the expected range. The ability to view the results within the program without sending them to a printer not only saves paper, but lets the user quickly try design alternatives.

2. Incorporating User-Defined Limits

When looking at the calculated results, it is often helpful to compare the results to established design guidelines. For example, if the fluid velocity in a pipe line is too great, that can be an indication that the pipe diameter is too small for the required flow rate. The ability to set design guides or limits and have the program indicate when these limits are exceeded makes it easier to identify and correct problem areas within the system.

3. Selecting the Results to Display

The program's printed results document how the system operates; and for large systems, the reports can be lengthy. Some of the listed programs give the user the option of choosing the reports the user wishes to print. Some programs give the user the opportunity to choose only a part of the results to print, thereby minimizing the size of the printed report. Finally, some programs let the user customize reports by selecting the fields to print, as well as the order in which they are listed.

4. Plotting the Piping Schematic

The piping schematic drawings for large systems can be very complex and require the ability to send the report to a large format printer or plotter. This is an important feature to consider if the user typically works on large piping systems. Some of the programs support tiled printing, the ability to print the piping schematic on multiple pages. With tiled printing, the piping schematic is divided into tiles, with the portion of the system on each tile printed to a single page. Once all the tiles are printed, they can be taped together, providing a large format drawing.

5. Exporting the Results

The ability to export the design data and calculated results to a text file makes it easy to share results between programs.

Special Hydraulic Considerations

For example, the results can be imported into a spreadsheet program to provide input information for external calculations, or to allow for the generation of a custom report using the page layout capability of the spreadsheet.

6. Sharing Results with Others

E-mail is rapidly becoming the preferred method of communicating between associates and customers. The ability to send piping system reports as e-mail attachments makes it easier to share results with others. Some programs create reports in a universal format such as Hypertext Markup Language (HTML) or Portable Document Format (PDF). These report files then become attachments to an e-mail message. The recipients then open the attached file with the appropriate reader program, where they can view and print the results. This advanced printing feature helps streamline the workflow.

7. Sharing Results Using a Viewer Program

To provide a greater degree of information sharing, some of the programs listed in Table 3.4 support proprietary viewer formats that contain all the calculated results and design information. The creator of the piping system model sends a viewer file that the recipient opens with a corresponding viewer program. The recipient can then view and print all the calculated reports with the viewer program.

G. Conclusion

Anyone who designs or analyzes pump systems or selects pumps should consider using computer software to assist in this effort. The capability of software solutions to save time, to reduce overall system costs, to handle extremely complex systems, and to permit optimization of the system are all important reasons to use these valuable tools.

In the past, fluid flow software was primarily used as an engineering and design tool. Because of improved ease of use, more people working with piping systems are using piping software to simulate the operation of the system. When searching for fluid piping software, it is important to consider the primary use of the software and which package best meets the user's needs. Any program being considered should have the basic features outlined in this section. It is left to the reader to make the determination of what enhanced features are important, and which fluid piping program will best meet the needs of the user company.

H. List of Software Vendors

Table 3.4 provides a list of fluid piping software, along with the company and contact information. Most programs listed have a demonstration disk that can be downloaded from their website or supplied on CD-ROM.

IV. PIPING LAYOUT

Improper piping layout, particularly on the suction side of the pump, can lead to several serious problems with pump operation, and to excessive pump maintenance costs. This section provides some guidelines on piping design and layout to help minimize operation and maintenance problems.

Suction piping should be as short and as straight as possible. Short suction lines minimize friction losses to help ensure adequate $NPSH_a$. Straight suction lines help the liquid enter the impeller in a straight line, to minimize uneven loading of the impeller and bearings. Keep suction line velocities in the 4 to 6-ft/sec range. Lower suction velocities are permissible if the pumped liquid is free of solids that must be kept in suspension.

Care should be taken if an elbow is located at the suction of a double suction pump. Chapter 4, Sections II.B and VI, describes double suction pumps. One of the primary benefits of a double suction pump is that, in single-stage configuration, the impeller produces almost no axial thrust. (See Chapter 4, Section II.D, for a discussion of axial thrust.) If an elbow is located at the suction of a double suction pump, the elbow should be in a vertical rather than horizontal orientation. Having an elbow in a horizontal orientation permits an

Special Hydraulic Considerations 197

uneven division of the flow between the two sides of the double suction impeller, which leads to excessive axial thrust. If the elbow must absolutely be located in the horizontal plane, it should be located several pipe diameters away from the pump. Also, several companies make flow straightening diffusers that can be located at the pump suction downstream of the elbow to help straighten the flow.

Check valves are notorious for having problems with sticking or not seating properly, so a check valve located at the discharge of a pump should always be able to be isolated from the system pressure for maintenance.

Allowing air to get into a pump can reduce head and flow, increase noise levels, and most importantly, lead to increased radial bearing loads, which can cause premature failure of bearings and seals, or shaft breakage. Sometimes, this *air entrainment* is confused with cavitation (Chapter 2, Section VI) because the symptoms are similar. Several aspects of pump system piping that can introduce air into a pump are described below.

Suction piping is generally one size larger than the suction flange of a pump to minimize the friction losses in the suction piping. This means that a reducer is required at the pump suction flange. If the pump is operating on a lift, the reducer should be an eccentric, rather than a concentric reducer (Figure 3.4). The reason for the eccentric reducer is that if the pump is operating on a lift, the pressure in the suction line is below atmospheric. Therefore, any leaks in the pipe flange joints can introduce air into the suction piping. If the reducer at the pump suction flange is concentric, it would have a small cavity at the top of the reducer next to the pump suction flange, which is higher than the suction flange itself. This is a cavity where air could accumulate and cause problems with pumping.

For systems with the pump operating on a suction lift, horizontal pipe runs on the suction line should be continuously sloping upward from the sump to the pump (Figure 3.4). The reason for this requirement is the same as in the previous paragraph. To do otherwise might create a cavity located higher than the top of the pump suction flange, where air

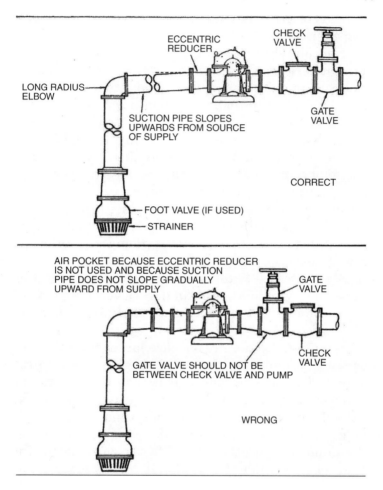

Figure 3.4 Use an eccentric rather than concentric reducer, and slope suction lines upward from the source if operating on a lift. Check valve should be able to be isolated for service. (Courtesy of Goulds Pumps, Inc., ITT Industries, Seneca Falls, NY.)

might accumulate and eventually pass into the pump or partially block flow into the pump.

Return lines back to a pump supply vessel should not be allowed to free fall into the vessel (Figure 3.5), as the turbulence caused by the falling water can produce air bubbles,

Special Hydraulic Considerations

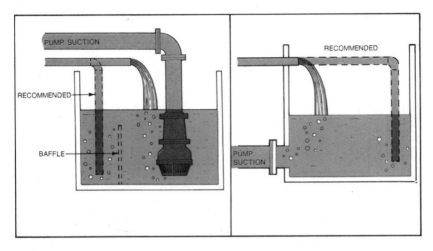

Figure 3.5 Return lines back to a pump suction vessel should not be allowed to free-fall into the vessel. (Courtesy of Goulds Pumps, Inc., ITT Industries, Seneca Falls, NY.)

which can move into the pump and accumulate. This guideline applies whether the pump is operating on a suction lift or on a suction head. Return lines should always be submerged below the liquid level, and located as far as practical in the vessel away from the pump suction connection. An additional measure is to separate the return line from the pump suction with a baffle, so that any air bubbles that are permitted to enter the system will float to the surface in the supply vessel, rather than being reintroduced into the pump.

Vortex formation is another phenomenon that can introduce air into a centrifugal pump pumping from an open sump. A vortex is a rotating swirl of water, often observed when the plug is pulled from a filled sink. A fully developed vortex in a supply vessel can draw air down into the pump suction. This can occur if the liquid drops below a prescribed minimum level, which is a function of the velocity of the liquid at the suction connection on the supply vessel. Although manufacturers are not in complete agreement as to the precise details, Table 3.5 shows suggested values of minimum submergence (liquid surface to top of suction pipe) to prevent vortices, as

Table 3.5 Minimum Submergence for Vortex Suppression

Suction Velocity (ft/sec)	Minimum Submergence (ft)
2.0	1.0
4.0	2.0
6.0	4.0
8.0	6.5
10.0	9.5

a function of velocity in the suction pipe. Where there is a bell at the inlet (e.g., a vertical turbine pump), velocity is calculated at the widest part of the bell, and submergence is measured from the liquid surface to the bell lip.

In some installations, it is impossible or impractical to keep the liquid level in the supply vessel above the recommended minimum level. In these situations, several steps can be taken to break up the vortex or keep it from entering the pump. One arrangement (Figure 3.6) uses a flat baffle plate located just above the pump suction connection. This plate helps prevent the vortex from fully forming and entering the pump suction, because the vortex would have to form in a horizontal plane, which is less likely to occur.

Another strategy, also shown in Figure 3.6, uses a cross-hatched baffle in the suction piping to break up the rotational swirling motion of the vortex. With vertical wet pit pumps, still another technique for breaking up vortices uses the suction screen normally attached to the bell of the pump for the purpose of preventing foreign materials above a certain size from entering the pump. The screen serves the same function as the cross-hatched plate just described, breaking up the vortex before it enters the pump.

V. SUMP DESIGN

A thorough treatment of this subject is beyond the scope of this book, but a few words deserve to be written on the subject

Special Hydraulic Considerations

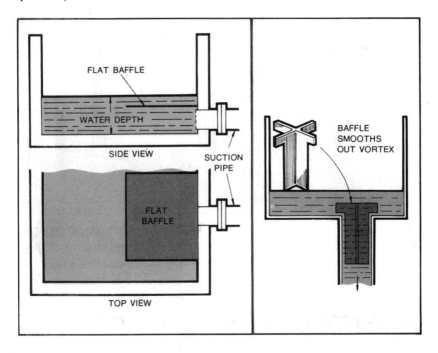

Figure 3.6 A flat baffle plate located just above the pump suction connection, or a cross-hatched baffle in the suction piping, minimize the possibility of vortex formation. (Courtesy of Goulds Pumps, Inc., ITT Industries, Seneca Falls, NY.)

of sump design. Many installations require multiple pumps in a common sump or intake structure. This is often the case for plant make-up water pumps, cooling water supply, and fire pumps, to name some common applications. The pumps are usually horizontal, split case, double suction pumps (see Chapter 4, Section VI) operating on a lift, or vertical turbine pumps (see Chapter 4, Section XI).

The primary objectives for the designer of an intake sump with multiple pumps are to ensure that each of the pumps in the sump is allowed to receive its intended flow rate, to make certain that the liquid entering each pump is moving in a straight line as it enters the pump, and to prevent the possibility of vortex formation that could lead to air introduction into the pump and system.

The Hydraulic Institute (Ref. [2]) offers guidelines to help meet these design objectives. These guidelines establish key dimensions for sumps containing multiple pumps, such as the optimal distance the pumps should be located from each other and from adjacent walls, acceptable and recommended angles for pipe openings to open channels, etc. Suction velocity in an open channel is recommended to be 1 ft/sec or less, and the flow should approach the pump in as straight a line as possible. The Hydraulic Institute offers some guidelines as to the configuration of multiple pumps. For example, multiple pumps should not be lined up in a narrow sump in the direction of the main flow path because of the likelihood that the final pump in the line would not receive as much flow as the first pump in the line.

If Hydraulic Institute Standards are used as a design guide for designing a multiple pump sump, a conservative design will result, one that will likely cause no pumping problems. (When asked to comment on proposed sump or intake structure designs, most pump manufacturers will advise the engineer or owner to follow Hydraulic Institute design guides.) While this approach is conservative, it may result in a more expensive intake structure. One possible alternative to explore early in the design phase for a plant intake system is a sump model test. This test, using dimensionless parameters such as the Froude number, can create a scale model of the sump and the profiles of the various pumps to be located in the sump. Then the flow to the sump can be modeled to allow for physical observation of the flow distribution to various parts of the sump and for observation of the formation of eddies and vortices. Also, the sump model design and pump orientation can easily be altered in a model test, by adding or moving baffles, rearranging the pumps in the sump, or changing the liquid level in the sump, until the best arrangement and design is achieved given the physical limitations of the sump. This model test approach can sometimes result in significant construction cost savings for an intake structure.

Sump model tests are available from some manufacturers of large intake pumps, from civil or mechanical engineering departments at major universities, and from a number of independent testing laboratories.

VI. FIELD TESTING

A. General

Field testing usually involves taking measurements of pump flow, total head, and power consumption from an operating pump. Using this test data, along with a knowledge of the motor efficiency, the pump efficiency can be computed. Thus, the complete operating characteristics of the pump can be produced at a range of flow values, allowing a complete new set of pump curves to be generated in the field.

Some people view a field test as a way to verify whether or not they were "cheated" by the pump supplier. This perceived benefit is often not realistically achievable for the simple reason that measurements taken in most field tests are less accurate than the tests the manufacturer performs in the test lab. Measuring flow in an installed pump is particularly difficult to do accurately without expensive flow measuring equipment.

A calculation of pump efficiency for an installed pump requires the measurement of flow, TH, and power draw to the motor (plus knowledge of the motor's efficiency at its operating load). When these three variables are measured to calculate efficiency, the accuracy of the result is equal to the square root of the sum of the squares of the accuracies of the individual variables tested. This is expressed by the following formula:

$$\%_E = \sqrt{\%_Q^2 + \%_H^2 + \%_P^2} \qquad (3.4)$$

where:
$\%_E$ = accuracy of pump efficiency
$\%_Q$ = accuracy of flow measurement
$\%_H$ = accuracy of head measurement
$\%_P$ = accuracy of power measurement

For example, suppose that pump TH and motor kilowatt draw can be computed to within 2% in a given installation, but pump flow can only be measured to within 5%. In this case, the expected accuracy of the computed efficiency, by Equation 3.4, is:

$$\%_E = \sqrt{5^2 + 2^2 + 2^2} = \sqrt{33} = 5.7\%$$

The preceding example illustrates that the accuracy of efficiency testing is no more precise than the accuracy of the least accurate variable measured, which in the majority of cases is flow. Very little benefit would be gained, in the preceding example, by modification of the test assembly to allow more accurate measurement of pump head or motor kilowatt draw.

Despite its shortcomings, field testing of pumps does have several major benefits. The primary benefit of field testing a pump is to establish a benchmark of head, flow, and power or efficiency at one or more points of operation on the pump curve when the pump is first installed in a new system. These benchmark data points can then be used to compare against pump performance in subsequent periodic field tests. These tests can be a big help in calling attention to a developing problem before significant deterioration can occur.

One pump malfunction that can be observed through field testing is operation of the pump at an unhealthy point on the H–Q curve (flow below the minimum recommended or above the maximum allowable). Another measurable malfunction is cavitation, causing the pump curve to drop off from where it should be at the point where cavitation commences, as shown in Figure 2.9. Another problem observable through a field test is a blocked or partially blocked suction line, which allows the pump to hit its factory tested curve only at shutoff, and falls below the new pump curve at all other points. Finally, a field test can show when a pump is experiencing excessive recirculation leakage, signifying the need to replace wear rings on a closed impeller and to reset clearances or replace an open impeller. The excessive recirculation is indicated by curve points falling low on head and capacity compared to what they should be, and by a reduction in efficiency. Early detection of these problems allows the pump to be returned to a safe operating point, ensure that system problems can be checked out, or permit better planning of a pump repair to restore recirculation clearances.

B. Measuring Flow

The first component of pump field testing discussed here is flow measurement. There are quite a few different methods to measure pump flow. They vary as to the accuracy of their measurement, their cost, and their complexity. Many are useful only for a specific range of flows or for certain types of liquids. Most are located at a fixed place in the piping system, while a few types are more portable. Most cause a pressure drop as liquid flows through the measuring device. Being exposed to the pumped liquid, many flowmeter types are subject to a loss of accuracy over time due to the effects of corrosion and/or erosion. The following are brief discussions on the major types of flow measurement systems.

1. Magnetic Flowmeter

Magnetic flowmeters are commonly used by pump manufacturers in their laboratory testing facilities. These instruments are very accurate and easily adaptable to computer-based data acquisition instruments. Their shortcomings are that they are expensive, not portable, and usable only for electrically conductive liquids. (They will not work on hydrocarbons, for example.) Also, the electrical contacts that are exposed to the liquid in a magnetic flowmeter can become contaminated, thus reducing accuracy.

2. Mass Flowmeter

Mass flowmeters work on the Coriolis principle. They are highly accurate and work on a broad range of liquid types. Disadvantages include the fact that they are expensive, are only available for fairly small pipe sizes (up to about 4 in.), and have a substantial pressure drop.

3. Nozzle

A nozzle is a special type of venturi (see Section VI.B.10 to follow), with a converging section and a throat, but no diverging section. With a nozzle, the throat empties to atmosphere.

Flow is a function of the pressure immediately upstream of the converging section, measured with a pressure gauge.

A nozzle has the same shortcomings as the venturi meter. Its accuracy can deteriorate rapidly if deposits form on its interior surfaces or when wear occurs. Additionally, because the nozzle must discharge to atmosphere, its use is limited to very small systems or pumps that can discharge freely to atmosphere.

4. Orifice Plate

The orifice plate is one of the most commonly used industrial flow measurement devices. Similar to a venturi (see Section VI.B.10 to follow), flow through an orifice plate is a function of the differential pressure across the device. Its chief advantages are its low initial cost and ease of relocation in a piping system. The main disadvantages of orifice plates include limited turndown ratio (range of flow rate which can be measured), high pressure drop, and loss of accuracy as the edges of the orifice wear.

5. Paddle Wheel

Practical only on quite small systems, paddle wheel flow instruments are the most common devices used for residential water meters. These paddle wheels simply count the rotations as flow goes past them, with the flow being proportional to the rotational speed. They are inexpensive, but not very accurate, and are usually only used on water at relatively low temperatures.

6. Pitot Tube

A pitot tube is a relatively simple flow measuring device and is especially useful for large systems handling ambient temperature water. The pitot tube in its simplest form consists of a tube with a right-angle bend which, when immersed with the bent part pointed directly into the flow, indicates flow velocity by the distance water rises in the vertical stem. It is

Special Hydraulic Considerations

often installed in a side opening of the pipe with packing to allow a traverse of the pipe to establish a velocity profile. This allows the computation of an average velocity and the resulting flow.

7. Segmental Wedge

With this flow measuring device, flow is a function of the differential pressure measured across a wedge located in the pipe line. While measuring flow quite accurately, these devices are expensive and not portable.

8. Turbine Meter

A sophisticated version of a paddle wheel, this type of flowmeter works on the positive displacement principle that flow is proportional to speed. Like the paddle wheel, the turbine meter, acting like a gear pump in reverse, measures rotational speed and correlates it to flow. These meters are highly accurate, and are sometimes used as "custody transfer" meters on pipe line systems.

Shortcomings of turbine meters include the fact that they are quite expensive, have a fairly high pressure drop, are subject to wear by abrasives in the liquid, and are not portable. Many also have the disadvantage of having limited turndown ratio, so that if the application requires flow measurement over a wide range of flows, it is possible that more than one meter will be required.

9. Ultrasonic Flowmeter

These flowmeters work with a pair of transducers mounted on the outside of the pipe, which send and receive ultrasonic signals. Taking into account the sonic velocity of the pipe material and the liquid, flow is proportional to the lag between the signal going with the flow and the one going against it.

The distinct advantage of ultrasonic flowmeters over most others types of flow measuring devices is that the meter is nonintrusive. That is, it is mounted on the outside of the pipe, and not inside the pipe. This means it is not subject to a reduction in accuracy due to wear, it can be worked on without penetrating the pipe, and it is portable so that it can be used in numerous locations. Accuracies to less than 1% are achievable with these devices. They are also capable of high turndown ratios.

Shortcomings of ultrasonic flowmeters are that they are somewhat tempermental with liquids containing gases, and the flow is subject to being distorted by upstream valves or fittings if they are closer than about ten pipe diameters. Also, the setup may have to be modified if different liquids are being measured at the same location.

10. Venturi

Venturis are commonly used for industrial flow measurement, and may be the most economical choice of flow instrument for water supply systems. A venturi has a converging section, a throat section, and a diverging section. The flow through the venturi is a function of the upstream and throat dimensions, and the differential pressure across the venturi.

The advantages of a venturi include that it requires very little maintenance and causes very little head loss. The major shortcoming of this method of measuring flow is the fact that while the venturi is quite accurate when newly installed, its accuracy can deteriorate if deposits form on its interior surfaces or when wear occurs.

11. Volumetric Measurement

Volumetric method of flow measurement is usually practical only for very small systems, and is only accurate if the pump flow rate is very stable. It consists simply of measuring how long a pump operating at steady state takes to empty or fill a known volume in a vessel. The measured volume, divided by the time, equals the average pump capacity.

12. Vortex Flowmeter

A vortex flowmeter has a protrusion in a pipe that causes a vortex. It has been likened in principle to a flag fluttering in

Special Hydraulic Considerations

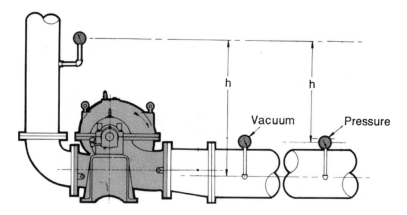

Figure 3.7 Measuring TH by using pressure or vacuum gauges. (Courtesy of Goulds Pumps, Inc., ITT Industries, Seneca Falls, NY.)

the wind, where the faster the wind flows, the higher the frequency of the fluttering of the flag. Similarly, the frequency of the vortex can be proportioned to flow.

These flowmeters are not recommended for applications with a Reynolds number less than about 10,000. This means they have limited turndown ratio, and also will not work at low flow velocities. Additionally, they lose their accuracy as the protruding body is worn or coated with deposits.

C. Measuring TH

The most commonly used method to measure total head is by means of pressure gauges (or vacuum gauges in the case of a vacuum in the suction). The gauge setup is shown on Figure 3.7. The formula for TH when there is a vacuum in the suction is:

TH = Discharge gauge reading (in ft)
      + Vacuum gauge reading (in ft)
      + Distance between point of attachment of vacuum gauge and centerline of discharge gauge, h (in ft)

$$+ \left(\frac{V_d^2}{2g} - \frac{V_s^2}{2g} \right) \quad (3.5)$$

When there is a positive pressure in the suction, a pressure gauge is used on the pump suction rather than a vacuum gauge, and the formula becomes:

TH = Discharge gauge reading (in ft)
      − Suction gauge reading (in ft)
      + Distance between centerlines of discharge and suction gauges, h (in ft)

$$+ \left(\frac{V_d^2}{2g} - \frac{V_s^2}{2g} \right) \tag{3.6}$$

The third term in each of these two formulae is positive, assuming that the discharge pressure gauge is located higher than the pressure or vacuum gauge on the suction. If not, the term is negative. Actually, the two gauges are often so close in elevation to each other that this term is ignored.

The final term in the above two equations is the change in velocity head that is caused by the difference in size of the suction and discharge lines at the gauge locations. It is found by first measuring flow (see Section VI.B above), or approximating it if flow is not being measured. Then $V^2/2g$ is calculated at the suction and discharge gauge locations, or is looked up in friction tables such as Table 2.1. As discussed in Chapter 2, this difference in velocity head is often small enough compared to the total pump head that it can safely be ignored.

Note that when using pressure gauges, distances are measured from the centerline of the gauge. With vacuum gauges, distances are measured from the point of attachment of the gauge to the pipe.

Make sure that gauges used to test pumps are in calibration, and that they are the proper pressure rating to get an accurate reading at the pressure levels expected in the pump system.

Differential pressure transmitters can also be used to measure pump head, converting differential pressure in psi to a 4 to 20-milliamp signal. These are substantially more expensive than gauges, but can be used to continuously monitor pump head, and are more accurate than most gauges.

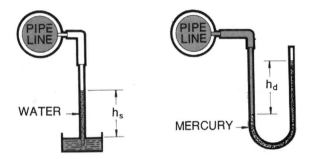

Figure 3.8 Manometers can be used to measure vacuum or low pressure. (Courtesy of Goulds Pumps, Inc., ITT Industries, Seneca Falls, NY.)

A manometer (Figure 3.8) can also be used to measure low pressure or vacuum. Manometers can be easily field fabricated using a liquid of a known specific gravity. Water and mercury are common liquids used in manometers.

D. Measuring Power

The power required by a pump and motor (input power) can be measured by a kilowatt transducer if one is available. More commonly, to obtain input KW, volts and amperes are measured using a voltmeter and an ammeter (for three-phase power, one uses the average of the amperage measured in each of the three legs). The following formula yields input KW:

$$KW_{in} = \frac{1.732 \times \text{Volts} \times \text{Amps} \times \text{Power factor}}{1000} \quad (3.7)$$

The constant 1.732 (square root of 3) in Equation 3.7 is used for three-phase power measurements. For single-phase, that term in Equation 3.7 becomes 1.0.

The power factor in Equation 3.7 can be obtained from the motor manufacturer, and varies with motor load.

Note that a power reading in KW can be converted to horsepower (HP) by the following equation:

$$HP = KW \times 1.34 \quad (3.8)$$

E. Measuring NPSH

Chapter 2, Section VI.E, described the test done by the manufacturer at the factory to produce the pump's $NPSH_r$ curve. This test can also be carried out in the field but, practically speaking, it seldom is, due to the complexity of the test setup and the difficulty in accurately measuring the required parameters.

The $NPSH_a$ for a system at the suction of an operating pump can be measured in the field by means of a gauge pressure reading taken at the pump suction, using the following formula:

$$NPSH_a = P_b \pm G - H_{vp} + H_v \qquad (3.9)$$

where:
P_b = barometric pressure (in ft)
G = gauge reading at pump suction (in ft) (plus if above atmospheric, minus if below atmospheric)
H_{vp} = vapor pressure (in ft)
H_v = velocity head in suction pipe at the gauge connection (in ft)

$NPSH_a$ can be measured by the above method for a system where the pump is suspected of cavitating. The measured value of $NPSH_a$ can then be compared with the value of $NPSH_r$ shown on the pump curve.

4

Centrifugal Pump Types and Applications

I. OVERVIEW

Now that the reader has been introduced to pumps in Chapter 1, and has developed a solid foundation in hydraulic theory related to centrifugal pumps from Chapters 2 and 3, it is time to begin looking at hardware. This chapter devotes itself to a discussion of the mechanical features and applications of the most important types of centrifugal pumps.

The chapter begins with a discussion on impeller types, because the impeller is the most important component of the pump. The chapter then describes the two most common types of impellers (open and closed), with a discussion on the physical features, types of leakage joint, and applications most suited to each type of impeller. Single suction and double suction impellers are described. The explanations of these major impeller types lead into discussions of impeller wear rings, suction specific speed, axial thrust and thrust balancing, filing of impeller vane tips, and solids handling impellers.

Following the introduction to impellers, the bulk of the remainder of Chapter 4 examines the most common types or configurations of centrifugal pumps. Each pump type is described as to its physical configuration and to its important hydraulic and mechanical design features. The applications for which each pump type is best suited are described. Note that discussions on sealing systems are presented separately in Chapter 5.

The reader is again referred to Appendix A at the end of this book, which lists the major suppliers of centrifugal and positive displacement pumps in the United States, along with an indication of the types of pumps they manufacture. As mentioned in Chapter 1, Appendix A should by no means be considered all-inclusive, but merely a listing of the manufacturers with whom the author is familiar. The reader should be able to use an Internet search, a *Thomas Register,* or similar directory to locate particular manufacturers shown in Appendix A. For any pump application being considered, the best advice is to obtain literature from a number of the manufacturers for the types of pumps being considered.

One of the early decisions that must be made in centrifugal pump selection is the type of pump configuration that will be used. The solution is not always obvious, and often there are a number of alternatives. Consider, for example, a very common pump application problem, that of emptying a liquid out of a sump where the liquid level in the sump is below the level where the pump would normally be located. There are at least a half dozen different centrifugal alternatives for this application problem. The choices could include a submersible pump, a pump with only the impeller and casing immersed in the liquid (either a vertical column sump pump or a vertical turbine pump), a self-priming centrifugal pump, a nonself-priming end suction pump with a priming system or with a foot valve, or an ejector system. Any one of these choices might work for the application, but the question is, which one is the best, given the particular conditions? Which one results in the lowest first cost for the pump, which has the lowest installed cost (not the same as first cost of the pump alone), which uses less energy to operate, and which has the lowest expected maintenance costs? The answers to these questions depend, among other factors, on the flow and head required, physical limitations of the installation, the corrosive nature of the pumped liquid, whether there are abrasives or solids in the liquid, the depth of the sump, and the liquid temperature. By a careful examination of the most common pump types, this chapter provides many of the tools to help make this type of application decision.

Chapter 4 includes a discussion of pump standards and design specifications, with particular emphasis on three important standards for centrifugal pumps: (1) American National Standards Institute (ANSI), (2) American Petroleum Institute (API), and (3) International Standards Organization (ISO).

Finally, Chapter 4 concludes with sections on two of the most important pump accessories: (1) couplings and (2) electric motors.

II. IMPELLERS

A. Open vs. Closed Impellers

There are two major classes of mechanical designs for centrifugal pump impellers: *open* and *closed*. These two types of impellers are illustrated in Figure 4.1. The distinction between these two impeller types has nothing to do with the

Figure 4.1 Open and closed impellers. (Courtesy of Goulds Pumps, Inc., ITT Industries, Seneca Falls, NY.)

impeller specific speed, head/flow characteristics, or number and shape of impeller vanes. Rather, the distinction has to do principally with the way the *leakage joint* for the impeller is designed. The leakage joint exists because the pressure on the discharge (or back side) of the impeller is higher than the pressure on the suction (or front side) of the impeller. Because the impeller is rotating within a stationary casing, there must naturally be a certain amount of running clearance between the impeller and casing. Therefore, a certain amount of liquid tends to leak from the discharge side of the impeller back to the suction side, through the leakage joint. It is desirable to minimize the amount of leakage across the leakage joint in a centrifugal pump. Recall from Chapter 2 that volumetric loss is one of the contributing factors to the inefficiency of a pump.

Figure 4.1 shows both an open and a closed impeller, both viewed from the suction side. The primary difference is that with the open impeller, the impeller vanes are clearly visible when viewed from the suction side of the impeller. The closed impeller, on the other hand, has a shroud covering the vanes on the suction or front side, and an axially oriented hub that provides the inlet for the liquid into the vane passageways.

The easiest way to understand how the leakage joints for the two types of impellers work is to look at cross sections of pumps with the two types of impellers. Figure 4.40 shows a pump with an *open impeller*. The leakage joint for the open impeller consists of the front edges of the impeller vanes, which are machined to a contour that is identical to the contour of the casing adjacent to the impeller. To minimize leakage across the leakage joint, the front edges of the impeller vanes must be kept very close to the mating face of the casing, while at the same time avoiding the possibility of the rotating impeller rubbing against the casing. A common axial clearance or impeller setting to achieve this is 0.012 to 0.015 in.

Open impeller pumps are generally considered a good choice for pumping liquids that contain solids or stringy materials. The exposed vanes reduce the likelihood of debris becoming trapped as it passes through the impeller. Most open impeller pumps have the capability to adjust the impeller axially in the field to account for wear on the front edges of the impeller vanes. This wear can occur as the liquid squirts past the leakage joint from discharge to suction, and is exacerbated if the liquid contains abrasives or is corrosive. For open impellers with *pump-out vanes* (see Section II.D below), adjustment of the axial clearance affects the resulting pressure in the *stuffing box* and the pump *axial thrust*, in addition to pump efficiency. (See Chapter 5, Section III, for a more complete discussion of stuffing boxes.)

The pump in Figure 4.40 carries out this axial adjustment by means of two sets of machine bolts at the coupling end of the pump, located on a common bolt circle. Half of the bolts are fitted into drilled and tapped openings in the bearing housing, while the other half act as jacking screws. When it is desired to adjust the impeller axial setting, the pump is shut down, and the bolts that go into the tapped openings are tightened, moving the entire rotating assembly toward the suction end of the casing. At the same time, the technician turns the pump by hand from the coupling end. When the front edges of the impeller vanes come in contact with the casing, it begins to rub, and the impeller axial setting is then essentially zero. The impeller is then adjusted away from the casing by turning the other set of machine bolts (the ones that act like jacking screws). This moves the whole rotating assembly away from the inlet side of the pump casing.

The amount of axial movement of the impeller during this adjustment is measured in one of two ways. One method requires the gap behind the plate where the bolts are located to be measured using feeler gauges, and then remeasured as the jacking screws widen this gap, until the desired axial setting is achieved. Another method uses a dial indicator attached to the pump base, with the pin of the indicator on the plate where the jacking screws are located. Either method allows the impeller to be located a precise distance from the front end of the pump casing.

A major benefit of this style of impeller is that its leakage rate can be reduced by resetting the impeller in the field as just described, without disassembling the pump. This tightening of the impeller clearance improves the pump's efficiency. However, this readjustment of clearance as the impeller wears can only be done a limited number of times without affecting pump performance. As the front edges of the impeller vanes wear due to the corrosive/erosive effect of the liquid squirting past the leakage joint, the width of the impeller vanes is actually being made smaller. Eventually, if the vanes wear enough, the reduced impeller width causes the capacity of the pump to begin to diminish. At that point, the only way to bring the pump back to its original hydraulic performance is to replace the impeller entirely. Also, as discussed in Section II.D below, axial adjustment of this impeller type affects stuffing box pressure and axial thrust.

Another open impeller type has no pump-out vanes and has an axially oriented (parallel to the shaft) leakage joint on the stuffing box side of the impeller. This design minimizes the effect of axial adjustment but gives up some of the solids handling and wear capability of the open impeller.

Both types of open impellers require a locked thrust bearing that limits axial movement to 0.001 in.

A very common type of *closed impeller* pump is illustrated in Figure 4.9 and Figure 4.11. With this type of impeller, the leakage joint is the axially oriented annular space between the rotating hub on the suction side of the impeller and the space next to the hub on the pump casing. Quite often, the rotating impeller hub, or the mating part on the pump casing, or both, will be fitted with *wear rings*. These are replaceable rings that, if wear occurs in the leakage joint from corrosion and/or erosive wear, can be more economically replaced than the alternative of replacing the entire impeller or casing. The pumps in Figures 4.9 and 4.14 are fitted with wear rings in the pump casing.

Because of the axial orientation of the leakage joint for the closed impeller described above, the axial setting on the closed impeller is not nearly as critical to the pump's performance as is the case with the axial setting on an open impeller. With a closed impeller, where the leakage joint hub may be nearly an inch long (varying with impeller size), moving the impeller axially a tenth of an inch or so in either direction has a very minor influence on the amount of leakage across the leakage joint. The amount of leakage across the joint is primarily a function of the differential pressure and the diametral clearance at the impeller front hub. Therefore, the axial setting is primarily to ensure that the impeller is not rubbing against the front side of the casing.

A shortcoming of this type of closed impeller construction is that, although the leakage joint can be renewed on a pump fitted with wear rings by replacing the rings, the pump must be taken out of service and disassembled to do this. This is in contrast to the open impeller pump described above, which can have its leakage joint adjusted without disassembling the pump.

A closed impeller pump may have a wear ring on either the impeller, the casing, or both *(dual wear rings)*. If there is only one wear ring, the ring material is selected to be softer than the part it is running against, so that the wear ring will wear, rather than the impeller or casing. Dual rings, if they are supplied in chrome stainless, are usually heat treated to maintain a 50 Brinell hardness difference between the impeller and casing rings.

It may be possible to reduce the initial cost of a pump by purchasing it without wear rings, and then adding them at the first maintenance outage of the pump. Normally, this is false economy because the pump casing and/or impeller would have to be remachined to accept the rings during the outage, and this operation would likely cost more when done as part of a pump repair than if supplied that way by the manufacturer.

Closed impeller wear ring clearances vary according to the diameter of the impeller, or the diameter of the leakage joint (impeller suction side hub diameter). One standard for wear ring clearance, from the American Petroleum Institute (API) Standard (see Section XIV.C) is shown in Table 4.1. Another guideline for wear ring clearances calls for a diametral clearance of 0.001 to 0.0015 in. per inch of impeller diameter. For example, a 10-in. diameter impeller would have a diametral wear ring clearance of 0.010 to 0.015 in. Wear ring clearances should be opened up from the recommended amounts by about 0.002 to 0.005 in., depending on the impeller size, for galling materials such as 316 stainless, for operating temperatures over about 500°F, and for multi-stage pumps.

Table 4.1 Recommended Wear Ring Clearances for Closed Impellers

Diameter of Rotating Member at Clearance (inches)	Minimum Diametral Clearance (inches)
<2	0.010
2.000–2.499	0.011
2.500–2.999	0.012
3.000–3.499	0.014
3.500–3.999	0.016
4.000–4.499	0.016
4.500–4.999	0.016
5.000–5.999	0.017
6.000–6.999	0.018
7.000–7.999	0.019
8.000–8.999	0.020
9.000–9.999	0.021
10.000–10.999	0.022
11.000–11.999	0.023
12.000–12.999	0.024
13.000–13.999	0.025
14.000–14.999	0.026
15.000–15.999	0.027
16.000–16.999	0.028
17.000–17.999	0.029
18.000–18.999	0.030
19.000–19.999	0.031
20.000–20.999	0.032
21.000–21.999	0.033
22.000–22.999	0.034
23.000–23.999	0.035
24.000–24.999	0.036
25.000–25.999	0.037

Source: API standard 610, 7th edition. (Courtesy of American Petroleum Institute, Washington, D.C.)

Generally speaking, a closed impeller pump is more efficient than an open impeller pump of the same size and specific speed. There are exceptions to this rule, however, and there are several reasons for this inconsistency. The area of the leakage joint for a closed impeller is smaller than the corresponding annular leakage space on an open impeller. This would seem to mean that closed impellers are more efficient. Many vertical turbine pumps (Section XI), which offer both open and closed impeller designs, have a more efficient open impeller design. One reason is the fact that vertical turbine pumps may have the impeller setting even tighter than the standard 0.012 to 0.015 in. discussed above, at least when the pump is new. Seizure of the rotor is less likely with an open impeller leakage joint than with an axially oriented one such as in the closed impeller described above. Another point is that liquid flowing through an open impeller is exposed to a machined surface on one side (the machined casing contour), while in a closed impeller the liquid is exposed to an as-cast surface on both sides.

An alternative wear ring design, know as *serrated rings* (Figure 4.2), relies on machined grooves in the rings to provide additional pressure breakdown as liquid leaks across the

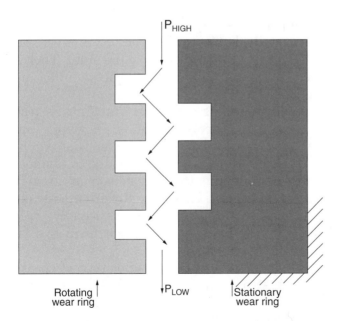

Figure 4.2 Serrated wear rings minimize leakage to reduce efficiency losses without tightening ring clearances.

rings. Volumetric losses can be minimized without reducing ring clearances to the point where galling is a concern.

A closed impeller generally produces lower axial thrust than an open impeller of roughly the same rating and specific speed. The difference in thrust (which is explained in more detail in Section II.D to follow) between closed and open impellers is generally quite significant, commonly 30 to 40%.

Closed impellers are almost always used with multi-stage pumps. The reason for this is that with most multi-stage pumps, each of the impellers is fixed in its location on the shaft by means of a split ring and groove arrangement that is used to transmit the axial thrust created by the impeller to the thrust bearing. If a multi-stage pump had open impellers and only one impeller were replaced, it would be impossible to reset all of the impellers to a tight leakage joint clearance. For this reason, most multi-stage pumps have closed impellers. Exceptions to this rule are vertical turbine pumps (Section XI) and some very small radially split multi-stage pumps (Section VII.C), both of which may have methods of attaching the impeller to the shaft other than the split thrust rings just described. On these types of pumps, the impeller location is not fixed with respect to the shaft. Therefore, these pumps are able to be offered by some manufacturers in multi-stage open impeller configuration.

The closed impeller designs discussed thus far are best suited to relatively clean, noncorrosive liquids. They are especially suited for high-temperature applications. The axially oriented leakage joint allows axial movement due to temperature expansion without affecting the rate of leakage through the leakage joint. On the downside, the axially oriented leakage joint may clog with stringy solids, wears rapidly with erosive solids, and is more likely to gall when furnished in stainless steel.

Many abrasive slurry pumps use another type of closed impeller, which is illustrated in Figure 4.32. This impeller type is designed for clogging solids and abrasive and corrosive slurries. The design uses a leakage joint that is radial to the shaft, more like an open impeller. The impeller may be rubber lined, made of hard metal for abrasion resistance, or made of

Centrifugal Pump Types and Applications

corrosion-resistant metal. See further discussion of slurry pumps in Section X to follow.

From a manufacturing perspective, open impellers require less machining because they do not have the front shroud. On the other hand, machining the front edges of the vanes is an interrupted cut, which may require slow machine tool speeds. Open impeller vanes can be hand-dressed and more easily underfiled (Section II.E) to improve performance, because the vanes are not covered by the shroud on the front side.

In summary, there are many factors affecting the right choice of impeller type for a given pump type and application. Because of their usual higher efficiencies and lower axial thrust, closed impellers are normally preferred, provided that the liquid is reasonably clean and free of solids and abrasives. If there are solids and abrasives in the pumped liquid, open impellers or closed impellers with radially oriented leakage joints may be more suitable.

Some open impellers are referred to by some pump manufacturers as *semi-open* or *semi-enclosed*. A semi-open impeller has a full or partial shroud on the back side of the impeller vanes. The open impeller shown in Figure 4.1 has such a partial shroud and thus, in the strictest sense, is a semi-open impeller. A pure open impeller, strictly speaking, would have only vanes joined together by a hub at the center. Practically speaking, most open impellers must have at least a partial back shroud to enable machining of the vanes without bending or breaking them off. The larger the shroud on the back side of the impeller, the greater the amount of thrust generated by the impeller. (Refer to the discussion on impeller thrust in Section II.D below.)

B. Single vs. Double Suction

Figure 4.3 illustrates the difference between single suction and double suction impellers. Single suction impellers, both open and closed, are by far the most common type of impeller. With this impeller type, liquid enters the impeller from one side only.

A double suction impeller is a special type of impeller that has liquid entering the impeller from both sides. Double

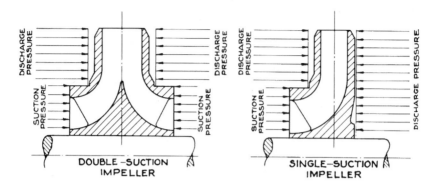

Figure 4.3 Single suction and double suction impellers and thrust development. (From *Pump Handbook*, I.J. Karassik et al., 1986. Reproduced with permission of McGraw-Hill, Inc., New York.)

suction impellers are usually associated with horizontally split case pumps, either in single stage (Section VI) or as the first stage in multi-stage (Section VII) configuration. They are also supplied in vertical configurations, either as a single stage, or as the first stage of vertical turbine pumps (Section XI). Double suction impellers are always closed impellers.

There are two primary benefits of double suction impellers compared to single suction impellers. The first benefit is that the axial thrust of a double suction impeller is much less than that of a single suction impeller, and very nearly zero. (Refer to further discussion of how impellers generate thrust in Section II.D to follow.) Use of a double suction impeller means that thrust bearing loads are lower, or that the pump may not even need a thrust bearing.

The second benefit of a double suction impeller is that the $NPSH_r$ of a double suction impeller is much lower than that of a comparably sized single suction impeller with the same type of inlet design. This is true because each of the two inlets on the double suction impeller has only one half of the flow going into it. (See Chapter 2, Section VI for discussion of NPSH.)

Double suction impellers are often used in conjunction with double volute casings. The double volute casing (see Chapter 1, Section V) is there to reduce the radial bearing loads on the pump, and is usually of most benefit in higher flow pumps where radial bearing loads are likely to be higher. The double suction impeller is present to reduce the $NPSH_r$, which is more likely to be a problem at higher flows.

C. Suction Specific Speed

Suction specific speed, a nondimensional index used to describe the geometry of the suction side of an impeller, was first explained in Chapter 2, Section VII. The suction specific speed is designated S, and its formula, very similar to the formula for specific speed (Chapter 2, Section VII), is:

$$S = \frac{N \times \sqrt{Q}}{NPSH_r^{3/4}} \quad (4.1)$$

As with specific speed, the terms in Equation 4.1 are taken at BEP, full diameter. The value of Q in Equation 4.1 for a double suction impeller (Section II.B) is taken as one half the total pump capacity.

Typical values of S for most standard designed impellers lie in the range of 7000 to 9000. The discussion of NPSH in Chapter 2, Section VI.C, mentions special types of single suction impellers that have enlarged inlets to reduce the $NPSH_r$. These *large eye* single suction impellers typically have a value of S in the range of 10,000 to 13,000. Caution should be used in applying high suction specific speed impellers because the acceptable range of operation on the pump curve may be more restricted with pumps having higher S values. Operating outside this safe range may cause the pump to operate unstably, have excessive recirculation, and may cause cavitation, noise, and vibration. The reasons for this phenomenon are described in Chapter 8, Section III.B.

Another example of a high suction specific speed impeller is an *inducer*. An inducer is a special type of impeller that looks almost like a tapered screw having only a few threads (Figure 4.4). It is usually used directly in front of another impeller to reduce the value of $NPSH_r$ of the second impeller. The value of suction specific speed for an impeller with an

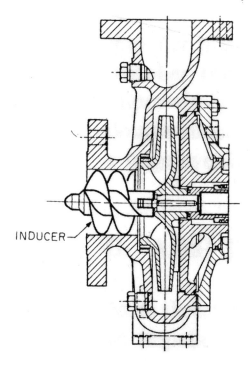

Figure 4.4 Inducers are used to reduce $NPSH_r$. (Reprinted by permission of *Hydrocarbon Processing*, December 1979, by Gulf Publishing. All rights reserved.)

inducer is about 18,000. This means, considering the discussion of the previous paragraph, that an inducer has an even more restrictive acceptable range of flow than a comparable impeller without one.

Section II.B above discusses the benefits of a double suction impeller, including the lowered value of $NPSH_r$. Because the flow into a double suction impeller is split in two, and only half the flow goes into each inlet of the impeller, this means that the double suction impeller is able to achieve the improved $NPSH_r$ without a higher suction specific speed. Therefore, the double suction impeller is not subject to the same restrictive flow range as is a large eye single suction impeller or an inducer.

D. Axial Thrust and Thrust Balancing

Axial thrust is caused by pressure acting against the cross-sectional area of an impeller. This thrust must be accommodated by the pump's thrust bearing. Figure 4.3 shows the thrust profile for a single suction closed impeller and a double suction impeller. With the single suction impeller, the back side of the impeller sees nearly full discharge pressure. The front side is subject to near-discharge pressure outside the leakage joint and near-suction pressure inside the leakage joint. Thus, there is a net thrust toward suction, the ordinary direction of axial thrust for a single stage pump having a single suction impeller.

In an open impeller, the pressure on the front side of the impeller breaks down from discharge to suction across the leakage joint. On the back side of the impeller, the pressure may be nearly discharge pressure across the entire back face, or may be lower than this closer to the hub if the impeller is fitted with pump-out vanes (described below). This results in a net axial thrust that is higher than that of a comparably sized closed impeller.

With the double suction impeller in Figure 4.3, because flow evenly splits between the two impeller inlets, the thrust forces on each side of the impeller cancel out each other. This results in a complete balancing of axial thrust, provided that the flow is evenly split between the two impeller inlets. Many larger double suction pumps have a thrust bearing to account for the possible uneven distribution of the flow between the two impeller inlets.

Because thrust in a centrifugal pump impeller is a function of the differential head as well as the area under pressure, thrust varies depending on where the pump is operating on its curve. The farther to the left of BEP a pump operates, the higher will be the thrust generated and the load the thrust bearing must carry. This may be a factor limiting minimum flow, as is discussed in Chapter 8, Section III.B.

There are a number of ways to balance or reduce the amount of axial thrust generated by a single suction impeller. One such method, very commonly used with enclosed impellers,

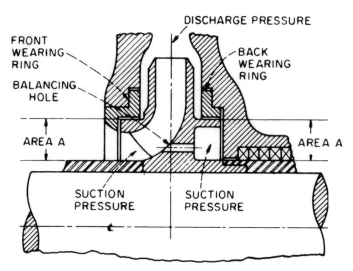

Figure 4.5 Axial thrust is balanced by the addition of a back wear ring and holes drilled in the impeller. (From *Pump Handbook*, I.J. Karassik et al., 1986. Reproduced with permission of McGraw-Hill, Inc., New York.)

is illustrated in Figure 4.5. The pump in this figure has a second wear ring located on the back side of the impeller. Additionally, several small holes have been drilled in the impeller, permitting a leak path from the back side of the impeller to suction. High-pressure liquid from the pump discharge leaks across the back wear ring, reducing in pressure as it throttles across the ring. Then the liquid, when it reaches the cavity at the back side of the impeller, is exposed to the leak path through the drilled holes, and leaks back to the suction side of the impeller. The net effect of these two features on the pump is that the pressure in the cavity on the back side of the impeller is reduced to nearly suction pressure, thus reducing the axial thrust as well as lowering the pressure that must be carried by the seal.

This thrust balancing arrangement has a cost, naturally. First, there is additional machining and an additional component (the wear ring on the back side of the impeller) required when the pump is built. Second, the clearance of the back wear ring must be regularly maintained if the thrust balance is to remain effective. Otherwise, if the wear ring clearance were allowed to open excessively, the liquid could not relieve through the drilled holes fast enough and the pressure on the back side of the impeller would increase, raising thrust and stuffing box pressure. Finally, this feature causes an additional loss of a point or two of efficiency because more liquid is leaked back to suction rather than being pumped through the pump. These costs are often deemed acceptable in return for the greatly reduced size of thrust bearing that must be used in the pump, as well as the reduction in stuffing box pressure.

Another means of reducing axial thrust, commonly found on open impellers, is through *pump-out vanes*. These vanes, shown on the back side of the impeller in Figure 4.40 have a width of only $1/16$ to $1/8$ in. and usually follow the contour of the impeller vanes. The pump-out vanes produce a pumping action as the pump impeller rotates, pushing liquid out of the area behind the impeller. This reduces the pressure on the back side of the impeller, lowering the axial thrust. An additional (and sometimes more important) benefit of back pump-out vanes is that, with the pressure on the back side of the impeller reduced, the packing or mechanical seal must seal against a lower pressure, which usually means a longer packing or seal life. Still another benefit of the back pump-out vanes is to keep solids and abrasives out of the packing or seal area, a common application for slurry pumps. (Refer to Section X of this chapter.)

Note that as the open impeller wears and is adjusted axially to reestablish higher efficiency, this reduces the effectiveness of the back pump-out vanes (because the clearance on the back side of the impeller would have increased from the impeller adjustment). This, in turn, results in higher stuffing box pressure and higher axial thrust. Therefore, prior to impeller adjustment, users should consider the trade-off between higher pump efficiency (lower pumping costs) on the one hand, vs. higher thrust load and seal pressure (possibly higher maintenance costs) on the other hand. Each situation is unique, of course, and due consideration should be given to other factors such as the conservatism in the bearing design and the pump operating point or duty cycle.

Under conditions of high suction pressure, especially with an impeller that has been thrust balanced as described above, the thrust may reverse and can result in a high thrust toward the stuffing box, opposite of the normal direction of thrust.

Another type of thrust balancing device, common with multi-stage pumps that do not have opposed impellers (see Section VII), is a *balance drum*. This device, shown on the pump in Figure 4.24, has a sleeve or drum attached to the shaft at the high-pressure end of the pump. The drum is exposed to full discharge pressure on the front side, and a much lower pressure on the back side, creating a thrust counter to the normal direction of thrust. The balance drum runs in a long throttle bushing that might be serrated or have a labyrinth design to provide the maximum amount of pressure breakdown. This large amount of pressure breakdown makes these devices rather high-maintenance devices, but this may be the only feasible way to balance axial thrust in some types of multi-stage pumps.

E. Filing Impeller Vane Tips

Impeller vane tips can be filed on both the inlet and the outlet edges of the vanes as a way to sometimes improve pump performance. Figure 4.6 illustrates the location of filing operations. Impeller outlet tips are underfiled by removing a certain amount of material from the under side of the vanes at the tip, and then filing back several inches to achieve a smooth contour. This has the effect of increasing the normal dimension between adjacent vanes (from d to d_F in Figure 4.6), as well as changing the angle of the liquid exiting from the impeller. This often results in a slight improvement of performance, including the possibility of an increase of several percent in flow, head, or efficiency of the pump.

The results of underfiling the vane outlets are not consistent from one type of pump to another, as tests by manufacturers on a variety of impeller sizes and specific speeds

Centrifugal Pump Types and Applications

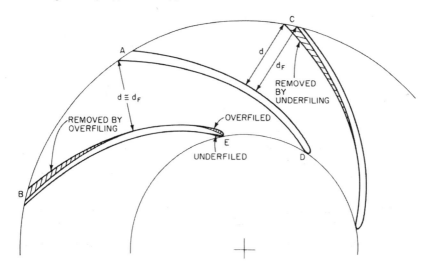

Figure 4.6 Underfiling of vane outlets may improve performance. Filing vane inlet edges may improve NPSH$_r$. (From *Pump Handbook*, I.J. Karassik et al., 1986. Reproduced with permission of McGraw-Hill, Inc., New York.)

have demonstrated. However, once underfiling has been shown to be effective for a particular impeller, future underfiling of impellers made of the same material and with the same drawings will have repeatable results. Therefore, underfiling is sometimes considered by a manufacturer at the time that a newly developed pump is being tested, especially if the pump's early test results do not come up to the manufacturer's expectations. In that case, if the underfiled impeller performs much better, the manufacturer may elect to add the underfile to the bill of material, so that all future impellers made to that same drawing number are underfiled as well. Sometimes, an underfile is done by a manufacturer to achieve quoted efficiencies. The decision to do an impeller underfile is not made lightly by the manufacturer, and is only done if the improvement of pump performance justifies the additional labor required to do the job.

Underfiling can also be done in the field if it is desired to slightly increase the flow or head of an impeller, or if the efficiency must be improved slightly to keep a motor from being overloaded. Before a user attempts an underfile, it is a good idea to consult with the manufacturer to see if the manufacturer has attempted an underfile on this impeller in the past and knows what to expect in the way of performance change.

Underfiling the impeller outlet vane tips must be done on all of the vanes, and on the under side only. Referring to Figure 4.6, overfiling of impeller outlet tips has not proven effective in improving pump performance. Also, the user must realize that the process of underfiling is destructive to the impeller. That is, if the user is not happy with the results, the process cannot be reversed, because it involves removing metal from the impeller. Following an underfile, the impeller must be rebalanced too.

Underfiling is normally done by hand, using a pencil grinder. It is not practical to underfile smaller, low specific speed, closed impellers because the vanes are so narrow that a pencil grinder cannot reach in to do the filing operation. The practicality of underfiling also depends on the impeller material because the removal of metal must be done by hand. For example, if the impeller material is stainless steel and the impeller is any size at all, an underfiling operation would take so long to perform that it might not be worth the effort.

Filing of impeller inlet vane tips, also shown on Figure 4.6, can sometimes be done to achieve a reduction in the $NPSH_r$ of the pump. This improvement can be achieved if the impeller vane inlet edges are blunt or very rough.

F. Solids Handling Impellers

Section II.A above discussed the fact that open impellers can handle liquids containing solids better than closed impellers. When solids sizes are larger than about 1 in., however, special impeller designs are used to handle these larger solids. These special impeller types include *nonclog* impellers and *vortex* impellers.

Nonclog impellers are designed to accommodate large solids without clogging the pump. The impeller is an enclosed

Centrifugal Pump Types and Applications

type, usually has a minimal number of vanes (often two or three), and the vanes are designed with a width and curvature to allow solids to pass through the pump. These impellers are usually in the mixed flow specific speed range, designed to handle high capacities with relatively low heads. The largest impellers of this style have capacities up to 50,000 gpm, heads up to several hundred feet, and are capable of handling solids up to 6 in. in diameter.

Nonclog impellers come in a number of pump configurations, depending on the details of the installation, many of which are shown later in this chapter. For sewage applications, they can be supplied in horizontal end suction or in vertical dry pit installations. They can also be supplied in a vertical wet pit column arrangement (Figure 4.26), or in a self-priming arrangement, where the pump is located above the wet well (Figure 4.15). Another nonclog option for sewage and industrial waste services that is growing in popularity is the submersible configuration (Figure 4.29).

The other impeller type designed to handle large solids and stringy material is a *vortex* impeller, the principle of operation for which is illustrated in Figure 4.7. Also called a *recessed* impeller, this impeller type is an open impeller, but with a large space between the front edges of the impeller vanes and the casing. The pumped liquid is induced into the pumping chamber by the vortex created as the impeller rotates. Thus, the liquid passes through the pump without actually coming in direct contact with the impeller. This impeller type is quite inefficient, but may be a good alternative for handling solids or stringy material. Like the nonclog impeller, the vortex impeller also comes in a variety of configurations (e.g., end suction, submersible, and vertical column type). A vertical column pump with a recessed impeller is shown in Figure 4.33.

III. END SUCTION PUMPS

A. Close-Coupled Pumps

Figure 4.8 and Figure 4.9 illustrate an end suction close-coupled centrifugal pump, by far the most common type of

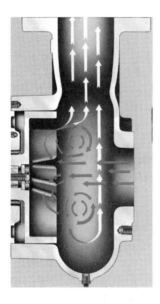

Figure 4.7 Vortex impeller: principle of operation. (Courtesy of Aurora Pumps, North Aurora, IL.)

Figure 4.8 Close-coupled end suction pump. (Courtesy of Crane Pumps & Systems, Inc., Piqua, OH.)

centrifugal pump. This pump type is called an end suction pump because the liquid enters the pump from the end, with the discharge being at a right angle from the shaft. The pump

Figure 4.9 Sectional view of close-coupled end suction pump. (Courtesy of Crane Pumps & Systems, Inc., Piqua, OH.)

type is referred to as close-coupled because of the fact that the impeller is directly connected to the motor shaft, rather than being separated by a shaft coupling. With a close-coupled pump, the pump casing (almost always a single volute casing) is directly attached to the end face of the motor (or separated by a connecting bracket). There are no separate bearings, either radial or thrust, in the pump. All radial and axial thrust loads must be supported by the bearings in the motor. Because there are no separate bearings for the pump, most makers of this type of configuration usually limit their offering to about 60 HP (although several manufacturers go up to 100 HP). Still, this covers a huge majority of centrifugal pumps produced.

The close-coupled end suction pump is popular because of its compactness, its simplicity, and the fact that it is the lowest cost configuration for a single stage pump. Over 50 U.S. companies make this type of pump in standard materials such as iron and bronze. A number of pump manufacturers offer the pump in stainless and other alloys, as well as in plastics. Close-coupled pumps are usually fitted with closed impellers because of the difficulty in adjusting axial clearances with this type of pump. The thrust bearing for the motors used with this pump type is often not limited to 0.001-in. axial movement, as is necessary with open impellers, but instead usually allows more than 0.005 in. movement. Close-coupled pumps are quite often fitted with a single mechanical seal. The pump may or may not have wear rings. Case wear rings are shown on the pump in Figure 4.9. Upper temperature limits are usually about 225°F, and the most common service is for ambient or relatively low temperature water.

Close-coupled pumps normally come in sizes up to 4 or 5 in. discharge, with flows up to about 2000 gpm and heads up to about 700 ft.

Note that when maximum flows and heads are given in this chapter for particular pump types, it should be understood that these maximum values are not available concurrently. A pump achieving the highest flow listed would normally be run at a slower speed than 3600 rpm, and would have a much lower head than the maximum head listed. The interested reader should look at several manufacturers' catalogs to ensure that a particular rating can be made.

Suction and discharge connections for end suction close-coupled pumps are often threaded up to about 2 in. and flanged in larger sizes. The usual 60-cycle operating speeds for this pump type are 3600, 1800, and 1200 rpm.

Close-coupled pumps require a C-Face motor, different from the foot-mounted motors used with most other pump types. C-Face motors have the inboard end machined to dimensions standardized by NEMA, the National Electrical Manufacturers' Association, so that pump manufacturers can easily adapt their pumps to them.

In addition to their relatively low cost, compactness, and simplicity, close-coupled pumps also enjoy the distinct advantage, compared to the frame-mounted pump design discussed in the following section, of not requiring coupling alignment. The rabbet fits of the motor and pump volute ensure the concentricity of these two components.

Centrifugal Pump Types and Applications 237

Because of the advantages cited above, close-coupled pumps are the most popular type of pump for general light-duty services. They are also favored by OEMs (original equipment manufacturers), companies that make a machine or system that incorporates a pump, and by many commercial users.

Industrial users of pumps have historically not favored close-coupled pumps, preferring instead the more common foot-mounted motor, which is used on frame-mounted pumps and on many other types of industrial rotating machines (blowers, mixers, etc.). Reasons for the lack of full acceptance of close-coupled pumps include the relatively low upper limitations on horsepower, flow, head, and temperature in the close-coupled configuration; and the fact that very few special features are available for this pump type. Also, 30 years ago, C-Face motors were not competitively available in the variety of enclosures available with foot-mounted motors. Today, C-Face motors are readily available in all types of enclosures and at prices competitive with foot-mounted motors.

B. Frame-Mounted Pumps

The end suction frame-mounted pump, illustrated in Figure 4.10 and Figure 4.11, differs from the close-coupled pump described in Section III.A above in that the pump and motor (which is foot mounted) are separated by a shaft coupling. The pump has its own bearing frame, with a radial and thrust bearing, and the pump and motor are usually mounted on a common cast or fabricated bedplate.

Frame-mounted pumps use motors that are the same as those used on other rotating equipment such as fans, blowers, mixers, etc. In a frame-mounted pump, shaft deflection, runout, and thrust carrying capability are controlled and designed by the pump manufacturer.

This type of pump, because it has a separate bearing frame, can be made in much larger sizes than close-coupled pumps, and is therefore more commonly used in industrial and heavy-duty commercial applications. Also, the end suction configuration is simpler and has a lower first cost than any other single stage alternative. Some manufacturers offer this

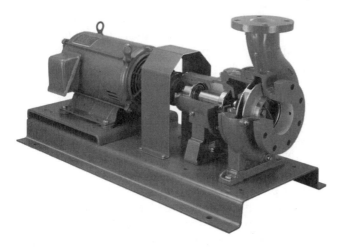

Figure 4.10 Frame mounted end suction pump. (Courtesy of Crane Pumps & Systems, Inc., Piqua, OH.)

pump configuration up to flows of about 5000 gpm, and several offer it in much higher flows (to about 25,000 gpm). The casing is usually offered in single volute up to about 2000 gpm, and may be double volute for higher flows.

Because of its use in many process applications that involve high pressures, high temperatures, and corrosive liquids, frame-mounted pumps are offered in a broad range of materials and with a variety of sealing options. Provisions are often made to cool the bearing housing and seal housing for high-temperature applications.

Many end suction frame-mounted pumps are equipped with shaft sleeves (Figures 4.9 and 4.11). The sleeve is usually sealed against the impeller with an O-ring or gasket. Its primary function is to isolate the shaft from the pumped liquid, so that the shaft is not exposed to the potentially corrosive and erosive liquid, and to protect the shaft from wear at the seal or packing. Therefore, the shaft can be made of a less-corrosion-resistant alloy. Also, if there is a corrosive or abrasive attack, it is the less complex (and therefore usually less expensive) sleeve that needs to be replaced, rather than the more complex and expensive shaft.

Centrifugal Pump Types and Applications   239

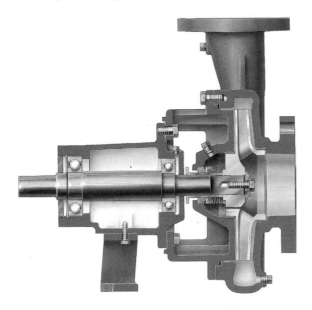

Figure 4.11 Sectional view of frame-mounted end suction pump. (Courtesy of Crane Pumps & Systems, Inc., Piqua, OH.)

Bearing lubrication for frame mounted end suction pumps is generally accomplished either by grease or by oil. Figure 4.11 shows grease lubricated bearings, which is the more common lubrication system employed in lighter-duty commercial pumps such as those used in commercial buildings for hot water and HVAC systems. The bearings are greased through grease fittings (not shown in figure 4.11).

Oil lubrication is the more common system for industrial frame-mounted centrifugal pumps (regardless of the configuration). The pump in Figure 4.40, shows an oil lubricated bearing housing, which typically includes a vented fill cap. A constant level oiler (Figure 4.12) maintains a consistent level of oil in the bearing housing, as either too much oil or too little oil can be detrimental to proper bearing lubrication. The oiler bubble glass holds an inventory of oil. If oil leaks out of the bearing housing through the lip seals on either end of the housing, an air path is exposed in the oiler, which allows oil to flow from the oiler bubble glass into the bearing housing

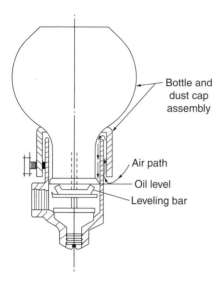

Figure 4.12 Constant level oiler maintains oil level in bearing housing. (Courtesy of Goulds Pumps, Inc., ITT Industries, Seneca Falls, NY.)

until the proper level is once again achieved. Not only does the oiler bubble glass hold an inventory of oil, but it also allows a visual determination when the bearing housing lip seals are leaking and need replacement.

IV. INLINE PUMPS

Inline pumps (Figure 4.13 and Figure 4.14), are usually oriented vertically, and may be close-coupled or frame mounted. The suction and discharge flanges are located in line with each other, and on opposite sides of the pump. There are three advantages of the inline configuration compared with a comparably sized end suction horizontal pump. First, an inline pump takes up less floor space than a horizontal end suction pump of the same size. This may be important if the installation must fit into a tight spot in a building or plant, on a ship or offshore platform, or in a skid-mounted assembly. Second, the piping coming into and out of the pump is simpler

Centrifugal Pump Types and Applications

Figure 4.13 Inline pump. (Courtesy of Crane Pumps & Systems, Inc., Piqua, OH.)

because the suction and discharge flanges are in line with each other. This eliminates the necessity for the piping changing directions as it must do with an end suction pump, and may also eliminate elbows or other fittings. Finally, compared to a frame-mounted horizontal pump, the inline configuration usually requires no field alignment, because the motor and pump are aligned by rabbet fits.

Offsetting these advantages are several shortcomings of this type of configuration. With the vertical orientation, the motor is supported by the pump rather than by a baseplate, so larger sizes may require external support. This design is inherently less stable from a structural standpoint because loads are not transferred via the bedplate to the foundation as is the case with horizontal frame-mounted pumps. This limits the size that most manufacturers offer this configuration to under 200 HP. Finally, leakage from the packing or mechanical seal does not always freely drip down to a collection cup

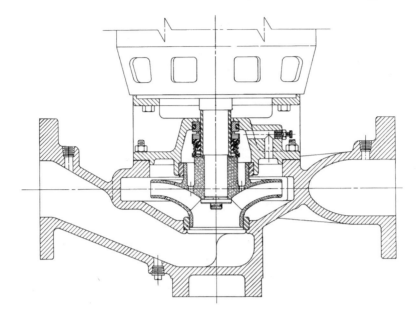

Figure 4.14 Sectional view of inline pump. (Courtesy of Crane Pumps & Systems, Inc., Piqua, OH.)

where it can be piped to a disposal system, as is the case with horizontally oriented pumps. Rather, it can collect at the stuffing box mounting area. Although the leakage can be drained from here to waste, it is more likely to cause corrosive damage to the pump or motor support.

V. SELF-PRIMING CENTRIFUGAL PUMPS

Chapter 1, Section V, explains that with a centrifugal pump, the pump's suction line and impeller inlet must be filled with liquid and vented of air before the pump can start satisfactorily. With a pump located in a system with a positive suction head, merely opening the valve at the pump suction and opening the vent valves at the top of the pump casing floods the suction line and vents the casing. If the pump is operating on a suction lift, however, the suction line must be filled before the pump can be started, a procedure known as "priming" the pump.

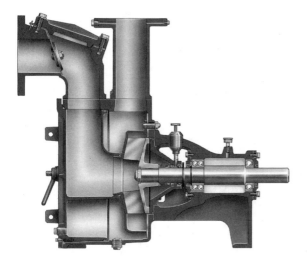

Figure 4.15 Self-priming centrifugal pump can operate on a suction lift without external priming. This particular pump is a nonclog style, designed to handle liquids containing solids. (Courtesy of Crane Pumps & Systems, Inc., Piqua, OH.)

Chapter 8, Section II.B, describes several alternative priming techniques. A self-priming centrifugal pump (Figure 4.15) automatically performs this priming procedure. This pump type can be close-coupled or frame mounted.

The ability of the pump to prime itself is due to the unique design of the pump casing. Figure 4.16 illustrates how the self-priming centrifugal pump works. The casing is double volute and acts just like any other double volute casing when the pump is operating. When the pump is shut down, liquid drains out of the pump and the suction line, but a quantity of liquid remains in the lower volute of the casing. When it is time for the pump to be started again, the lower volute acts as an intake for the impeller. As liquid moves from the lower volute chamber to the impeller, it does two things. First, the liquid keeps the seal parts wetted during the priming cycle so that the seal faces do not run dry. Second, the liquid moves into the upper volute, where it forces air out the discharge of the pump. This elimination of air from the pump creates a

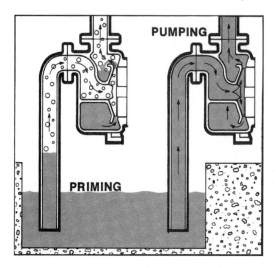

Figure 4.16 Illustration of how the self-priming centrifugal works. (Courtesy of Goulds Pumps, Inc., ITT Industries, Seneca Falls, NY.)

vacuum in the pump suction, which begins to draw liquid up the suction line. The process, known as the priming cycle, continues until the liquid moves up the suction line and into the impeller eye. Then, priming is complete and the pump once again acts like any other centrifugal pump.

The priming cycle generally takes from 3 to 10 minutes to complete, depending on the size of the pump and the volume of the suction line. Maximum lift capability is determined by the maximum vacuum that the pump will pull during the priming cycle. Most self-priming centrifugals are restricted to suction lifts of no more than 20 to 25 ft of water. These pumps are usually restricted to flows of about 3000 gpm, although several manufacturers offer flows up to about 6000 gpm.

Given the dilemma of emptying a sump filled with dirty, corrosive, and abrasive liquid, a self-priming centrifugal pump is probably one of the best solutions, and is most likely the alternative that minimizes maintenance headaches. That is because with this configuration, the only components in contact with the sump liquid are the suction pipe, pump

Centrifugal Pump Types and Applications

casing, impeller, and stuffing box (or seal housing and wetted seal parts). These are all components that can readily handle nasty liquids. There are no moving parts at all located down in the sump, and no bearings in contact with the liquid. The only drawbacks to the self-priming pump are that it may be more expensive than an alternative such as a column sump pump (especially in exotic alloys) due to the large casing, and the limitations on maximum flow and suction lift. The benefits often outweigh the shortcomings, and so this pump is popular as a tough service sump pump. It is also very common as a dewatering pump in industrial and construction arenas. For the latter, there are versions of this pump that are portable, skid mounted, gasoline engine driven, and some with impellers designed to handle large solids.

VI. SPLIT CASE DOUBLE SUCTION PUMPS

Figure 4.17 illustrates a horizontally split case double suction pump, with a similar pump being shown in Figure 4.18 in a cross-sectional view. The casing configuration is called *horizontally split case*, but a more general term is *axially split*

Figure 4.17 Horizontally (axially) split case, double suction pump. (Courtesy of Crane Pumps & Systems, Inc., Piqua, OH.)

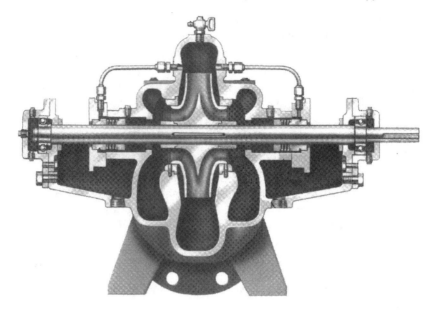

Figure 4.18 Sectional view of horizontally (axially) split case, double suction pump. (Courtesy of Crane Pumps & Systems, Inc., Piqua, OH.)

case, that is, with the casing split along the axis of the pump shaft. In fact, some manufacturers also make this design in a vertical orientation as shown in Figure 4.19 (for savings of floor space, similar to the benefit of a vertical inline configuration). The most common arrangement of the axially split case pump is with a horizontal orientation.

The benefit of the horizontally split case configuration is that the mechanical design of the casing is more structurally stable because the impeller is supported by bearings on either side of the impeller. The pump is sometimes called a *between the bearings* design, as opposed to *overhung* designs (Figures 4.8, 4.9, 4.10, 4.11, 4.13, and 4.14), where the impeller is supported by bearings located on one side of the pump only. Also, with an axially split case configuration as shown in Figures 4.17 through 4.19, the suction and discharge flanges are directly in line with each other, and on opposite sides of the pump, often simplifying the piping arrangement.

Figure 4.19 Vertically oriented, axially split case, double suction pump. (Courtesy of Crane Pumps & Systems, Inc., Piqua, OH.)

The axially split casing design allows the pump casing cover to be unbolted, the bearing covers to be removed, and the pump rotating element (shaft, impeller, sleeves, bearings, and seals) plus the casing wear rings to be completely removed as a unit for maintenance without having to unbolt the pump suction and discharge flange connections, which are located in the lower half of the casing.

In Figure 4.18, the double suction impeller is clearly shown. The horizontally split case design is the most common one for horizontal, single stage pumps with double suction impellers. As previously discussed in Sections II.B, II.C, and II.D in this chapter, the major benefits of the double suction impeller are that the pump has a lower $NPSH_r$ than a comparable single suction impeller (without resorting to a high suction specific speed design for the inlet), and the fact that in single stage configuration, the thrust loads are eliminated with a double suction impeller. These are two very compelling benefits. When combined with the previously mentioned benefits of the axially split casing configuration, they are the reasons for the popularity of the axially split case double suction pump. Because problems with NPSH and high thrust loads are often associated with larger pumps, these pumps are most commonly used with higher flow applications. These pumps commonly employ a double volute casing for flows higher than about 1500 gpm, to reduce radial bearing loads and to permit smaller-diameter shafts.

The applications for the horizontally split case double suction pump include such services as plant raw water supply, cooling water supply, cooling tower pumps, fire water pumps, pipe line pumps, and bilge and ballast pumps. They are offered in flows up to 70,000 gpm, and heads up to about 2000 ft. Material options usually include all iron, bronze fitted (cast iron casing with bronze impellers, sleeves, and wear rings), and all 316 stainless steel. Some manufacturers also offer other higher alloy impeller, sleeve, and wear ring materials.

Because of the balanced axial thrust, many smaller double suction pumps have no thrust bearing at all, but merely radial bearings on each side of the impeller. Larger sizes often have a thrust bearing on one end. This is to accommodate brief periods during start-up when thrust is developed by the pump, and also to account for the possibility that the flow is not evenly split between the two impeller inlets.

Many double suction horizontally split case pumps have two shaft sleeves that serve to locate the impeller at the

Centrifugal Pump Types and Applications

proper point on the shaft (the sleeves are usually threaded on at either side of the impeller), as well as to isolate the shaft from the pumped liquid to protect it against corrosive and abrasive attack.

Figure 4.18 shows case wear rings on the pump. The design shown holds the case ring against rotation by a pin that fits into a groove in the lower half of the casing. Note that the case ring in Figure 4.18 is contoured to guide the flow into the impeller inlet.

Double suction axially spit case pumps must have two sets of packing, or two mechanical seals, because the shaft penetrates the casing on both sides. As Figure 4.18 illustrates, however, the packing or seals are subjected to suction pressure rather than discharge pressure. This is normally lighter duty for the packing or seals, as the lower the pressure being sealed against, the longer the service life of the packing or seals. This can present problems, however, when the pump is operating on a suction lift. In that situation, the pressure inside the casing on the suction side of the impeller is below atmospheric pressure. This presents an opportunity for air to leak into the pump across the packing or seal (more likely to be a problem with packing). Chapter 3, Section IV, deals with the detrimental consequences of introducing air into a centrifugal pump. This possibility is eliminated on a double suction pump by means of the bypass piping, which circulates high-pressure liquid from the pump discharge around to the stuffing box area, where it creates a liquid seal preventing air from leaking into the pump. This introduction of liquid under pressure into the stuffing box also ensures that the packing or seal faces are lubricated at all times when the pump is operating, which, as discussed in Chapter 5 to follow, is essential for long life of the packing or mechanical seal.

Note that several manufacturers make a *radially split case* version of the single stage double suction pump, with a configuration quite different from that shown in Figure 4.18. The radially split case pump design is primarily used for high-temperature process applications, and is limited to flows of about 10,000 gpm. The configuration may be top suction, top discharge; or side suction, side discharge.

VII. MULTI-STAGE PUMPS

A. General

Multi-stage pumps generate the highest heads of any centrifugal pump at normal operating pump speeds. (A special type of high-speed centrifugal pump is discussed in Chapter 7, Section VII.) Multi-stage pumps have multiple impellers that operate in series. The flow moves through the pump from one stage to the next, with a volute or diffuser section following each impeller, so that the head is increased as the liquid moves through the pump. In most axially split case designs, the pumps have from two to as many as fifteen stages. Some radially split case multi-stage pumps are available with many more stages than that. Vertical turbine pumps are a special type of multi-stage pump, and they are discussed separately in Section XI to follow.

Multi-stage pumps achieve much higher heads than can be obtained from even large-diameter single stage impellers. Also, compared to single stage pumps at the same head and capacity, multi-stage pumps can achieve higher efficiencies than single stage pumps. (Refer to Chapter 6, Section II, for a more detailed explanation of why this is true.) Compared to positive displacement pump alternatives, multi-stage pumps may not be as efficient, but are often lower priced and are almost always smoother operating and with lower maintenance costs than positive displacement alternatives. (And as Chapter 1 indicates, positive displacement pumps cannot achieve as high a flow rate as multi-stage centrifugals.)

Applications for multi-stage pumps include boiler feed, high-pressure process applications, spraying systems, de-scaling, pressure boosters for high-rise buildings, reverse osmosis, and pipe line. The following two sections describe the most important types of multi-stage pumps.

B. Axially Split Case Pumps

Axially split case multi-stage pumps are usually horizontally oriented (so the two terms "horizontally split case" and "axially split case" are used interchangeably in this section). This pump type may have from as few as two stages (Figure 4.20) to as many as fifteen stages (Figure 4.21 shows five stages but could have more). Some manufacturers make horizontally split case multi-stage pumps in flows up to about 15,000 gpm, and these pumps can develop heads up to about 7000 ft. Applications with hot water are typically limited to about 1800 psi with this pump type, while pumps carrying ambient temperature liquids can handle considerably higher pressures.

The horizontally split case offers the same advantages as discussed for axially split casing configurations in Section VI, except that the suction and discharge flanges, while on opposite sides of the pump, are usually not directly in line with each other.

Axially split multi-stage pumps are made with both volute and diffuser casing designs. Volute designs with more than two stages are usually dual volute. A two-stage pump may have single volute construction, with the volute for one stage 180° opposed from the other. This achieves nearly the same effect of balancing radial loads as a dual volute casing on a single stage pump. The pump shown in Figure 4.21 is a double volute design.

Thrust loads in a horizontally split case pump can be minimized by orienting half of the impellers in one direction, and the other half in the opposite direction. The pumps shown in Figures 4.20 and 4.21 illustrate this. The flow passes through half of the stages (or nearly half in the case of an odd number of stages), then crosses over to the opposite side of the pump and goes through the remaining stages. The net effect, if there is an even number of stages, and if the impellers are all trimmed to the same diameter, is that thrust loads are balanced. Most multi-stage pumps still have a thrust bearing to accommodate thrust imbalances at start-up, to accommodate designs with an odd number of impellers, or if the impellers are not all trimmed to the same diameter.

Horizontally split case multi-stage pumps have a center case bushing located between the two impellers that are back-to-back in the center of the pump. This bushing must maintain a tight running clearance because there is a high differential

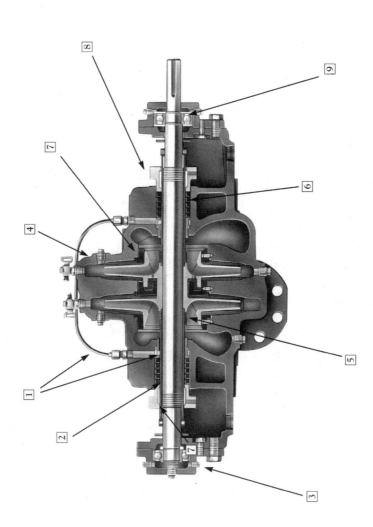

Figure 4.20 Horizontally split case two-stage pump. (Crane Pumps & Systems, Inc., Piqua, OH.)

Centrifugal Pump Types and Applications   253

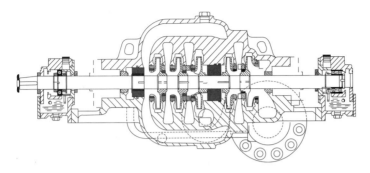

Figure 4.21 Horizontally (axially) split case multi-stage pump. (Courtesy of Sulzer Pumps, Portland, OR.)

pressure across the bushing, and the leakage across it should be minimized to maintain the pump's efficiency and preserve thrust balance.

There are two stuffing boxes or seals on this type of pump, but ordinarily one would see suction pressure and the other would see some intermediate pressure. Therefore, there is usually a connecting line to equalize the pressure that the two seals must seal against, to equalize the maintenance interval for the two sets of seals.

Many multi-stage pumps are offered with optional large eye, first stage impellers for improved $NPSH_r$ of the first stage. (See Section II.C for a discussion of suction specific speed and enlarged inlets for impellers.) The pump shown in Figure 4.21 shows such a special first stage impeller. Some designs also offer a double suction impeller for the first stage.

Material options for horizontally split case multi-stage pumps usually include cast iron (for lower pressure) or steel (for higher pressure), 12% chrome, bronze, or 316 stainless steel.

The horizontal multi-stage pump with opposed impellers is a difficult casing to seal with the complex geometry required for the crossover described above. This relies heavily on the gasket between the casing halves doing its job. Particularly in hot water services, a very slight gasket imperfection could result in leakage across the casing parting plane from high pressure areas to low pressure areas, a condition known as

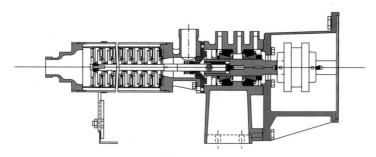

Figure 4.22 Radially split case multi-stage (tube design) TONKA-FLO pump. (Courtesy of GE Infrastructure Water Technologies, Minnetonka, MN.)

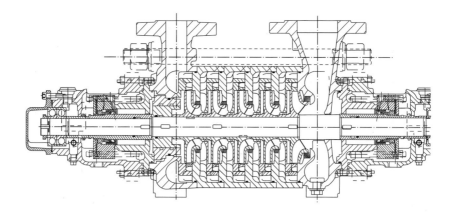

Figure 4.23 Radially split case multi-stage pump (tie-rod design). (Courtesy of Sulzer Pumps, Portland, OR.)

wire drawing, which can very quickly cause serious damage to the casing. The radially split case diffuser pump described in the next section does not have this case gasket complexity and sealing problem.

C. Radially Split Case Pumps

There are several types of radially split case multi-stage pumps. Figure 4.22 and Figure 4.23 show two types of radially split case pumps that have impellers and diffusers stacked

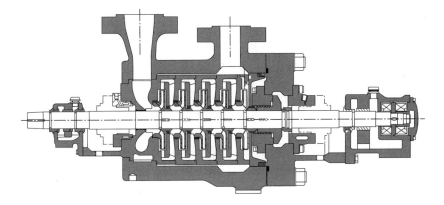

Figure 4.24 Barrel pump. (Courtesy of HydroTex Dynamics, Inc., Houston, TX.)

together, with the entire assembly held together by a tube (Figure 4.22) or tie-rods (Figure 4.23). These pumps, also called *ring-joint* or *ring-section* pumps, are generally considered in the United States to be lighter duty than the axially split case pumps described in Section VII.B above. They are available with flows to about 3500 gpm, and with heads to about 5000 ft. Several manufacturers make large, heavily engineered versions of this pump for process applications, but the most common applications for this configuration are high-pressure water booster systems, reverse osmosis, and small boiler feed services. Smaller sizes and lighter-duty versions of this pump are available vertically oriented.

Figure 4.24 shows a heavier-duty, radially split case pump known as a *double barrel*, *barrel*, or *double case* pump. This pump is used for electric utility boiler feed and other very heavy duty process services. Pumps of this configuration can achieve flows to 50,000 gpm, heads to 10,000 ft, and can handle temperatures up to about 700°F. These pumps for boiler feed service generally operate at 5000 to 7000 rpm, and often require as much as 100 ft of NPSH. Accordingly, they usually require a booster pump ahead of them to provide sufficient $NPSH_a$.

The impellers and diffusers of the double barrel pump shown in Figure 4.24 are enclosed in a pressure-containing outer barrel, with all of the impellers oriented in the same direction. This orientation of the impellers in one direction produces high thrust loads; and some form of hydraulic thrust balancing is provided, such as the balance drum introduced in Section II.D and shown in Figure 4.24 just to the right of the last impeller. Kingsbury tilting pad thrust bearings with external pressure lubrication are commonly used for this pump type.

The first stage impeller of a radially split multi-stage pump may be a special low-$NPSH_r$ design, either large eye or double suction. The pump shown in Figure 4.23 shows a first-stage impeller with a larger eye area than the other impellers.

Another version of the double case configuration uses volutes instead of diffusers, and provides a crossover so that half of the impellers are opposed to each other. This configuration is generally more costly than the diffuser design just described, but it eliminates the high maintenance balance drum.

VIII. VERTICAL COLUMN PUMPS

Figures 4.25, 4.26, 4.27, and 4.28 show a vertical column pump, one of the most common pumps used in sump pump service, or in transferring or circulating liquid from a tank or sump. The pump is a single stage design and is often installed as a duplex unit as shown in Figure 4.25. A thrust bearing is located at the top of the pump. The pump discharges through a discharge column pipe, and the shaft is enclosed in a central vertical column pipe. This pump type is actually just a single stage end suction pump that has been oriented vertically, with the addition of the long column pipe enclosing the shaft and the discharge pipe. The impeller is immersed below the liquid level, so that it has adequate $NPSH_a$ and submergence. This pump style can have an open impeller design (typical for sump pump service), or a nonclog style impeller (for sewage and other liquids containing large solids) as shown in Figure 4.26. The suction inlet on the casing is usually opened wider because there is no suction piping, and the suction inlet is often covered with a screen to keep out solids that are too large

Figure 4.25 Vertical column pump, duplex unit. (Courtesy of Crane Pumps & Systems, Inc., Piqua, OH.)

to pass through the pump. Some manufacturers offer an adapter to the casing that allows the pump to be located next to a tank, rather than immersed in the tank, and with the suction piping running from the tank to the suction adapter. This configuration is often referred to as a *dry pit* configuration.

As a sump pump, this pump has the advantage of being relatively inexpensive, of rather simple construction, and easy to take apart for maintenance. Because it has been adapted from horizontal designs, and with the flows being usually less than 3000 gpm, the casing is usually single volute. This is one of the weaknesses of this design in its basic configuration, because the pump has no radial bearings except a sleeve bearing just above the impeller. Because the single volute

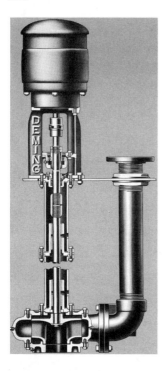

Figure 4.26 Sectional view of vertical column pump. (Courtesy of Crane Pumps & Systems, Inc., Piqua, OH.)

construction causes radial loads, this sleeve bearing in larger pumps is often observed to wear on one side because the radial load tends to push the impeller to one side.

The lubrication of the sleeve bearing mentioned in the preceding paragraph is the other weakness of this pump style. There are usually other sleeve bearings supporting the shaft at intervals of 3 ft. (This pump type is usually limited to a length of about 20 ft.) There are several alternatives for lubricating these sleeve bearings. In one design style, the pumped liquid is allowed to get into the shaft column to lubricate the bearings. This may not be a good solution if the pumped liquid is corrosive or contains abrasives, although it is acceptable if the liquid is clean and noncorrosive. Also, the shaft can be hardened at the sleeve bearing locations to minimize abrasive wear. Another approach has tubing directed down the outside

Centrifugal Pump Types and Applications

of the shaft column, so that the bearings can be lubricated externally by water, oil, or grease. While this approach isolates the bearing from the pumped liquid, it adds to the amount of water or oily waste that must be processed in a sump pump system.

Despite the shortcomings discussed above, this pump design is popular because of its relative low cost and ease of maintenance. It also has another distinct advantage of not requiring a seal. The shaft column pipe usually has holes open to the sump (when used in its normal wet-pit configuration), so that the pipe does not build up pressure, but relieves to the sump instead. There is often a lip seal or packing at the top of the column pipe to keep vapors from leaking out at this point.

This pump type is available in iron, bronze, and higher alloys for more corrosive services. There are several makers of smaller versions of this pump type in a cantilevered configuration (eliminating bearings in the pumped liquid), with sealless pumping being one of their major selling features. These smaller pumps (see Figure 4.27) are supplied in nonmetallic materials such as CPVC, polypropylene, and PVDF, and are popular in the plating industry. (See Chapter 7, Section V for discussion of nonmetallic pumps.)

Section X in this chapter on slurry pumps discusses an alternative design of the cantilevered configuration in larger sizes. These pumps eliminate the bearings in the shaft column, and some have special impeller types to handle solids and abrasives.

Still another variation of vertical column pump is the *drum pump*, shown in Figure 4.28. Drum pumps are designed to fit the bung of a standard 55-gallon drum, for emptying the drum of chemicals or other liquids. There are also designs that can accommodate pumping out of totes, vats, intermediate bulk containers, carboys, and open vessels up to 10 ft deep. Drum pumps are portable, and can be supplied with a wide range of AC and DC motors and compressed air drives. The usual construction has centrifugal-style impellers, although several manufacturers also have offerings where the rotor is a progressing cavity design for higher viscosity liquids. Typical

Figure 4.27 Nonmetallic cantilevered submerged column pump. (Courtesy of CAMAC Industries, Sparta, NJ.)

drum pump wetted materials of construction include polypropylene, PVDF, aluminum, and stainless steel.

IX. SUBMERSIBLE PUMPS

Submersible pumps (Figure 4.29) use a motor that is designed to operate submerged in the pumped liquid. The motor is often encapsulated and filled with oil, which is separated from the pumped liquid by a mechanical seal. These pumps are usually designed to pump sewage or industrial wastewater from a pit or a tank, with the pumped liquid often containing solids. The

Figure 4.28 Drum pump. (Courtesy of Lutz Pumps, Norcross, GA.)

simplest version of this design is the residential sump pump, located in the basements of homes where the water table is such that water will accumulate in the basement. There are also home sump pumps of the column type as described in Section VIII above, but this older design is less popular for residential sump pumps, and the submersible configuration now dominates this application. There are much larger submersible models, of course, for handling the higher flows and heads and the larger solids sizes required for commercial, municipal, and industrial applications. Also, refer to Section XI for a discussion of submersible versions of vertical turbine pumps.

Figure 4.29 Submersible pump. (Courtesy of Crane Pumps & Systems, Inc., Piqua, OH.)

Submersible pumps such as shown in Figure 4.29 are single stage, and the impeller is usually a *nonclog* style, described in Section II.F.

Another feature often found on these pumps is a slide rail system and quick connect discharge, which allow the pump to be able to be lowered into a waste sump and coupled to the discharge piping with minimal time spent in the sump by the mechanic.

The submersible nonclog pump for waste handling applications is a relatively recent pump configuration, replacing an earlier generation of vertical bottom suction dry pit nonclog pumps. Submersible motors have improved in reliability over the past twenty years, making this configuration more acceptable for most applications. This pump configuration is available normally only in iron and bronze, although there is limited availability of stainless steel for corrosive industrial wastes. Residential sump pumps often have the impeller, volute, and motor housing made largely of plastic.

Centrifugal Pump Types and Applications

Figure 4.30 Grinder pump. (Courtesy of Crane Pumps & Systems, Inc., Piqua, OH.)

Another type of submersible pump, called a *grinder* pump (see Figure 4.30), is used to grind and pump sewage in pressurized sewer systems. This relatively new application allows for smaller sewer lines in new housing and commercial developments, because the sewage is ground up and pumped under pressure, rather than having to drain by gravity to a collection station or treatment plant. Use of the grinder pump also means that sewage lines in new housing developments can follow the contour of the land, rather than being continuously sloping to a collection station. Grinder pumps have higher heads than sump and effluent pumps, and have cutting blades to grind up the sewage.

As the next section indicates, submersible pumps can also be used in abrasive applications, although with special materials and impeller types to tolerate the abrasive nature of the liquids pumped.

X. SLURRY PUMPS

Slurry pumps have many industrial applications where liquids containing abrasive particles must be pumped. Applications for slurry pumps include mine dewatering and transporting of mine slurries and mine tailings, fly ash and bottom ash sluicing in coal-fired power plants, mill scale handling, sand and gravel sluicing in quarry and dredging operations, paper mill waste processing, and clay slurry pumping.

The major concern in pumping these and other abrasive slurries is the abrasive wear on the pump impeller, casing, and other wetted parts. Many of these slurries are corrosive as well. The major considerations are the selection of the best pump materials for the service, and the design of the pump components to resist wear or be isolated from the abrasive liquid.

The two most common material choices for slurry pumps are rubber-lined pumps and hard metal pumps. Rubber-lined pumps (Figure 4.31) are normally chosen if the abrasives are fine, and particularly if the abrasive particles are more rounded in shape, or are corrosive as well as abrasive. If the abrasive solids are larger in size, or if they have a more jagged or irregular shape, hard metal pumps are the more likely choice. Most hard metal pumps are made of a high-nickel iron, heat treated to hardness levels above 600 on the Brinell scale. These metal pumps are so hard that the components cannot be machined using regular machine tools, but rather must be machined using grinders. For this reason, flanges are often cast with slots to eliminate the necessity to drill holes. In addition to high-nickel iron, other material choices for slurries that are more corrosive than abrasive are 316 stainless steel and CD4MCu.

Impellers, whether rubber lined or hard metal, are often supplied in one size only because it is impractical or impossible to trim them to achieve varying hydraulics. Therefore, many of these pumps are set up to run at one of several

Figure 4.31 Rubber-lined end suction slurry pump. (Courtesy of Goulds Pumps, Inc., ITT Industries, Seneca Falls, NY.)

speeds, with belt drives being the most common device to achieve this. On the family of curves for this type of pump, the curves are usually shown at a single impeller diameter but operating at several possible speeds. Slurry pumps are generally run at speeds of 1200 rpm and slower in an effort to reduce excessive erosion in the high-velocity areas.

There are several configurations offered for slurry pump applications. There are end suction pumps, as exemplified by the rubber-lined pump in Figure 4.31. A variation on this configuration is a side suction design, shown in Figure 4.32 in a hard metal design. In this configuration, the impeller is reversed, with the primary benefit of this feature being that the stuffing box is exposed to suction pressure rather than discharge pressure, for longer packing and sleeve life. Both versions use enclosed impellers with a radially oriented leakage joint.

For wet pit applications, some manufacturers offer a column sump pump, similar to the one discussed in Section VIII above, except that the pump is cantilevered to eliminate any bearings in the pumped liquid. This pump configuration, shown in Figure 4.33, is usually limited to a length of about

Figure 4.32 Hard metal side suction slurry pump. (Courtesy of Goulds Pumps, Inc., ITT Industries, Seneca Falls, NY.)

12 ft, and must have an oversized shaft to eliminate submerged bearings.

Finally, some manufacturers offer a submersible configuration of a hard metal pump, similar in its basic configuration to Figure 4.29 except having the wetted parts constructed of hard metal alloys.

Slurry pumps come with a variety of special features, in addition to the ones already discussed, to help them withstand the effects of abrasive wear. Some of these pumps are equipped with *vortex* or *recessed impellers*, as shown in Figure 4.33 and described in Figure 4.7. This impeller is recessed so that it increases the liquid velocity while remaining outside the main liquid flow path. This results in a very inefficient pump, but this is often offset by the much greater resistance to abrasion and the much larger solids and stringy material that can be handled with this design.

Other features commonly found on slurry pumps include pump-out vanes on the back side of the impeller (described

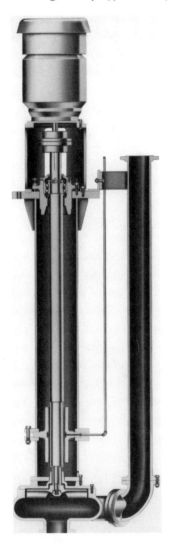

Figure 4.33 Vertical cantilevered slurry pump with recessed impeller. (Courtesy of Goulds Pumps, Inc., ITT Industries, Seneca Falls, NY.)

in Section II.D), replaceable wear plates on both the suction and the discharge sides of the casing, extra heavy bearings to withstand the shock load of pumping solids, and hard-faced sleeves to resist abrasion at the packing area.

Figure 4.34 Vertical turbine pump. (Courtesy of Johnston Pump Company, Brookshire, TX.)

XI. VERTICAL TURBINE PUMPS

The vertical turbine pump (Figure 4.34) is a special class of pump, in a category all its own. It was first designed as a well pump to bring groundwater to the surface for irrigation and drinking water. In addition to these two major applications, the vertical turbine pump has a great many industrial applications such as transfer, booster, fire pump, cooling water, and makeup water supply. The term "turbine" is somewhat misleading, but is used for the sake of convention.

For many industrial applications, the vertical turbine pump offers quite a few advantages over alternative choices (such as split case double suction pumps). The way that they can be staged allows designs with optimum efficiency to be obtained for most applications, and they may be the only choice for some applications. Other benefits include lower installed cost in many cases, no foot valve or priming required, and more material options. There are several offsetting shortcomings, however, which can lead to maintenance problems if these pumps are not chosen carefully and maintained properly.

As shown in Figure 4.34, liquid enters the lower end of the vertical turbine pump through the *suction bell*. The flow passes through one or more stages that have either open or closed impellers and diffuser cases called *bowls*. Because of its deep-well origin, economics dictate that the pump be made as narrow as possible, to minimize the required well diameter. Consequently, the diffuser bowls are located more or less in line with the impellers rather than outside them, as is the case with horizontal diffuser pumps where the diameter is not critical but the shaft length affects rotor dynamics.

Vertical turbine pumps can have anywhere from one to about twenty-five stages, providing a wide range of flow and head for this pump design. Bowls range in size from a diameter of 4 in. to about 100 in., with flow rates greater than 200,000 gpm being offered by several manufacturers. Multistage vertical turbines can achieve heads of up to about 5000 ft.

The vertical lineshaft to which the pump impellers are attached runs up the length of the pump column, which is as long as necessary to get the bowl assembly down far enough in the liquid to achieve the minimum required submergence and to have adequate $NPSH_a$. No priming is required. The ordinary industrial application has the pump taking suction from a pond, lake, river, tank, or intake structure, and requires a pump length of less than 50 ft. However, there are many vertical turbine pumps installed in wells for irrigation services worldwide with settings over 1000 ft deep.

At the top of the pump, above the column section, the discharge head supports the vertical motor, mounts the pump at grade, and turns the direction of flow 90°. Lighter-duty vertical turbine pumps have threaded and coupled column sections and cast iron discharge heads, while heavier-duty industrial pumps use flanged column sections and fabricated discharge heads. Pumps for industrial services are offered in a variety of alloys for corrosion resistance, while the standard materials for water services are iron bowls, iron or bronze impellers, iron or steel discharge heads, and steel column pipe.

There are several alternative configurations to the one shown in Figure 4.34. If it is taking suction from a source other than atmospheric pressure, the pump can be mounted in a barrel with a special double-shelled discharge head, as shown in Figure 4.35. This canned configuration can have a suction pressure greater than atmospheric pressure (used as a booster pump in this application), or below atmospheric pressure (for example, in hotwell condensate applications). A vertical turbine in a barrel may be a very economical choice for a condensate pump, where $NPSH_a$ is usually minimal. Use of this pump type requires that the barrel be only long enough to have the first stage impeller low enough to achieve adequate $NPSH_a$. Referring to the formula for $NPSH_a$ in Chapter 2, Equation 2.18, H for a vertical turbine installation is the distance from the surface of the suction vessel to the inlet of the first stage impeller, and $H_f = 0$. Therefore, it may be only necessary to drill a cavity large enough to accommodate the pump barrel, as an alternative to excavating and building an entire pump room for a horizontal pump below the condenser.

Another configuration of vertical turbine pumps uses a submersible motor mounted to the bottom of the pump as shown in Figure 4.36. Usually, the motor is a specially designed long and narrow submersible motor, capable of fitting down a well casing. The major advantage of this configuration is that it eliminates the need of any shaft and bearings above the bowl assembly. On a very deep setting, this can save a great deal of money, as well as eliminate a major source

Centrifugal Pump Types and Applications

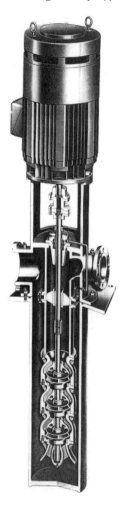

Figure 4.35 Vertical turbine pump mounted in a barrel can be used for high-pressure suction sources, or for pumping from vessels below atmospheric pressure. (Courtesy of Goulds Pumps, Inc., ITT Industries, Seneca Falls, NY.)

of maintenance problems, as discussed below. Submersible motors are commonly used with bowl sizes up to 8 and 10 in., and are available in much larger sizes as well. In Europe, submersible designs are used for most vertical turbine applications. In the United States, this configuration is becoming

Figure 4.36 Vertical turbine pump with a submersible motor eliminates column shaft and bearings. (Courtesy of Johnston Pump Company, Brookshire, TX.)

more popular as the reliability and affordability of submersible motors continue to improve in larger sizes. Submersible vertical turbines have made significant inroads in the industrial markets. In addition to clean services, they are used in seawater (e.g., fire pumps and seawater lift services) for platform and dockside applications. A special application is the horizontal installation of a submersible turbine for booster service.

One of the shortcomings of vertical turbine pumps is the fact that bearings to hold radial loads are sleeves mounted in each bowl, located in retainers at intervals of every 3 to 5 ft in the column assembly, and at the stuffing box just below the discharge head. The diffuser casing design that these pumps employ does minimize radial loads, but the pump still requires these guide bearings at regular intervals. With the configuration shown in Figures 4.34 and 4.35, the bearings in the bowls and column must be lubricated by the product being pumped. This means they are likely to wear if the pumped liquid is abrasive and/or corrosive. To help minimize this, a variety of material combinations are available for bearings, as well as coatings for the shaft at bearing journal locations.

The bearings can also be worn if any of the shaft sections are not straight, which should be carefully checked at the time of assembly of the pump. A typical straightness tolerance is 0.0005 in. per foot of shaft length.

If the many registered fits on the pump (at each bowl and column joint) are not machined concentrically, or are assembled with the looseness of the fit all stacked up on the same side, this can also result in bearings wearing as the pump rotates. To preclude this, the bowl assembly, which is usually assembled horizontally on a bench, should be rotated 90° on the bench every stage or two during assembly.

As a bearing wears at a specific location in the pump, it tends to cause the shaft to whip as it rotates, which further exacerbates the problem and causes the bearing wear to spread up and down the pump to other bearings. This is difficult to observe if it happens far down in the liquid, out of sight and hearing of the observer, thus increasing the likelihood of a problem going undetected. This progressive bearing wear can eventually cause wear rings or impellers to rub and, in the worst case, can cause the bowl assembly to seize up, break the shaft, or cause other extensive damage to the pump. At the least, it can cause excessive vibration, which can shorten seal and motor bearing life, and reduce hydraulic performance as recirculation increases.

Another potential problem with vertical turbine pumps has to do with the fact that they are much more flexible than most close-coupled or frame-mounted centrifugal pumps, making them quite a bit more susceptible to *resonance* than other pump types. Resonance is a condition that can occur with pumps if the operating speed is too close to the natural frequency of the equipment as installed in the field. Resonance produces extremely high vibration levels in the equipment and should be avoided. Most pumps are so rigid that their operating speed never approaches the natural frequency of the equipment as they come up to full speed from a stopped position. Vertical turbine pumps, on the other hand, are much more flexible, and they generally pass through a natural frequency on the way up to full speed. The natural frequency of the installed vertical turbine pump is a function of many variables, the most important of which are pump length, weight of the bowl assembly, weight and natural frequency of the electric motor, column diameter and wall thickness, discharge head dimensions, shaft diameter, foundation mounting system, and piping support system. Some of these variables, such as piping and foundation details and motor natural frequency, must be communicated to the pump manufacturer by outside parties. There occasionally arises a field problem where the pump, when first started up, ends up operating at a speed too close to its natural frequency. This phenomenon, called resonance, causes the pump to vibrate excessively. When the problem exists, it is usually related to the portion of the pump from the foundation up, rather than the portion below the pump mounting flange. The only possible solutions to a resonance problem are to change the natural frequency of the pump by acting to stiffen or soften the pump (by substituting certain components, welding stiffeners onto the pump, externally supporting the pump, or changing the foundation support system); or to change the operating speed of the pump if that is possible (such as on a variable-speed system).

Note that the foregoing discussion of resonance and natural frequency is separate from the issue of *shaft rotating critical speed*. This is a design issue involving the selection by the manufacturer of correct bearing spacing and shaft diameter for given operating speeds and thrust values to ensure that the pump rotating speed is a safe margin from the rotating critical speed of the shaft/bearing system.

Axial thrust loads generated by the vertical turbine pump must be accommodated by a thrust bearing. Designs of U.S. pump manufacturers use a high-thrust vertical motor to handle the thrust loads. The normal thrust is downward, and the thrust loads must be supplied to the motor manufacturer with the motor specification. Axial thrust increases at higher values of TH, so the thrust loads should generally be specified as a continuous-duty load at the normal pump operating point, as well as a short-term load based on the pump running at or near shut-off.

Vertical turbine pumps mounted in barrels with high suction pressure may be subject to upthrust at some operating points, with this most likely to occur at runout flow. This condition should be carefully checked and avoided if possible by adding hydraulic balancing devices or making system modifications. This condition should be avoided for several reasons, the most important being that operating a vertical pump in upthrust means that the shaft is in compression rather than tension. This makes it much more susceptible to problems with alignment and imbalance, as well as changing the shaft rotating critical speed. Also, some thrust bearings are not designed for continuous operation in reversed thrust.

European designs of vertical turbine pumps typically use a separate thrust bearing assembly mounted between the motor and pump head to carry the thrust, allowing the use of normal thrust vertical motors. In North America, external thrust assemblies are used in only a few applications, most notably in the automotive manufacturing industry.

If a submersible motor is undesirable or unavailable, pumps longer than 50 ft to water level must rely on something other than the pumped liquid to lubricate the upper sleeve bearings in the pump column and discharge head. Otherwise, when the pump is first started, it will take too long for the pumped liquid to reach the upper bearings, and these bearings will overheat and damage the bearing and/or shaft. A design option to consider as an alternative to a submersible

Figure 4.37 An enclosing tube isolates the column bearing from the pumped liquid, and lubricates the bearings inside the tube with water or oil. (Courtesy of Goulds Pumps, Inc., ITT Industries, Seneca Falls, NY.)

configuration for settings over 50 ft in length uses an enclosing tube around the column shaft (Figure 4.37) that isolates the column shaft and bearings from the pumped liquid. Lubricating liquid (in older installations primarily oil, now more commonly

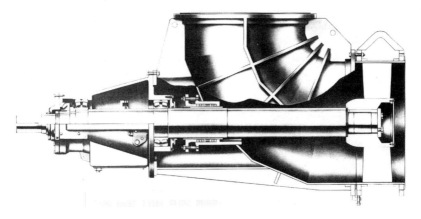

Figure 4.38 Axial flow (propeller) pump. (Courtesy of Goulds Pumps, Inc., ITT Industries, Seneca Falls, NY.)

water due to environmental release restrictions on oil) is allowed to enter the enclosing tube at the top to provide lubrication for the column bearings. This enclosed lineshaft construction also protects the lineshaft bearings from abrasive wear.

XII. AXIAL FLOW PUMPS

This category of pump is discussed in Chapter 2, Section VII, in the discussion of specific speed. An axial flow pump is designed to deliver a very high flow rate with a low head. Flows greater than 200,000 gpm are not uncommon, with heads usually less than 50 ft. These pumps normally come in one of two configurations. One type is similar to the vertical turbine pump discussed in Section XI. With a single impeller and a vertically oriented shaft, this pump type is used in an open body of water for such services as condenser cooling water for power plants.

Another configuration of axial flow pumps is shown in Figure 4.38. This configuration, with the impeller mounted in a cast or fabricated elbow, is used in closed systems where a very high flow rate must be achieved, with the only head requirement being the piping friction losses in the closed loop. An example of such an application is evaporator recirculation.

Figure 4.39 Regenerative turbine pump. (Courtesy of Crane Pumps & Systems, Inc., Piqua, OH.)

XIII. REGENERATIVE TURBINE PUMPS

This style of pump was first mentioned in Chapter 1, Section IV.B, as a special type of pump in the kinetic category, also known as a *peripheral* pump. Figure 4.39 shows a regenerative turbine pump, the impeller of which is clearly seen as markedly different from a traditional centrifugal pump impeller. Instead of having the traditional backward-curved vanes, a regenerative turbine impeller has radially oriented teeth or buckets, having an increasing depth with increasing diameter. The flow path of the liquid as it passes through the rotating impeller's teeth is illustrated in Figures 4.39. As the impeller rotates, it increases the liquid's velocity. As the liquid moves past the teeth, the expanding area from the increasing depth causes the liquid velocity to decrease, achieving the change to pressure energy.

The regenerative turbine pump is capable of achieving quite high heads in single stage construction at 3600 rpm (up to about 700 ft), and several manufacturers make this pump type with more than one stage. Flows in this pump type are limited to about 150 gpm. The head–capacity curve of this pump type is quite steep, with shutoff head being two to three times higher than the head at the best efficiency point. Also, the horsepower curve of a regenerative turbine pump behaves opposite to most centrifugal pumps, with the BHP of a turbine pump increasing as flow decreases. Because of the steep head–capacity curve and the shape of the horsepower curve, it is usually recommended that regenerative turbine pumps be protected against over-pressurization by means of a pressure relief valve, either incorporated into the pump or external to the pump.

The most common application for the regenerative turbine pump is small commercial boilers. For applications where this pump type can achieve the hydraulics, it is often the lowest cost and most compact alternative, although not as efficient as a multi-stage pump. Another characteristic that sets the regenerative turbine apart from other centrifugal pumps is that the pump can handle up to 20% vapor or noncondensable gases in the pumped liquid.

The greatest drawback of this pump type is that very tight clearances must be maintained within the pump to prevent the pump hydraulic performance from rapidly deteriorating. Consequently, the pumped liquid must be very clean and the pump will tolerate no abrasives. Most regenerative turbines have no capability to make adjustments to account for wear in the pump, although one manufacturer offers an adjustable impeller.

Iron casings and bronze impellers are the most common material options for this pump type. Several manufacturers offer regenerative turbines in all bronze or all 316 stainless steel.

XIV. PUMP SPECIFICATIONS AND STANDARDS

A. General

One of the most commonly asked questions by engineers and others who specify or purchase pumps is: How detailed should a pump specification be, and what is the minimum information it should contain? As with most of the questions raised in this book, there is no absolute answer, and any answer that is given must start with, "It depends." As an example, for an application requiring a relatively low flow and head (such as 100 gpm at 50 ft), where the liquid pumped is clean water at ambient temperature, and the pump configuration will be the manufacturer's most standard close-coupled offering, this information, along with the motor voltage and enclosure type, is pretty much all that is required. At the other end of the spectrum, many larger and more complex pumps for municipalities, other government entities, or electric utilities must be specified in great detail to ensure that the buying agency will not be forced to buy equipment that is not satisfactory for the service merely because it is the lowest bid submitted.

With regard to the operating conditions to which the pump is exposed, it is safe to say that it is nearly impossible to supply too much information to the pump manufacturer. The following information regarding operating conditions should be provided in the specification in as much detail as possible.

1. Liquid Properties

The important liquid properties to specify include the liquid type, operating temperature, specific gravity, pH, viscosity, vapor pressure, amount and type of suspended and dissolved solids, and amount and type of abrasives or other solids. Normal conditions and the expected range of these properties should be specified.

2. Hydraulic Conditions

Design parameters that should be specified include design capacity, total head, $NPSH_a$ (at maximum continuous flow, and at a specified reference point such as floor elevation), suction lift or suction head (and how these will vary), and maximum suction pressure.

In addition to design conditions, the specification should indicate the minimum and maximum flow the pump will be required to deliver, and the duty cycle for the different flows (i.e., other expected pump flow rates and approximate percentage of operating time at each flow rate). Keep in mind, however, that many manufacturers only guarantee the hydraulic conditions at a single operating point on the pump curve unless other points are specified and specifically agreed upon by the supplier. In some cases, asking other operating points to be guaranteed may increase the cost of the equipment, or restrict the number of potential suppliers.

For severe services or critical installations, a system head curve should be developed to determine the pump duty cycle. Or, software such as described in Chapter 3, Section III, can be used to develop the pump operating duty cycle.

3. Installation Details

Important installation details that should be specified include the location of the installation, elevation above sea level, unusual ambient conditions (such as extremely high humidity or salt-laden atmosphere), type of suction vessel or sump, and space or weight limitations.

When it comes to design details, the buyer (or specifying engineer) should, to the extent possible, allow the bidders to offer their recommendations for the best equipment to meet the specified service conditions. Of course, there will be circumstances where the buyer or specifying engineer may want to detail more specific design requirements. One reason might be if a specific design standard is to be invoked, such as ANSI or API, which are described in the following sections, or FDA requirements for sanitary pump services. Or, the buyer may wish to specify a particular pump configuration, materials of construction, or design details, based on the buyer's experience with pumps operating in similar services. Wear rings may be considered optional by the manufacturer on certain pumps, and so they should be specified if required.

Lacking experience with pumps in similar services, it may be best to allow the manufacturers to offer their recommendations for the specified service, leaving the door open to explore alternatives if the evaluation process produces them. Where the manufacturers can offer their standard construction,

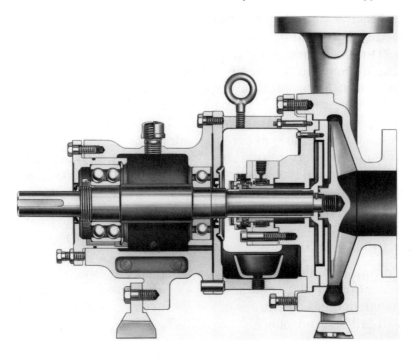

Figure 4.40 ANSI pump. (Courtesy of Goulds Pumps, Inc., ITT Industries, Seneca Falls, NY.)

the pump's cost is usually less than that for uniquely built equipment. The bidding manufacturers should be asked to describe their offerings, as to configuration type, materials of construction, and design details, so that a proper evaluation can be carried out and alternatives considered if necessary.

The type of driver for the pump should be specified, with details of construction covered if the buyer is purchasing the driver separately.

B. ANSI

The American National Standards Institute (ANSI) Standard B73.1 (Ref. [7], also known as ASME B73.1) for chemical process pumps was developed more than 30 years ago. Figure 4.40 shows a pump built to the ANSI standard.

Centrifugal Pump Types and Applications

The standard was written to ensure that chemical process pumps built by any manufacturer that met the standard would use the same criteria for such design details as required casing thickness, allowable shaft deflection, bearing life, etc. Another very important goal of the standard was to standardize the configuration and envelope dimensions of the pump for common sizes. Thus, the standard configuration for the ANSI pump is end suction, frame mounted, with the discharge of the pump in line with the centerline of the casing. No matter which manufacturer is making the pump, the key envelope dimensions of each company's size $1 \times 1½ - 6$ pump, for instance, are identical. The key envelope dimensions here are the shaft centerline to the discharge flange face; the suction flange face to the coupling end of the shaft; flange diameters; flange bolt circle; size, number, and location of flange bolts; and shaft and keyway dimensions. This standardization of dimensions allows an engineering consultant or user company that is designing a plant containing hundreds of ANSI pumps to complete all details of piping and foundation design without necessarily having chosen a supplier of the pumps. It also allows a user to switch suppliers when replacing a pump, with making piping or foundation chagnes.

Although it is known as a chemical process pump, the ANSI pump is one of the most versatile and widely used pumps in a variety of industrial applications. The pump can achieve flows up to 5000 gpm, with heads as high as 750 ft at lower flow rates. Because the pump is often used with corrosive chemicals, it is readily available in an extremely broad range of material options, probably the widest array of material options of any pump type. Standard construction material is ductile iron, but the ANSI pump is available in 316 stainless steel, bronze, CD4MCu, hastelloy, titanium, and several other exotic metals. A number of nonmetallic material options are also available, including FRP and Teflon® PTFE lined versions.

While most companies that make ANSI pumps offer as many as 25 sizes of impellers and casings for their complete line, the makers of this pump type have modularized the design so that there are no more than six or eight stuffing box sizes, and no more than two or three bearing frames to cover the entire ANSI line. This modularization minimizes the amount of spare parts that must be maintained by the user in a chemical plant, which may have hundreds or even thousands of ANSI pumps.

The ANSI pump is most commonly supplied with an open impeller, adjustable for wear as described in Section II.A above. The pump has a sleeve to isolate the shaft, and can be equipped with a variety of mechanical seal types. Bearings are normally oil lubricated, and the bearing housing can be externally cooled to allow pumping liquids up to 500°F. The pump is normally *back pull-out* design, which means that the impeller, sleeve, seal, and bearings can be removed from the back of the pump for maintenance without disturbing the suction and discharge piping connections.

C. API

The American Petroleum Institute (API) Standard 610 for refinery pumps was written to develop a stringent, detailed design standard for process pumps in refinery services and other petrochemical applications. The need for this standard stems from the fact that the liquids being handled in refineries and petrochemical plants are often high pressure, high temperature, and usually flammable or volatile, so safety is a very important consideration. Furthermore, the cost of lost production time in a refinery can be considerable, as can the cost of pump maintenance. Therefore, every effort is made in the API 610 Standard to maintain pressure integrity and reliability of the pump, and to reduce maintenance expense. This attention to design detail extends to such diverse subjects as casing thickness criteria, nozzle loading criteria, allowable shaft deflection and runout, bearing design, and mechanical seals.

The API 610 Standard is not written around a single configuration as is the ANSI standard for the end suction configuration. Instead, the API 610 Standard covers a variety of pump types, including end suction, axially and radially split

Centrifugal Pump Types and Applications

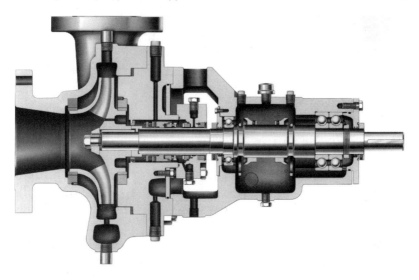

Figure 4.41 API 610 covers a number of pump configurations, including the end suction single stage frame-mounted design shown here. (Courtesy of Goulds Pumps, Inc., ITT Industries, Seneca Falls, NY.)

case multi-stage, vertical turbine, and others. Because they encompass such a variety of pump types, API 610 pumps are available over a very wide range of capacities and heads.

The 610 Standard references a number of other API specifications, such as API 671 for couplings and API 677 for gear units, as well as standards from the American Society for Testing and Materials (ASTM) and other organizations. Also, there are several specific API standards covering special pump types (e.g., API 685 for sealless pumps).

Figure 4.41 shows a horizontal end suction single stage pump designed to API 610. Note that the pump casing is centerline supported, rather than being supported by pump feet as is the ANSI pump discussed in Section XIV.B. Centerline support is a requirement of API 610 for horizontal pumps with a pumped liquid temperature of 350°F and higher. The centerline support requires pedestals on the bedplate upon which support "wings" on the casing of the pump are mounted.

This minimizes the amount of movement of the pump centerline due to thermal expansion when pumping hot liquids, so it is especially suitable for high-temperature applications. It also allows the pump to operate with a relatively large temperature swing without having to realign the pump because of thermal growth.

Pump impellers are specified in API 610 as closed impellers, to be fitted with wear rings. The wear rings must be locked against rotation by a more positive locking arrangement than a simple interference or press fit. Some manufacturers use set screws or pins to secure the wear rings, while a few companies weld the rings in place. API 610 has its own recommendations for wear ring clearances (see Table 4.1).

Material options for API 610 do not include as many exotic alloys for corrosion resistance as found in the ANSI standard, because most hydrocarbon products are not corrosive. Material options include cast iron, bronze, carbon steel, 12% chrome, and 316 stainless steel.

The API 610 Standard is considered one of the most thorough and exacting of pump specifications. Because of this, it is sometimes used as a design specification for other heavy-duty process pumps outside the refinery and petrochemical industries.

D. ISO

The International Standards Organization (ISO) standards are making significant inroads in the United States, particularly among engineering contractors and pump manufacturers involved in major projects outside the United States. The ISO standards are broader in scope than the ANSI or API standards, and they include such subjects as certification of manufacturers' quality assurance programs in addition to equipment design standards.

ISO Technical Committee 115 (TC115) is responsible for international pump standards. The secretariat for the U.S. Technical Advisory Group to this committee is the Hydraulic Institute. ISO TC115 has two primary working subcommittees. Subcommittee 1 (SC-1) covers dimensions and technical specifications, and the secretariat is the British Standards

Institute. Subcommittee 2 (SC-2) covers methods of testing pumps, and the secretariat is UDMA, the German Standards Organization.

The ISO standards reference metric units exclusively, and are based on 50-cycle electric current. Metric dimensions are generally the preferred dimensions outside the United States. However, most developing countries have a great deal of installed equipment such as ANSI pumps built with inch dimensions (USCS units).

The global trend is moving toward metric dimensions, and most U.S. manufacturers with global marketing organizations supply pumps with both metric and inch dimensions. An indication of the growing importance of ISO in some U.S. industries is the fact that API 610 uses ISO metric standards for mechanical seals.

XV. COUPLINGS

The primary objectives of the shaft coupling in a pump are to:

- Connect the pump and driver shafts.
- Transmit the power (torque at a given speed) between the separate rotating shafts of the pump and driver. (This assumes the pump is not close-coupled.) This causes both shafts to rotate in unison, and at identical speeds.
- Compensate for minor amounts of misalignment and movement of the shaft due to vibration or thermal effects.

For vertical pumps, because the alignment of the components is assured through registered fits, the coupling can be a rigid type that does not tolerate misalignment, as shown in Figure 4.42. For most horizontal pumps, it is nearly impossible to maintain perfect concentricity of centerlines of the two shafts. *Flexible couplings* are used to transmit the torque, while allowing for some amount of angular and parallel misalignment. (Refer to Chapter 7, Section IX, for a discussion of these types of shaft misalignment.)

Figure 4.42 Rigid adjustable coupling. (Courtesy of Lovejoy, Inc., Downers Grove, IL.)

In addition to transmitting the torque and allowing for angular and parallel misalignment, flexible couplings also accept torsional shock and dampen torsional vibration, minimize lateral loads on bearings from misalignment, and allow for axial movement of the shafts even under misaligned conditions, without transferring thrust loads from one machine element to another.

One of the confusing things about couplings is that there are so many alternatives from which to choose. Most of the coupling types discussed in this section could be applied successfully with pumps below 200 HP in size. The selection of the optimum coupling type for a given application must take into consideration a lot of factors, including the cost and importance of the pump and the coupling, the pump's duty cycle, torque to be transmitted, shaft size, permissible misalignment, temperature, and required maintenance. This section should help bring the reader a better understanding of

Figure 4.43 Gear coupling. (Courtesy of The Falk Corporation, Milwaukee, WI.)

the various flexible coupling types, to help pick the most suitable one for a given application.

Flexible couplings are broadly divided into two major categories: *metallic* and *elastomeric* couplings. Metallic couplings have mechanical elements such as gears (Figure 4.43), grid springs (Figure 4.44) or chains, or metallic elements such as discs (Figure 4.45) or diaphragms. The elements allow for some movement and misalignment of the two connected shafts. One significant feature of the mechanical element type is that they usually require lubrication.

Elastomeric couplings are further divided into two subcategories, those that operate using compression (e.g., Lovejoy jaw type, Figure 4.46) and those that operate using shear (e.g., Falk Torus corded tire type, Figure 4.47). There are also some that use some combination of shear and compression. In general, the elastomeric couplings are lower in cost and have lower upper limits of torque than metallic couplings. Elastomeric couplings do not require lubrication.

Figure 4.44 Grid spring coupling. (Courtesy of The Falk Corporation, Milwaukee, WI.)

Figure 4.45 Disc coupling. (Courtesy of The Falk Corporation, Milwaukee, WI.)

To help make the optimal pump coupling selection, refer to Table 4.2 (Flexible Coupling Functional Capabilities Chart) and Table 4.3 (Flexible Coupling Evaluation Factors Chart), which have been adapted from charts produced by a major coupling manufacturer. These charts provide the specification

Figure 4.46 Jaw type coupling. (Courtesy of Lovejoy, Inc., Downers Grove, IL.)

Figure 4.47 Corded tire type coupling. (Courtesy of The Falk Corporation, Milwaukee, WI.)

limits of the major coupling types, and rate them in the context of the most important application criteria.

XVI. ELECTRIC MOTORS

The electric motor (Figure 4.48) is one of, if not the most common machine used in industrial and commercial settings.

Table 4.2 Flexible Coupling Functional Capabilities Chart

Flexible Coupling Type	Maximum Continuous Torque (in.-lb)	Maximum Bore Diameter (inches)	Maximum Angular Misalignment (degrees)	Maximum Parallel Offset (inches)	Maximum Axial Freedom (inches)
Elastomeric Types					
Jaw	170,000	7.000	1.00	0.015	0.028–0.310
Curved jaw (1)	88,500	5.688	0.90–1.30	0.008–0.086	0.023–0.181
Shear type donut – sleeve (2)	72,480	5.500	1.00	0.010–0.062	0.125
Corded tire (3)	450,000	11.000	4.00	0.125	0.250
Bonded tire – urethane (4)	175,000	8.000	4.00	0.188	0.125
Rubber in shear – flywheel (5)	177,000	6.700	0.50	0.020	0.080–0.866
Metallic Types					
Gear	54,000,000	45.000	1.50	0.005–0.320	0.125–1.00
Grid spring	2,700,000	16.313	0.25	0.012–0.022	0.125–0.250
Disc	4,000,000	15.500	0.50	0.009	0.009–0.400
Diaphragm (6)	6,000,000	19.670	0.50	(7)	0.100–0.875

Notes:
(1) European standard for Jaw type couplings.
(2) For example, Woods Sureflex, Lovejoy S-Flex.
(3) For example, Dodge Paraflex, Falk Torus.
(4) For example, Rex Omega.
(5) Used for engine driven vertical turbines.
(6) High-speed applications.
(7) Manufacturers' catalogs do not give ratings.

Courtesy of Lovejoy, Inc., Downers Grove, IL.

Table 4.3 Flexible Coupling Evaluation Factors Chart

Flexible Coupling Type	Axial Forces Generated	Torque Capacity to O.D.	Torsional Stiffness	Backlash	Relative Cost	Reactionary Loads Due to Misalignment	High Speed Capacity	Misalignment Capacity	Inherent Balance	Ease of Assembly	Dampening Capacity	Lubrication Required	Fail Safe	Elements Field Replaceable	Estimated Service Life (Yr)
Elastomeric Types															
Jaw	M	M	M	L–M	L	H	F	G	F	E	G	N	Y	Y	3–5
Curved jaw (1)	M	M	M	L–M	L	H	G	G	F	E	G	N	Y	Y	3–5
Shear type donut – sleeve (2)	L	L	L	L–M	L	L	F	G	F	E	E	N	N	Y	2–3
Corded tire (3)	H	L	L	L	L–M	L	F	E	F	G	E	N	N	Y	3–5
Bonded tire – urethane (4)	L	L	L	L	L–M	L	F	E	F	G	G	N	N	Y	2–3
Rubber in shear – flywheel (5)	L	L	L	M	M–H	L	F	E	F–G	E	E	N	N	Y	3–5
Metallic Types															
Gear	M–H	H	H	M–H	M	H	E	F–G	G	F	None	Y	Y	N	3–5
Grid spring	M	H	M	M	M	M	G	F	G	F	F–G	Y	Y	Y	3–5
Disc	L–M	H	H	None	H	L–M	E	F	E	F	None	N	N	Y	4–8
Diaphragm (6)	L	M–H	M–H	None	H	L–M	E	F	E	F	None	N	N	Y	5

Note: H = High, M = Medium, L = Low; E = Excellent, G = Good, F = Fair; Y = Yes, N = No.

(1) European standard for Jaw type couplings.
(2) For example, Woods Sureflex, Lovejoy S-Flex.
(3) For example, Dodge Paraflex, Falk Torus.
(4) For example, Rex Omega.
(5) Used for engine-driven vertical turbines.
(6) High-speed applications.

Courtesy of Lovejoy, Inc., Downers Grove, IL.

Figure 4.48 Three-phase electric motor. (Courtesy of Baldor Electric Company, Fort Smith, AK.)

Many books have been written on the subject of electric motors, but the following section should provide an overview of electric motors for readers who are interested to learn more about the driver type used for the vast majority of pumps.

The following materials are excerpted from the *Cowern Papers*, written by and with the permission of Edward Cowern, P.E., North Haven, CT.

A. Glossary of Frequently Occurring Motor Terms

Below is a glossary of some of the most frequently occurring terms related to electric motors.

1. Amps

Full Load Amps — The amount of current the motor can be expected to draw under full load (torque) conditions is called Full Load Amps. It is also known as *nameplate amps*.

Locked Rotor Amps — Also known as *starting inrush*, this is the amount of current the motor can be expected to draw under starting conditions when full voltage is applied.

Service Factor Amps — This is the amount of current the motor will draw when it is subjected to a percentage of overload equal to the service factor on the nameplate of

Centrifugal Pump Types and Applications

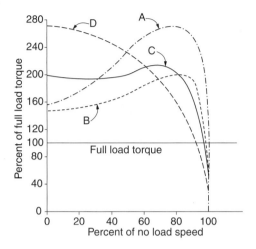

Figure 4.49 Torque speed curves for three-phase electric motors. (Courtesy of Edward Cowern, P.E., North Haven, CT.)

the motor. For example, many motors will have a service factor of 1.15, meaning that the motor can handle a 15% overload. The service factor amperage is the amount of current that the motor will draw under the service factor load condition.

2. Code Letter

The code letter is an indication of the amount of inrush or locked rotor current that is required by a motor when it is started.

3. Design Letter

The design letter is an indication of the shape of the torque speed curve, as defined by the National Electrical Manufacturers Association (NEMA). Figure 4.49 shows the typical shape of the most commonly used three-phase design letters, Design A, B, C, and D. Design B is the standard industrial duty motor which has reasonable starting torque with moderate starting current and good overall performance for most industrial applications. The other designs are only used on fairly specialized applications. Design A motors are not commonly specified but specialized motors used on injection molding applications have characteristics similar to Design A. The most important characteristic of Design A is the high pullout torque. Design C is a high starting torque motor that is usually confined to hard-to-start loads, such as conveyors that are going to operate under difficult conditions. Design D is a so-called high slip motor and is normally limited to applications such as cranes, hoists, and low-speed punch presses where high starting torque with low starting current is desirable. Generally, the efficiency of Design D motors at full load is rather poor and thus they are normally used on those applications where the torque characteristics are of primary importance.

4. Efficiency

Efficiency is the percentage of the input power that is actually converted to work output from the motor shaft. Efficiency is stamped on the nameplate of most domestically produced electric motors. (See Section XVI.I for more details.)

5. Frame Size

Motors come in various frame sizes to match the requirements of the application. In general, the frame size gets larger with increasing horsepower or with decreasing speed. To promote standardization in the motor industry, NEMA (National Electrical Manufacturers Association) prescribes standard frame sizes for certain dimensions of standard motors. For example, a motor with a frame size of 56 will always have a shaft height above the base of 3½ inches. (See Section XVI.E for more details.)

6. Frequency

This is the frequency for which the motor is designed. The most commonly occurring frequency in the United States is 60 cycles, but on an international basis, other frequencies such as 50 cycles can be found.

7. Full Load Speed

An indication of the approximate speed that the motor will run when it is putting out full rated output torque or horsepower is called full load speed.

8. High Inertial Load

These are loads that have a relatively high flywheel effect. Large fans, blowers, punch presses, centrifuges, commercial washing machines, and other types of similar loads can be classified as high inertial loads.

9. Insulation Class

The insulation class is a measure of the resistance of the insulating components of a motor to degradation from heat. Four major classifications of insulation are used in motors. They are, in order of increasing thermal capabilities, A, B, F, and H. (See Section XVI.D for more details.)

10. Load Types

Constant Horsepower — The term "constant horsepower" is used in certain types of loads where the torque requirement is reduced as the speed is increased and vice versa. The constant horsepower load is usually associated with metal removal applications such as drill presses, lathes, milling machines, and other similar types of applications.

Constant Torque — Constant torque is a term used to define a load characteristic where the amount of torque required to drive the machine is constant regardless of the speed at which it is driven. For example, the torque requirement of positive displacement pumps and blowers is constant.

Variable Torque — Variable torque is found in loads having characteristics requiring low torque at low speeds and increasing values of torque as the speed is increased. Typical examples of variable torque loads are centrifugal fans and centrifugal pumps.

Table 4.4 AC Motor Synchronous Speeds

# Poles	Synchronous Speed	
	60 Cycles	50 Cycles
2	3600	3000
4	1800	1500
6	1200	1000
8	900	750
10	720	600

11. Phase

Phase is the indication of the type of power supply for which the motor is designed. Two major categories exist; single phase and three phase. There are some very spotty areas where two-phase power is available but this is very insignificant.

12. Poles

This is the number of magnetic poles that appear within the motor when power is applied. Poles always come in sets of two (a north and a south). Thus, the number of poles within a motor is always an even number such as 2, 4, 6, 8, 10, etc. In an alternating current (AC) motor, the number of poles work in conjunction with the frequency to determine the synchronous speed of the motor. At 50 and 60 cycles, the common arrangements are as given in Table 4.4.

13. Power Factor

Percent power factor is a measure of a particular motor's requirements for magnetizing amperage. (See Section XVI.I for more details.)

14. Service Factor

The service factor is a multiplier that indicates the amount of overload a motor can be expected to handle. For example, a motor with a 1.0 service factor cannot be expected to handle more than its nameplate horsepower on a continuous basis.

Centrifugal Pump Types and Applications 299

Similarly, a motor with a 1.15 service factor can be expected to safely handle intermittent loads amounting to 15% beyond its nameplate horsepower. (See Section XVI.C for more details.)

15. Slip

Slip is used in two forms. One is the slip rpm, which is the difference between the synchronous speed and the full load speed. When this slip rpm is expressed as a percentage of the synchronous speed, then it is called percent slip or just "slip." Most standard motors run with a full load slip of 2 to 5%.

16. Synchronous Speed

This is the speed at which the magnetic field within the motor is rotating. It is also approximately the speed that the motor will run at under no load conditions. For example, a 4 pole motor running on 60 cycles would have a magnetic field speed of 1800 rpm. The no load speed of that motor shaft would be very close to 1800, probably 1798 or 1799 rpm. The full load speed of the same motor might be 1750 rpm. The difference between the synchronous speed and the full load speed is called the slip rpm of the motor.

17. Temperature

Ambient Temperature — Ambient temperature is the maximum safe room temperature surrounding the motor if it is going to be operated continuously at full load. In most cases, the standardized ambient temperature rating is 40°C (104°F). This is a very warm room. Certain types of applications, such as on board ships and boiler rooms, may require motors with a higher ambient temperature capability such as 50° or 60°C.

Temperature Rise — Temperature rise is the amount of temperature change that can be expected within the winding of the motor from nonoperating (cool condition) to its temperature at full load continuous operating condition. Temperature rise is normally expressed in degrees centigrade.

18. Time Rating

Most motors are rated for continuous duty, which means that they can operate at full load torque continuously without overheating. Motors used on certain types of applications, such as waste disposal, valve actuators, hoists, and other types of intermittent loads, will frequently be rated for short-term duty such as 5 minutes, 15 minutes, 30 minutes, or 1 hour. Just like a human being, a motor can be asked to handle very strenuous work as long as it is not required on a continuous basis.

19. Voltage

This refers to the voltage rating for which the motor is designed.

B. Motor Enclosures

The most reliable piece of electrical equipment in service today is a transformer. The second most reliable is the three-phase induction motor. Properly applied and maintained, three-phase motors will last many years. One key element of motor longevity is proper cooling. Motors are generally classified by the method used to dissipate the internal heat.

Several standard motor enclosures are available to handle the range of applications from "clean and dry" such as indoor air handlers, to the "wet or worse" as found on roofs and wet cooling towers. The most common enclosure types are summarized below.

1. Open Drip Proof

Open Drip-Proof (ODP) motors are good for clean and dry environments. As the name implies, drip-proof motors can handle some dripping water provided it falls from overhead or no more than 15° off vertical. These motors usually have ventilating openings that face down. The end housings can frequently be rotated to maintain "drip-proof" integrity when the motor is mounted in a different orientation. These motors are cooled by a continuous flow of the surrounding air through the internal parts of the motor.

Centrifugal Pump Types and Applications

2. Totally Enclosed Fan Cooled

Totally Enclosed Fan Cooled (TEFC) motors are cooled by an external fan mounted on the end opposite the motor output shaft. The fan blows ambient air across the outside surface of the motor to carry heat away. Air does not move through the inside of the motor, so TEFC motors are suited for dirty, dusty, and outdoor applications. There are many special types of TEFC motors, including *corrosion protected*, *chemical duty*, and *washdown* styles. These motors have special features to handle difficult environments. TEFC motors generally have "weep holes" at their lowest points to prevent condensation from puddling inside the motor. As in open drip-proof motors, if the TEFC motor is mounted in a position other than horizontal, the end housings can generally be repositioned to keep the weep holes at the lowest point.

3. Totally Enclosed Air Over

Totally Enclosed Air Over (TEAO) motors are applied on machines such as vane axial fans where the air moved by a direct connected fan passes over the motor and cools it. TEAO motors frequently have dual HP ratings, depending on the speed and temperature of the cooling air. Typical ratings for a motor might be 10 HP with 750 feet per minute (fpm) of 104°F air, 10 HP with 400 fpm of 70°F air, or 12.5 HP with 3000 fpm of 70°F air. TEAO motors are usually confined to Original Equipment Manufacturer (OEM) applications because the air temperature and flows need to be predetermined.

4. Totally Enclosed Non-Ventilated

Totally Enclosed Non-Ventilated (TENV) motors are generally confined to small sizes (usually under 5 HP) where the motor surface area is large enough to radiate and convect the heat to the outside air without an external fan or air flow. They have been popular in textile applications because lint cannot obstruct cooling.

5. Hazardous Location

Hazardous Location motors are a special form of totally enclosed motor. They fall into different categories, depending upon the application and environment, as defined in Article 500 of the National Electrical Code. The two most common hazardous location motors are Class I, *Explosion Proof*, and Class II, *Dust Ignition Resistant*. The term "explosion proof" is commonly but erroneously used to refer to all categories of hazardous location motors. Explosion proof applies only to Class I environments, which are those that involve potentially explosive liquids, vapors, and gases. Class II is termed Dust Ignition Resistant. These motors are used in environments that contain combustible dusts such as coal, grain, flour, etc.

C. Service Factor

Some motors carry a service factor other than 1.0. This means the motor can handle loads above the rated HP. A motor with a 1.15 service factor can handle a 15% overload, so a 10 HP motor with a 1.15 service factor can handle 11.5 HP of load. Standard open drip-proof motors have a 1.15 service factor, and fractional HP and sub-fractional ODP motors can have substantially higher service factors, in the range of 1.5 or even higher. Standard TEFC motors have a 1.0 service factor, but most major motor manufacturers now provide TEFC motors with a 1.15 service factor.

The question often arises whether to use service factor in motor load calculations. In general, the best answer is that for good motor longevity, service factor should not be used for basic load calculations. By not loading the motor into the service factor, the motor can better withstand adverse conditions that occur. Adverse conditions include higher than normal ambient temperatures, low or high voltage, voltage imbalances, and occasional overload. These conditions are less likely to damage the motor or shorten its life if the motor is not loaded into its service factor in normal operation. That being said, however, there are a good many lighter duty pumps whose motors are typically sized to run in the service factor. Examples would include light intermittent duty residential pumps such as utility sump pumps, or small OEM products where low cost is paramount.

D. Insulation Classes

The electrical portions of every motor must be insulated from contact with other wires and with the magnetic portion of the motor. The insulation system consists of the varnish that jackets the magnet wire in the windings along with the slot liners that insulate the wire from the steel laminations. The insulation system also includes tapes, sleeves, tie strings, a final dipping varnish, and the leads that bring the electrical circuits out to the junction box.

Insulation systems are rated by their resistance to thermal degradation. The four basic insulation systems normally encountered are Class A, B, F, and H. Class A has a temperature rating of 105°C (221°F), and each step from A to B, B to F, and F to H involves a 25°C (45°F) jump. The insulation class in any motor must be able to withstand at least the maximum ambient temperature plus the temperature rise that occurs as a result of continuous full load operation. Selecting an insulation class higher than necessary to meet this minimum can help extend motor life or make a motor more tolerant of overloads, high ambient temperatures, and other problems that normally shorten motor life.

A widely used rule of thumb states that every 10°C (18°F) increase in operating temperature cuts insulation life in half. Conversely, a 10°C decrease doubles insulation life. Choosing a one step higher insulation class than required to meet the basic performance specifications of a motor provides 25°C of extra temperature capability. The rule of thumb predicts that this better insulation system increases the motor's thermal life expectancy by approximately 500%.

E. Motor Frame Size

1. Historical Perspective

Industrial electric motors have been available for nearly a century. In that time there have been a great many changes.

One of the most obvious has been the ability to pack more horsepower in a smaller physical size. Another important achievement has been the standardization of motors by the National Electrical Manufacturers Association (NEMA).

A key part of motor interchangeability has been the standardization of frame sizes. This means that the same horsepower, speed, and enclosure will normally have the same frame size from different motor manufacturers. Thus, a motor from one manufacturer can be replaced with a similar motor from another company provided they are both in standard frame sizes.

The standardization effort over the last 50 years has resulted in one original grouping of frame sizes called "original." In 1952, new frame assignments were made. These were called "U frames." The current "T frames" were introduced in 1964. "T" frames are the current standard and most likely will continue to be for some time in the future.

Although "T" frames were adopted in 1964, there are still a great many "U" frame motors in service that will have to be replaced in the future. Similarly, there are also many of the original frame size motors (pre-1952) that will reach the end of their useful life and will have to be replaced. For this reason, it is desirable to have reference material available on frame sizes and some knowledge of changes that took place as a part of the so-called rerate programs.

Tables 4.5 and 4.6 show the standard frame size assignments for the three different eras of motors, broken down for open drip-proof (Table 4.5) and totally enclosed fan cooled (Table 4.6). For each horsepower rating and speed, there are three different frame sizes. The first is the original frame size, the middle one is the "U frame" size, and the third one is the "T frame." These are handy reference tables because they give general information for all three vintages of three phase motors in integral horsepower frame sizes.

One important item to remember is that the base mounting hole spacing ("E" and "F" dimensions) and shaft height ("D" dimension) for all frames having the same three digits regardless of vintage, will be the same.

Centrifugal Pump Types and Applications

Table 4.5 Frame Size Reference Table — Open Drip-Proof Motors

RPM NEMA Program HP	3600			1800			1200			900		
	Orig.	1952 Rerate	1964 Rerate	Orig.	1952 Rerate	1964 Rerate	Orig.	1952 Rerate	1964 Rerate	Orig.	1952 Rerate	1964 Rerate
1	—	—	—	203	182	143T	204	184	145T	225	213	182T
1.5	203	182	143T	204	184	145T	224	184	182T	254	213	184T
2	204	184	145T	224	184	145T	225	213	184T	254	215	213T
3	224	184	145T	225	213	182T	254	215	213T	284	254U	215T
5	225	213	182T	254	215	184T	284	254U	215T	324	256U	254T
7.5	254	215	184T	284	254U	213T	324	256U	254T	326	284U	256T
10	284	254U	213T	324	256U	215T	326	284U	256T	364	286U	284T
15	324	256U	215T	326	284U	254T	364	324U	284T	365	326U	286T
20	326	284U	254T	364	286U	256T	365	326U	286T	404	364U	324T
25	364S	286U	256T	364	324U	284T	404	364U	324T	405	365U	326T
30	364S	324US	284TS	365	326U	286T	405	365U	326T	444	404U	364T
40	365S	326US	286TS	404	364U	324T	444	404U	364T	445	405U	365T
50	404S	364US	324TS	405S	365US	326T	445	405U	365T	504	444U	404T
60	405S	365US	326TS	444S	404US	364T	504	444U	404T	505	445U	405T
75	444S	404US	364TS	445S	405US	365T	505	445U	405T	—	—	444T
100	445S	405US	365TS	504S	444US	404T	—	—	444T	—	—	445T
125	504S	444US	404TS	505S	445US	405T	—	—	445T	—	—	—
150	505S	445US	405TS	—	—	444T	—	—	—	—	—	—
200	—	—	444TS	—	—	445T	—	—	—	—	—	—
250	—	—	445TS	—	—	—	—	—	—	—	—	—

Courtesy of Edward Cowern, P.E., North Haven, CT.

Table 4.6 Frame Size Reference Table — Three-Phase Totally Enclosed Fan Cooled Motors

RPM	3600			1800			1200			900		
NEMA Program HP	Orig.	1952 Rerate	1964 Rerate	Orig.	1952 Rerate	1964 Rerate	Orig.	1952 Rerate	1964 Rerate	Orig.	1952 Rerate	1964 Rerate
1	—	—	—	203	182	143T	204	184	145T	225	213	182T
1.5	203	182	143T	204	184	145T	224	184	182T	254	213	184T
2	204	184	145T	224	184	145T	225	213	184T	254	215	213T
3	224	184	182T	225	213	182T	254	215	213T	284	254U	215T
5	225	213	184T	254	215	184T	284	254U	215T	324	256U	254T
7.5	254	215	213T	284	254U	213T	324	256U	254T	326	284U	256T
10	284	254U	215T	324	256U	215T	326	284U	256T	364	286U	284T
15	324	256U	254T	326	284U	254T	364	324U	284T	365	326U	286T
20	326	286U	256T	364	286U	256T	365	326U	286T	404	364U	324T
25	365S	324U	284TS	365	324U	284T	404	364U	324T	405	365U	326T
30	404S	326US	286TS	404	326U	286T	405	365U	326T	444	404U	364T
40	405S	364US	324TS	405	364U	324T	444	404U	364T	445	405U	365T
50	444S	365US	326TS	444S	365US	326T	445	405U	365T	504	444U	404T
60	445S	405US	364TS	445S	405US	364T	504	444U	404T	505	445U	405T
75	504S	444US	365TS	504S	444US	365T	505	445U	405T	—	—	444T
100	505S	445US	405TS	505S	445US	405T	—	—	444T	—	—	445T
125	—	—	444TS	—	—	444T	—	—	445T	—	—	—
150	—	—	445TS	—	—	445T	—	—	—	—	—	—

Courtesy of Edward Cowern, P.E., North Haven, CT.

2. Rerating and Temperature

The ability to rerate motor frames to get more horsepower in a frame has been brought about mainly by improvements made in insulating materials. As a result of this improved insulation, motors can be run much hotter. This allows more horsepower in a compact frame. For example, the original NEMA frame sizes ran at very low temperatures. The "U" frame motors were designed for use with Class A insulation, which has a rating of 105°C. The motor designs were such that the capability would be used at the hottest spot within the motor. "T" frame motor designs are based on utilizing Class B insulation with a temperature rating of 130°C. This increase in temperature capability made it possible to pack more horsepower into the same size frame. To accommodate the larger mechanical horsepower capability, shaft and bearing sizes had to be increased. Thus, you will find that the original 254 frame (5 HP at 1800 rpm) has a 1⅛-in. shaft. The 254U frame (7½ HP at 1800 rpm) has a 1⅜-in. shaft, and the current 254T frame (15 HP at 1800 rpm) has a 1⅝-in. shaft. Bearing diameters were also increased to accommodate the larger shaft sizes and heavier loads associated with the higher horsepowers.

3. Motor Frame Dimensions

Tables 4.7 and 4.8 show NEMA and IEC motor dimensions. Most of the motor dimensions are standard dimensions that are common to all motor manufacturers. One exception to this is the "C" dimension (overall motor length) which will change from one manufacturer to another.

4. Fractional Horsepower Motors

The term "fractional horsepower" is used to cover those frame sizes having two digit designations, as opposed to the three digit designations that are found in Tables 4.5 and 4.6. The NEMA frame sizes that are normally associated with industrial fractional horsepower motors are 42, 48, and 56. In this case, each frame size designates a particular shaft height,

Table 4.7 NEMA Motor Frame Dimensions (Courtesy of Baldor Electric Company, Fort Smith, AR.)

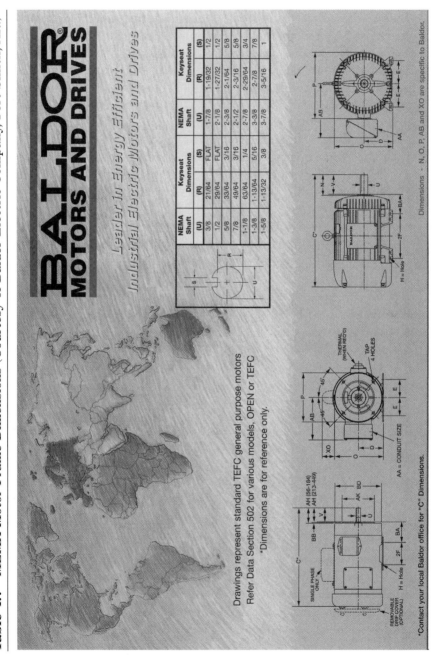

Centrifugal Pump Types and Applications

NEMA FRAME	D	E	2F	H	N	O	P	U	V	AA	AB	AH	AJ	AK	BA	BB	BD	XO	TAP
42	2-5/8	1-3/4	1-11/16	9/32 SLOT	1-1/2	5	4-11/16	3/8	1-1/8	3/8	4-1/32	1-5/16	3-3/4	3	2-1/16	1/8	4-5/8	1-9/16	1/4-20
48	3	2-1/8	2-3/4	11/32 SLOT	1-7/8	5-7/8	5-11/16	1/2	1-1/2	1/2	4-3/8	1-11/16	3-3/4	3	2-1/2	1/8	5-5/8	2-1/4	1/4-20
56	3-1/2	2-7/16	3	11/32 SLOT	2-7/16	6-7/8	6-5/8	5/8	1-7/8	1/2	5	2-1/16	5-7/8	4-1/2	2-3/4	1/8	6-1/2	2-1/4	3/8-16
56H			5		2-1/8														
143T	3-1/2	2-3/4	4	11/32 SLOT	2-1/2	6-7/8	6-5/8	7/8	2-1/4	3/4	5-1/4	2-1/8	5-7/8	4-1/2	2-1/4	1/8	6-1/2	2-1/4	3/8-16
145T			4																
182	4-1/2	3-3/4	4-1/2	13/32	2-11/16	8-11/16	7-7/8	7/8	2-1/4	3/4	5-7/8	2-1/8	5-7/8	4-1/2	2-3/4	1/8	6-1/2	2-3/8	3/8-16
184			5-1/2		2-11/16			7/8	2-1/4			2-1/8	5-7/8	4-1/2		1/8	6-1/2		3/8-16
182T			4-1/2		3-9/16			1-1/8	2-3/4			2-5/8	7-1/4	8-1/2		1/4	9		1/2-13
184T			5-1/2		3-9/16			1-1/8	2-3/4			2-5/8	7-1/4	8-1/2		1/4	9		1/2-13
213	5-1/4	4-1/4	5-1/2	13/32	3-1/2	10-1/4	9-9/16	1-1/8	3	3/4	7-3/8	2-3/4	7-1/4	8-1/2	3-1/2	1/4	9	2-3/4	1/2-13
215			7		3-7/8			1-3/8	3-3/8			3-1/8							
213T			5-1/2		3-1/2			1-1/8	3			2-3/4							
215T			7		3-7/8			1-3/8	3-3/8			3-1/8							
254U	6-1/4	5	8-1/4	17/32	4-1/16	12-7/8	12-15/16	1-3/8	3-3/4	1	9-5/8	3-1/2	7-1/4	8-1/2	4-1/4	1/4	10	—	1/2-13
256U			10		4-5/16			1-5/8	4			3-3/4							
254T			8-1/4		4-1/16			1-5/8	3-3/4			3-1/2							
256T			10		4-5/16			1-5/8				3-3/4							
284U	7	5-1/2	9-1/2	17/32	5-1/8	14-5/8	14-5/8	1-5/8	4-7/8	1-1/2	13-1/8	4-5/8	9	10-1/2	4-3/4	1/4	11-1/4	—	1/2-13
286U			11		5-1/8			1-5/8	4-7/8			4-5/8							
284T			9-1/2		4-7/8			1-7/8	4-5/8			4-3/8							
286T			11		4-7/8			1-7/8	4-5/8			4-3/8							
284TS			9-1/2		3-3/8			1-5/8	3-1/4			3							
286TS			11		3-3/8			1-5/8	3-1/4			3							
324U	8	6-1/4	10-1/2	21/32	5-7/8	16-1/2	16-1/2	1-7/8	5-5/8	2	14-1/8	5-3/8	11	12-1/2	5-1/4	1/4	13-3/8	—	5/8-11
326U			12		5-7/8			1-7/8	5-5/8			5-3/8							
324T			10-1/2		5-1/2			2-1/8	5-1/4			5							
326T			12		5-1/2			2-1/8	5-1/4			5							
324TS			10-1/2		3-1/2			1-7/8	3-3/4			3-1/2							
326TS			12		3-15/16			1-7/8	3-3/4			3-1/2							
364U	9	7	11-1/4	21/32	6-3/4	18-1/2	18-1/4	2-1/8	6-3/8	2-1/2	15-1/16	6-1/8	11	12-1/2	5-7/8	1/4	13-3/8	—	5/8-11
365U			12-1/4		6-3/4			2-1/8	6-3/8			6-1/8							
364T			11-1/4		5-5/8			2-3/8	5-7/8			5-5/8							
365T			12-1/4		6-1/4			2-3/8	5-7/8			5-5/8							
364TS			11-1/4		4			1-7/8	3-3/4			3-1/2							
365TS			12-1/4		4			1-7/8	3-1/2			3-1/2							
404U	10	8	12-1/4	13/16	7-3/16	20-5/16	20-1/8	2-3/8	7-1/8	3	18	6-7/8	11	12-1/2	6-5/8	1/4	13-7/8	—	5/8-11
405U			13-3/4		7-3/16			2-3/8	7-1/8			6-7/8							
404T			12-1/4		7-5/8			2-7/8	7-1/4			7							
405T			13-3/4		7-5/8			2-7/8	7-1/4			7							
404TS			12-1/4		4-1/2			2-1/8	4-1/4			4							
405TS			13-3/4		4-1/2			2-1/8	4-1/4			4							
444U	11	9	14-1/2	13/16	8-5/8	22-7/8	22-3/8	2-7/8	8-5/8	3	19-9/16	8-3/8	14	16	7-1/2	1/4	16-3/4	—	5/8-11
445U			16-1/2		8-5/8			2-7/8	8-5/8			8-3/8							
444T			14-1/2		8-1/2			3-3/8	8-1/2			8-1/4							
445T			16-1/2		8-1/2			3-3/8	8-1/2			8-1/4							
447T			20		8-15/16	22-15/16	23-3/4	3-3/8	8-1/2		21-11/16	8-1/4							
449T			25		8-15/16	22-15/16	23-3/4	3-3/8	8-1/2		21-11/16	8-1/4							
444TS			14-1/2		5-3/16	22-7/8	22-3/8	2-3/8	4-3/4		19-9/16	4-1/2							
445TS			16-1/2		5-3/16	22-7/8	22-3/8	2-3/8	4-3/4		19-9/16	4-1/2							
447TS			20		4-15/16	22-15/16	23-3/4	2-3/8	4-3/4	4NPT	21-11/16	4-1/2							
449TS			25		4-15/16	22-15/16	23-3/4	2-3/8	4-3/4	4NPT	21-11/16	4-1/2							

NEMA C-Face	BA Dim	NEMA C-Face	BA Dim
143-5TC	2-3/4	213-5TC	4-1/4
182-4TC	3-1/2	254-6TC	4-3/4

Table 4.8 IEC Motor Frame Dimensions (Courtesy of Baldor Electric Company, Fort Smith, AK.)

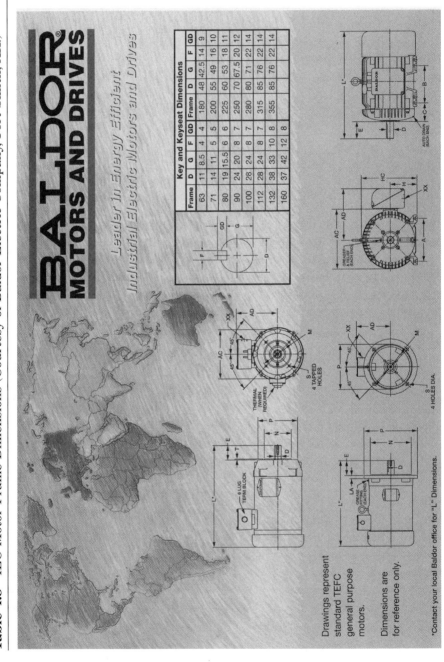

Centrifugal Pump Types and Applications

IEC Frame	Type	Foot Mounting A	B	C	H	Shaft D	E	LA	B5 Flange M	N	P	S	T	B14 Face M	N	P	S	T	L	General AC	AD	HC	XX
63	300	100 3.937	80 3.150	40 1.570	63 2.480	11 .433	23 .906	8 .313	115 4.528	95 3.740	140 5.512	9 .354	3 .118	75 2.953	60 2.362	90 3.540	M5	2.5 0.98	*	119 4.690	102 4	121 4.760	13 .500
																					116 4.567d	136d 5.375d	22d .880d
71	300 400	112 4.409	90 3.543	45 1.770	71 2.800	14 .551	30 1.181	8 .313	130 5.118	110 4.331	160 6.299	10 .393	3.5 .138	85 3.347	70 2.756	105 4.130	M6	2.5 .098	*	119 4.690 145 5.690d	102	131 5.140 149d 5.880d	18 .690 21d .844d
80	400 500	125 4.921	100 3.937	50 1.969	80 3.150	19 .748	40 1.575	13 .500	165 6.496	130 5.118	200 7.874	11 .430	3.5 .138	100 3.937	80 3.150	120 4.724	M6	3 .118	*	145 5.690 168d 6.614d	116 4.510 130 5.120	152 6 162d 6.380d	22 .880 21d .844d
90	S L	140 5.511	100 3.937 125 4.921	56 2.205	90 3.543	24 .945	50 1.969	13 .500	165 6.496	130 5.118	200 7.874	12 .472	3.5 .138	115 4.530	95 3.740	140 5.512	M8	3 .118	*	168 6.614 144d 5.687d	130 5.120 107d 4.250d	173 6.810 165d 6.531d	22 .880 21d .844d
100	S L	160 6.300	112 4.409 140 5.512	63 2.480	100 3.937	28 1.102	60 2.362	14 .562	215 8.465	180 7.087	250 9.840	14 .560	4 .160	130 5.108	110 4.331	160 6.299	M8	3.5 .138	*	200 7.875	149 5.875 153d 6.060d	180 7.906 239d 9.440d	27 1.062
112	S M	190 7.480	114 4.488 140 5.512	70 2.760	112 4.409	28 1.102	60 2.362	14 .562	215 8.465	180 7.087	250 9.840	14 .560	4 .160	130 5.108	110 4.331	160 6.299	M8	3.5 .138	*	243 7.875	149 5.875	214 8.437	27 1.062
132	S M	216 8.504	140 5.512 178 7.008	89 3.504	132 5.197	38 1.496	80 3.150	14 .562	265 10.433	230 9.055	300 11.811	14 .560	4 .160	165 6.496	130 5.118	200 7.874	M8	3.5 .138	*	243 9.562	187 7.375	256 10.062	27 1.062
160	M L	254 10	210 8.268 254 10	108 4.252	160 6.299	42 1.654	110 4.331	20 .787	300 11.811	250 9.842	350 13.780	19 .748	5 .200	215 8.465	180 7.087	250 9.840	M12	4 .160	*	329 12.940	242 9.510	329 12.940	35 1.375
180	M L	279 10.984	241 9.488 279 10.984	121 4.764	180 7.087	48 1.890	110 4.331	20 .787	300 11.811	250 9.842	350 13.780	19 .748	5 .200						*	395 15.560	333 13.120	372 14.640	51 2.008
200	L M	318 12.520	267 10.512 305 12.008	133 5.236	200 7.874	55 2.165	110 4.331	27 1.062	350 13.780	300 11.811	400 15.748	19 .748	5 .200						*	441 17.375	359 14.125	416 16.375	63 2.500
225	S M	356 14.016	286 11.260 311 12.244	149 5.866	225 8.858	60 2.362	140 5.512	19 .748	400 15.748	350 13.780	450 17.716	19 .748	6 .236						*	495 19.488	383 15.079	483 19.016	63 2.500
250	S M	406 15.984	311 12.244 349 13.740	168 6.614	250 9.843	70 2.756	140 5.512												*	520 20.472	457 17.992	513 20.197	63 2.500
280	S M	457 17.992	368 14.488 419 16.496	190 7.485	280 11.025	80 3.150	170 6.693												*	616 24.252	497 19.567	581 22.874	63 2.500
315	S M	508 20	406 16 457 18	216 8.500	315 12.400	85 3.346	170 6.693												*	759 29.900	683 26.880	682 26.840	102 4
355	S L	610 24	500 19.690 630 24.800	254 10	355 13.980	85 3.346	170 6.693												*	759 29.900	683 26.880	719 28.320	102 4

LEGEND

Metric (MM) Dimensions on top
Inch Dimensions below
d = DC Motors
1 mm = .03937"   1" = 25.40 mm

shaft diameter, and face or base mounting hole pattern. In these motors, specific frame assignments have not been made by horsepower and speed, so it is possible that a particular horsepower and speed combination might be found in three different frame sizes. In this case, for replacement it is essential that the frame size be known as well as the horsepower, speed, and enclosure. The derivation of the two digit frame number is based on the shaft height in sixteenths of an inch. A 48 frame motor would have a shaft height of 48 divided by 16, or 3 inches. Similarly, a 56 frame motor would have a shaft height of 3½ inches. The largest of the current fractional horsepower frame sizes is a 56 frame, which is available in horsepowers greater than those normally associated with fractional motors. For example, 56 frame motors are built in horsepowers up to 3 HP, and in some cases, 5 HP. For this reason, calling motors with two digit frame sizes "fractionals" is somewhat misleading.

5. Integral Horsepower Motors

The term "integral horsepower" generally refers to those motors having three digit frame sizes such as 143T or larger. When dealing with these frame sizes, one handy rule of thumb is that the centerline shaft height ("D" dimension) above the bottom of the base is the first two digits of the frame size divided by four. For example, a 254T frame would have a shaft height of $25 \div 4 = 6.25$ inches. Although the last digit does not directly relate to an "inch" dimension, larger numbers do indicate that the rear bolt holes are moved further away from the shaft end bolt holes (the "F" dimension becomes larger).

6. Frame Designation Variations

In addition to the standard numbering system for frames, there are some variations that will appear. These are itemized below along with an explanation of what the various letters represent.

C   This designates a *C-face* (flange) mounted motor. This is the most popular type of face mounted motor, and is the type used for close-coupled pumps. The C-face motor has a specific bolt pattern on the shaft end to allow mounting. The critical items on C-face motors are the "bolt circle" (AJ dimension), register (also called rabbet) diameter (AK dimension), and the shaft size (U dimension). C-face motors always have threaded mounting holes in the face of the motor.

D   The "D" flange has a special type of mounting flange installed on the shaft end. In the case of the "D" flange, the flange diameter is larger than the body of the motor and it has clearance holes suitable for mounting bolts to pass through from the back of the motor into threaded holes in the mating part. "D" flange motors are not as popular as "C" flange motors, and are almost never used on pumps.

H   Used on some 56 frame motors, "H" indicates that the base is suitable for mounting in either 56, 143T, or 145T mounting dimensions.

J   This designation is used with 56 frame motors and indicates that the motor is made for *jet pump* service with a threaded stainless steel shaft and standard 56C face.

JM The letters "JM" designate a special pump shaft originally designed for a mechanical seal. This motor also has a C face.

JP Similar to the JM style of motor having a special shaft, the JP motor was originally designed for a "packing" type of seal. The motor also has a C face.

S   The use of the letter "S" in a motor frame designates that the motor has a "short shaft." Short shaft motors have shaft dimensions that are smaller than the shafts associated with the normal frame size. Short shaft motors are designed to be directly coupled to a load through a flexible coupling. They are not supposed to be used on applications where belts are used to drive the load.

T   A "T" at the end of the frame size indicates that the motor is of the 1964 and later "T" frame vintage.

U   A "U" at the end of the frame size indicates that the motor falls into the "U" frame size assignment (1952 to 1964) era.

Y   When a "Y" appears as a part of the frame size, it means that the motor has a special mounting configuration. It is impossible to tell exactly what the special configuration

Table 4.9 Single Phase Motor Types and Characteristics

Category	Approximate HP Range	Relative Efficiency
Shaded pole	$1/100$–$1/6$ HP	Low
Split phase	$1/25$–$1/2$ HP	Medium
Capacitor	$1/50$–15 HP	Medium to High

is, but it does denote that there is a special nonstandard mounting.

Z Indicates the existence of a special shaft which could be longer, larger, or have special features such as threads, holes, etc. "Z" indicates only that the shaft is special in some undefined way.

F. Single Phase Motors

Three phase motors start and run in a direction based on the "phase rotation" of the incoming power. Single phase motors are different. They require an auxiliary starting means. Once started in a direction, they continue to run in that direction. Single phase motors are categorized by the method used to start the motor and establish the direction of rotation. The major three types of single phase motors, *shaded pole*, *split phase*, and *capacitor* motors are summarized in Table 4.9, and are described below.

Shaded pole motors employ the simplest of all single phase starting methods. These motors are used only for small, simple applications such as bathroom exhaust fans. In the shaded pole motor, the motor field poles are notched and a copper shorting ring is installed around a small section of the poles as shown in Figure 4.50.

The altered pole configuration delays the magnetic field build-up in the portion of the poles surrounded by the copper shorting rings. This arrangement makes the magnetic field around the rotor seem to rotate from the main pole toward the shaded pole. This appearance of field rotation starts the

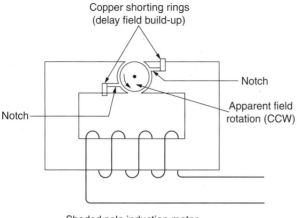

Figure 4.50 Shaded pole is the simplest of all single-phase starting methods. (Courtesy of Edward Cowern, P.E., North Haven, CT.)

rotor moving. Once started, the motor accelerates to full speed.

Split phase motors have two separate windings in the stator (stationary portion of the motor). See Figure 4.51. The winding shown in black is only for starting. It uses a smaller wire size and has higher electrical resistance than the main winding. The difference in the start winding location and its altered electrical characteristics causes a delay in current flow between the two windings. This time delay coupled with the physical location of the starting winding causes the field around the rotor to move and start the motor. A centrifugal switch or other device disconnects the starting winding when the motor reaches approximately 75% of rated speed. The motor continues to run on normal induction motor principles.

Split phase motors are generally available from 1/25 to 1/2 HP. Their main advantage is low cost. Their disadvantages are low starting torque and high starting current. These disadvantages generally limit split phase motors to applications where the load needs only low starting torque and starts are infrequent.

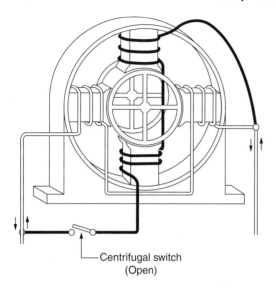

─ Centrifugal switch
(Open)

Figure 4.51 Split phase motors have two separate windings in the stator. (Courtesy of Edward Cowern, P.E., North Haven, CT.)

Table 4.10 Capacitor Motor Types and HP Ranges

Category	Usual HP Range
Capacitor Start–Induction Run	⅛–3 HP
Single Value Capacitor (Permanent Split Capacitor)	1/50–1 HP
Two Value Capacitor (Capacitor Start–Capacitor Run)	2–15 HP

Capacitor motors are the most popular single phase motors on pumps. They are used in many agricultural, commercial, and industrial applications where three-phase power is not available. Capacitor motors are available in sizes from subfractional to 15 HP. Capacitor motors fall into three types, summarized in Table 4.10 and described in the following paragraphs.

Capacitor Start–Induction Run motors form the largest group of general-purpose single phase motors. The winding and centrifugal switch arrangement is similar to that in a split phase motor. However, a capacitor start–induction run

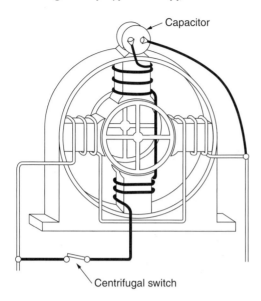

Figure 4.52 Capacitor start–induction run motors have a capacitor in series with the starter winding. (Courtesy of Edward Cowern, P.E., North Haven, CT.)

motor has a capacitor in series with the starter winding. Figure 4.52 shows the capacitor start–induction run motor. The starting capacitor produces a time delay between the magnetization of the starting poles and the running poles, creating the appearance of a rotating field. The rotor starts moving in the same direction. As the rotor approaches running speed, the starting switch opens and the motor continues to run in the normal induction motor mode.

This moderately priced motor produces relatively high starting torque (225–400% of full load torque) with moderate inrush current. Capacitor start motors are ideal for hard-to-start loads such as refrigeration compressors. Due to its other desirable characteristics, it is also used in applications where high starting torque may not be required. The capacitor start motor can usually be recognized by the bulbous protrusion on the frame that houses the starting capacitor.

In some applications it is not practical to install a centrifugal switch within the motor. These motors have a relay

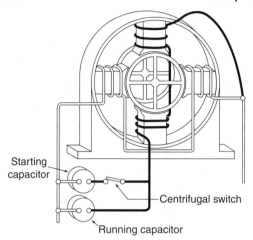

Figure 4.53 Two valve capacitor motors are used in large horsepower single phase motors. (Courtesy of Edward Cowern, P.E., North Haven, CT.)

operated by motor inrush current. The relay switches the starting capacitor into the circuit during the starting period. When the motor approaches full speed, the inrush current decreases and the relay opens to disconnect the starting capacitor.

Single Value Capacitor motors, also called *Permanent Split Capacitor (PSC) motors,* utilize a capacitor connected in series with one of the two windings. This type of motor is generally used on small sizes (less than 1 HP). It is ideally suited for small fans, blowers, and pumps. Starting torque on this type of motor is generally 100%, or less, of full load torque. A PSC motor would look the same as the capacitor start–induction run motor shown in Figure 4.52 except without the centrifugal switch.

Two Value Capacitor motors are utilized in large horsepower (2–15 HP) single phase motors. Figure 4.53 shows this motor. The running winding, shown in white, is energized directly from the line. A second winding, shown in black, serves as a combined starting and running winding. The black winding is energized through two parallel capacitors. Once the motor has started, a switch disconnects one of the capacitors, letting the motor operate with the remaining capacitor in series with the winding of the motor.

The two value capacitor motor starts as a capacitor start motor but runs as a form of a two phase or PSC motor. Using this combination, it is possible to build large single phase motors having high starting torques and moderate starting currents at reasonable prices.

The two value capacitor motor frequently uses an oversized conduit box to house both the starting and running capacitors.

G. Motors Operating on Variable Frequency Drives

Variable Frequency Drives (VFDs) are discussed in more detail in Chapter 6, Section IV. In the infancy of variable frequency drives, a major selling point was that VFDs could adjust the speed of "standard" three-phase induction motors. This claim was quite true when the variable frequency drives were "six-step" designs. The claim is still somewhat true, although pulse width modulated (PWM) VFDs have somewhat changed the rules. PWM drives are electrically more punishing on motor windings, especially for 460 and 575 volt drives.

"Standard" motors can still be used on many VFDs, especially on commercial pump, fan, and blower applications, as long as the motors are high-quality, conservative designs that use Inverter Spike Resistant (ISR) magnet wire. On these variable torque loads, a relatively small speed reduction results in a dramatic reduction in the torque required from the motor. For example, a 15% reduction in speed reduces the torque requirement by over 25%, so these motors are not stressed from a thermal point of view. Also, variable torque loads rarely need a wide speed range. Since the performance of centrifugal pumps, fans, and blowers falls off dramatically as speed is reduced, speed reduction below 40% of base speed is rarely required.

The natural question is, "What is meant by a high quality, conservative design?" Basically, this means that the motor must have phase insulation, should operate at a relatively low temperature rise (as in the case with most premium efficiency motors), and should use a high class of insulation (either F or H).

In addition, it is frequently desirable to have a winding thermostat in the motor that will detect any motor overheat conditions that may occur. Overheating could result from overload, high ambient temperature, or loss of ventilation.

Inverter Duty motors being offered in the marketplace today incorporate *premium efficiency* designs along with oversized frames or external blowers to cool the motor regardless of its speed. These motors are primarily designed for constant torque loads where the affinity laws do not apply. Inverter duty motors usually have winding thermostats that shut down the motor through the VFD control circuit in case of elevated temperature inside the motor. Inverter duty motors also have high-temperature insulating materials operated at lower temperatures. This reduces the stress on the insulation system. Although some of the design features of inverter duty motors are desirable for centrifugal pump applications using VFDs, these applications usually do not require inverter duty motors, which typically cost a good deal more than regular premium efficiency motors.

Some cautions should be observed. Generally speaking, the power coming out of a VFD is somewhat rougher on the motor than power from a pure 60-cycle source. Thus, it is not a good idea to operate motors on VFDs into their service factors.

In addition, when an old motor (one that has been in service for some time) is to be repowered from a variable frequency drive, it may be desirable to add a load reactor between the VFD and the motor. The reactor reduces the stress on the motor windings by smoothing out current variations, thereby prolonging motor life.

Reactors are similar to transformers with copper coils wound around a magnetic core. Load reactors increase in importance when the VFDs are going to run in the "quiet" mode. In this mode, the very high carrier frequency can create standing waves that potentially double the voltage peaks applied to the motor. The higher voltage can stress the motor insulation enough to cause premature failure.

H. NEMA Locked Rotor Code

The "NEMA Code Letter" is an additional piece of information on the motor nameplate. These letters indicate a range of *inrush* (*starting* or *locked rotor*) currents that occur when a motor starts across the line with a standard magnetic or manual starter. Most motors draw 5 to 7 times rated full load (nameplate) amps during the time it takes to go from standstill up to about 80% of full load speed. The length of time the inrush current lasts depends on the amount of inertia (flywheel effect) in the load. On centrifugal pumps with very low inertia, the inrush current lasts only a few seconds. On large, squirrel cage blowers, the inrush current can last considerably longer.

The locked rotor code letter quantifies the value of the inrush current for a specific motor. The lower the code letter, the lower the inrush current. Higher code letters indicate higher inrush currents. Table 4.11 lists the NEMA locked rotor code letters and their parameters.

Table 4.11 NEMA Locked Rotor Code Letters and Their Parameters

NEMA Code Letter	Locked Rotor KVA/HP	NEMA Code Letter	Locked Rotor KVA/HP
A	0–3.15	L	9.0–10.0
B	3.15–3.55	M	10.0–11.2
C	3.55–4.0	N	11.2–12.5
D	4.0–4.5	O	not used
E	4.5–5.0	P	12.5–14.0
F	5.0–5.6	Q	not used
G	5.6–6.3	R	14.0–16.0
H	6.3–7.1	S	16.0–18.0
I	not used	T	18.0–20.0
J	7.1–8.0	U	20.0–22.4
K	8.0–9.0	V	22.4 and up

I. Amps, Watts, Power Factor, and Efficiency

1. Introduction

There seems to be a great deal of confusion among the users of electric motors regarding the relative importance of power factor, efficiency, and amperage, particularly as related to operating cost. The following information should help to put these terms into proper perspective.

At the risk of treating these items in reverse order, it might be helpful to understand that in an electric bill, commercial, industrial, or residential, the basic unit of measurement is the kilowatt-hour. This is a measure of the amount of energy that is delivered. In many respects, the kilowatt-hour could be compared to a ton of coal, a cubic foot of natural gas, or a gallon of gasoline, in that it is a basic energy unit. The kilowatt-hour is not directly related to amperes, and at no place on an electric bill will you find any reference to the amperes that have been utilized. It is vitally important to note this distinction. The user is billed for kilowatt-hours, not necessarily for amperes.

2. Power Factor

Perhaps the greatest confusion arises due to the fact that early in our science educations, we were told that the formula for watts was amps times volts. This formula, watts = amps × volts, is perfectly true for direct current circuits. It also works on some AC loads such as incandescent light bulbs, quartz heaters, electric range heating elements, and other equipment of this general nature. However, when the loads involve a characteristic called inductance, the formula has to be altered to include a new term called *power factor*. Thus, the new formula for single phase loads becomes watts are equal to amps × volts × power factor. The new term, power factor, is always involved in applications where AC power is used and inductive magnetic elements exist in the circuit. Inductive elements are magnetic devices such as solenoid coils, motor windings, transformer windings, fluorescent lamp ballasts, and similar equipment that have magnetic components as part of their design.

Looking at the electrical flow into this type of device, we find that there are, in essence, two components. One portion is absorbed and utilized to do useful work. This portion is called the real power. The second portion is literally borrowed from the power company and used to magnetize the magnetic portion of the circuit. Due to the reversing nature of AC power, this borrowed power is subsequently returned to the power system when the AC cycle reverses. This borrowing and returning occurs on a continuous basis. Power factor then becomes a measurement of the amount of real power that is used, divided by the total amount of power, both borrowed and used. Values for power factor will range from zero to 1.0. If all the power is borrowed and returned with none being used, the power factor would be zero. If on the other hand, all of the power drawn from the power line is utilized and none is returned, the power factor becomes 1.0. In the case of electric heating elements, incandescent light bulbs, etc., the power factor is 1.0. In the case of electric motors, the power factor is variable and changes with the amount of load that is applied to the motor. Thus, a motor running on a work bench, with no load applied to the shaft, will have a low power factor (perhaps 0.1 or 10%); and a motor running at full load, connected to a pump or a fan might have a relatively high power factor (perhaps 0.78 or 78%). Between the no load point and the full load point, the power factor increases steadily with the horsepower loading that is applied to the motor. These trends can be seen on the typical motor performance data plots which are shown for a 15 HP, 1725 rpm, three-phase motor in Figure 4.54.

3. Efficiency

Now, let's consider one of the most critical elements involved in motor operating cost, motor efficiency. Efficiency is the measure of how well the electric motor converts the power that is purchased into useful work. For example, an electric heater such as the element in an electric stove converts 100% of the power delivered into heat. In other devices such as motors, not all of the purchased energy is converted into

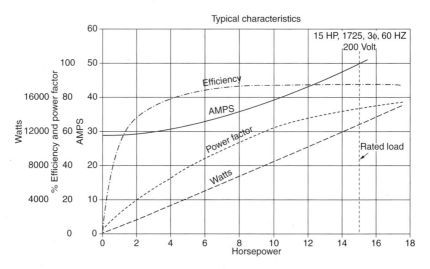

Figure 4.54 Typical motor performance data. (Courtesy of Edward Cowern, P.E., North Haven, CT.)

usable energy. A certain portion is lost and is not recoverable because it is expended in the losses associated with operating the device. In an electric motor, these typical losses are the copper losses, the iron losses, and the so-called friction and windage losses associated with spinning the rotor and the bearings and moving the cooling air through the motor.

In an energy-efficient (premium efficiency) motor, the losses are reduced by using designs that employ better grades of material, more material, and better designs to minimize the various items that contribute to the losses in the motor. For example, on a 10 HP motor, an energy efficient design might have a full load efficiency of 91.7%, meaning that, at full load (10 HP), it converts 91.7% of the energy it receives into useful work. A less efficient motor might have an efficiency of 82%, which would indicate that it only converts 82% of the power into useful work.

In general, larger motors can be expected to be found with higher efficiencies than smaller ones. For example, a 1 HP, 1800 rpm, premium efficiency motor might have an efficiency of 85.5%, while a 100 HP, 1800 rpm, premium efficiency motor might have an efficiency of 95.5%. In general, the efficiency of most motors will be relatively constant from 50 to 100% of rated load, although there may be a few percentage points fluctuation in this range.

4. Amperes

Now let's discuss amperes (abbreviated *amps*). Amperes are an indication of the flow of electric current into the motor. This flow includes both the borrowed as well as the used power. At low load levels, the borrowed power is a high percentage of the total power. As the load increases on the motor, the borrowed power becomes less and less of a factor and the used power becomes greater. Thus, there is an increase in the power factor as the load on the motor increases. As the load continues to increase beyond 50% of the rating of the motor, the amperage starts to increase in a nearly straight-line relationship. This can be seen in Figure 4.54.

5. Summary

Figure 4.54 shows significant items that have been discussed as plots of efficiency, power factor, amps, and watts, as they relate to horsepower. The most significant factor of all these is the watts requirement of the motor for the various load levels, because it is the watts that will determine the operating cost of the motor, not the amperage.

The user having an extremely low power factor in the total plant electrical system may be penalized by the utility company because the user is effectively borrowing a great deal of power without paying for it. When this type of charge is levied on the customer, it is generally called a *power factor penalty*. In general, power factor penalties are levied only on large industrial customers and rarely on smaller customers, regardless of their power factor. In addition, there are a great many types of power customers such as commercial establishments, hospitals, and some industrial plants that inherently run at very high power factors. Thus, the power factor of individual small motors that are added to the system will not have any significant effect on the total plant power factor.

It is for these reasons that the blanket statement can be made that increasing motor efficiency will reduce the kilowatt hour consumption and the power cost for all classes of power users, regardless of their particular rate structure or power factor situation. This same type of statement cannot be made relative to power factor.

The following formulas are useful for calculating operating costs for electric motors.

$$\text{Kilowatt-hours} = (\text{HP}^{**} \times 0.746 \times \text{Hrs. of operation})/\text{Motor Eff. (as a decimal)} \quad (4.2)$$

** Average Load HP (may be lower than motor nameplate HP)

$$\text{Kilowatt-hours} = (\text{Watts} \times \text{Hours of operation})/1000 \quad (4.3)$$

5

Sealing Systems and Sealless Pumps

I. OVERVIEW

This chapter addresses sealing in pumps. The chapter begins with a discussion of O-rings, which are widely used as sealing elements in pumps.

The chapter then moves on to shaft sealing. Most pumps require sealing of the shaft where it penetrates the pump casing. The casing at the point of shaft penetration is subjected to either a positive pressure or a vacuum. Some pump types such as horizontal split case are sealed at both ends, while other types such as end suction have only one sealing point.

A very basic method for shaft sealing is the packed stuffing box. The various components of the stuffing box and packing assembly are explained, along with limitations on their use.

Mechanical shaft seals are described as to their benefits, general function, description, and application. Particular types of mechanical seals (inside vs. outside, unbalanced vs. balanced, single vs. double and tandem) are explained and illustrated. A relatively new type of noncontacting, gas-lubricated seal is introduced in this chapter.

Finally, several types of centrifugal pumps that require no packing or seal are introduced. Sealless pumping is an

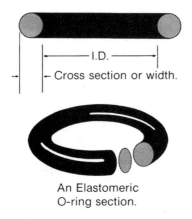

Figure 5.1 Anatomy of an O-ring. (Courtesy of Parker Hannifin Corporation, Seal Group.)

important technology not only because it guarantees zero emissions, but also because it eliminates the seal, a component that may require frequent maintenance. The two major types of sealless centrifugals, *magnetic drive* and *canned motor* pumps, are described and compared.

II. O-RINGS

For those readers who are unfamiliar with O-ring sealing technology, this section should aid in the understanding of the O-ring, its basic design principles, and the many sealing functions an O-ring can be expected to perform when properly specified and installed. The following material on O-rings is printed with permission of the Parker Hannifin Corporation, Seal Group. See also discussion of the various O-ring materials in Chapter 7, Section VI.

A. What Is an O-Ring?

An O-ring (Figure 5.1) is a torus, or a doughnut shaped object, generally made from an elastomer, although such materials as plastics and metal are sometimes used. This section will deal entirely with elastomeric O-rings used for sealing purposes.

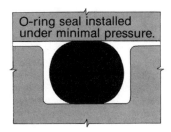

Figure 5.2 O-ring seal installed under minimal pressure. (Courtesy of Parker Hannifin Corporation, Seal Group.)

B. Basic Principals of the O-Ring Seal

An O-ring seal is a means of closing off a passageway and preventing an unwanted loss or transfer of fluid. The classic O-ring seal consists of two elements, the O-ring itself, and a properly designed gland or cavity to contain the elastomeric material. Prevention of the fluid loss or transfer may be obtained by several methods: welding, soldering, brazing, or the yielding of a softer material wholly or partially confined between two mating surfaces. This latter method best describes the design principle behind the operation of an O-ring seal.

C. The Function of the O-Ring

The elastomer is contained in the gland and forced to flow into the surface imperfections of the glands and any clearance available to it, creating a condition of "zero" clearance and thus effecting a positive block to the fluid being sealed (Figure 5.2). The pressure that forces the O-ring to flow is supplied by mechanical pressure or "squeeze," generated by proper gland design and material selection, and by system pressure transmitted by the fluid itself to the seal element. In fact, the classic O-ring seal may be said to be "pressure assisted" (Figure 5.3), in that the more system pressure, the more effective the seal, until the physical limits of the elastomer are exceeded and the O-ring begins to extrude into the clearance gap (Figure 5.4). This condition can usually be avoided by proper gland design and material selection.

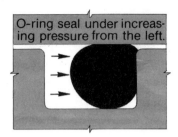

Figure 5.3 A pressure-assisted O-ring seal. (Courtesy of Parker Hannifin Corporation, Seal Group.)

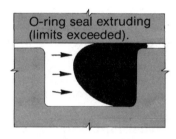

Figure 5.4 O-ring seal with pressure limited exceeded. (Courtesy of Parker Hannifin Corporation, Seal Group.)

D. Static and Dynamic O-Ring Sealing Applications

O-ring seals are generally divided into two main groups, *static seals*, in which there is little or no relative motion between the mating surfaces; and *dynamic seals*, which must function between surfaces with definite relative motion, such as the seal on the piston of a hydraulic cylinder. Of the two types, dynamic sealing is the more difficult and requires more critical design work and materials selection.

The most common type of dynamic motion utilizing an O-ring sealing system is *reciprocal motion* as found in hydraulic cylinders, actuators, and the like.

E. Other Common O-Ring Seal Configurations

Aside from reciprocating seals, there are other types of motion where an O-ring seal may be utilized. For example:

Sealing Systems and Sealless Pumps

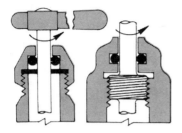

Figure 5.5 Oscillating seal. (Courtesy of Parker Hannifin Corporation, Seal Group.)

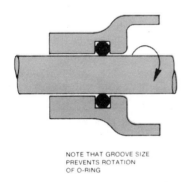

NOTE THAT GROOVE SIZE
PREVENTS ROTATION
OF O-RING

Figure 5.6 Rotary seal. (Courtesy of Parker Hannifin Corporation, Seal Group.)

Oscillating seals (Figure 5.5), in which the inner or outer member of the assembly moves in an arc relative to the other, rotating one of the members in relation to the O-ring.

Rotary seals (Figure 5.6), where an inner or outer member of the sealing assembly revolves around the shaft axis in only one direction. The direction may be reversed. If there are multiple brief arcs of motion, the designer should refer to parameters for oscillating seals.

Seat seals (Figure 5.7) utilize an O-ring to close a flow passage by distorting the face of the O-ring against the opposite contact face. Closing the passage distorts the seal element to create the closure.

Crush seals (Figure 5.8), a variation of the static seal, literally "crush" an O-ring into a space with a cross section

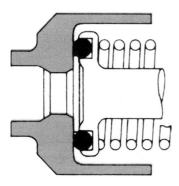

Figure 5.7 Seat seal. (Courtesy of Parker Hannifin Corporation, Seal Group.)

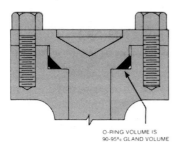

Figure 5.8 Crush seal. (Courtesy of Parker Hannifin Corporation, Seal Group.)

different from the standard gland. Although often an effective seal, the O-ring is permanently deformed and must be replaced if the unit is opened.

Pneumatic seals may be any of the previously described types, but are given a different classification because they seal a gas or vapor as opposed to a liquid. Thus, other design factors such as adequate lubrication (for dynamic seals), temperature increases due to compression of gases, and permeability of the seal element must be considered when the application is pneumatic.

Vacuum seals also may be any of the foregoing types (except pneumatic), and are classified separately because of special design considerations and the unusually low leakage requirements of vacuum systems.

F. Limitations of O-Ring Use

Although O-rings offer a dependable and reasonably economical approach to hydraulic sealing, they are not a "universal" seal, applicable to all sealing problems. Certain limitations must be imposed upon their use; among them, high temperatures, high frictional (rubbing) speeds, cylinder ports over which the seal element must pass, and excessive clearances. O-rings therefore may be considered for just about all sealing applications except:

- Rotary speeds exceeding 1500 feet per minute contact speed
- Environments (temperature and media) incompatible with any elastomeric material
- Insufficient structural support for anything but a flat gasket

III. STUFFING BOX AND PACKING ASSEMBLY

Figure 5.9 shows a stuffing box and packing assembly, the oldest and one of the most common shaft sealing systems for

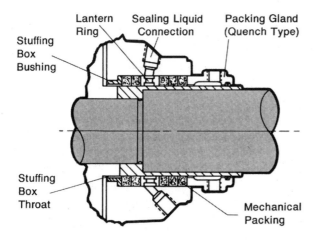

Figure 5.9 Stuffing box and packing assembly. (Courtesy of Goulds Pumps Inc., ITT Industries, Seneca Falls, NY.)

pumps. In Figure 5.1, the inside of the pump casing is on the left side of the figure, and the environment is on the right side. There are five components of the stuffing box and packing assembly, as described below.

A. Stuffing Box

The stuffing box houses the packing assembly and is the location where the shaft penetrates the casing that is under pressure or vacuum. In some pump configurations, the stuffing box is actually part of the pump casing, while for other pump types, the stuffing box is bolted to the casing. The stuffing box is machined so that the throat bore is concentric with the shaft. Its bore size and depth are designed to accommodate a specific size of packing and number of packing rings. Many stuffing boxes are fitted with drilled connections that allow the introduction of liquid into the packing area, or provide a path for liquid to escape to reduce the pressure on the packing, as discussed in Section III.E.

B. Stuffing Box Bushing

The stuffing box bushing is located at the bottom of the stuffing box, below the first ring of packing. It serves several functions. The bushing has a close running clearance between itself and the shaft or sleeve. This allows some amount of pressure breakdown as liquid throttles across it. In general, the lower the pressure against which the packing is sealing, the longer the expected service life of the packing. The bushing also serves to keep larger solids from entering the packing area where they might embed in the packing and cause undue wear on the shaft or sleeve. The bushing is replaceable so that the close clearance can be maintained throughout the pump's life.

The stuffing box bushing keeps wear from occurring in the stuffing box itself, a much more expensive component to replace than the bushing.

The stuffing box bushing serves as a landing or shoulder for the lowest ring of packing. If the clearance here is larger than provided by the stuffing box bushing, the soft, pliable

Sealing Systems and Sealless Pumps

packing could extrude into the clearance as the packing is tightened down by the *gland* (see Section III.D below), reducing the effectiveness of the packing and making its removal more difficult.

The stuffing box bushing also restricts the amount of liquid that might leak out of the stuffing box in the event of a complete failure of the packing rings and gland.

Inexpensive lighter-duty pumps may not be supplied with a stuffing box bushing, but it is a good idea to have such a bushing on any packed pump if one is available, for the reasons just mentioned.

C. Packing Rings

The mechanical packing rings themselves constitue the heart of the sealing device. Packing rings are usually formed of a pliable braided fibrous material, often impregnated with a lubricating medium. The packing material is square in cross section, and is formed into rings that are compressed into the stuffing box bore and around the shaft. These rings provide a seal between their outside diameter and the stuffing box (where there is no relative movement when the pump is operating) so that no liquid can leak out this way.

The packing rings also compress against the rotating shaft or sleeve, causing the liquid to break down in pressure as it leaks between the packing and the shaft or sleeve. To operate properly, *packing must leak*. As the liquid leaks between the packing and the shaft or sleeve, it cools and lubricates this interface. A common rule of thumb is that a packing assembly should be allowed to leak around 60 drops per minute to provide adequate lubrication and cooling of the packing and the surface against which it is running. This is not always properly understood by equipment owners or maintenance technicians, who sometimes keep tightening up on the packing gland to minimize the "mess" of the leaking packing assembly. If too little liquid is allowed to leak across the packing area, the packing can overheat due to the friction between itself and the shaft or sleeve. This can cause the packing to degrade quickly, and can cause wear on the shaft or sleeve.

Both of these degradations, in turn, cause the packing area to begin to leak excessively. It should be noted that there are several manufacturers that offer drip-less, dry running, self-lubricated packing; however, the author has limited experience with this packing.

Most packed pumps have from four to seven rings of packing, depending on the amount of pressure to be sealed. Packing cross-sectional width varies in size, depending on the size of the pump, with common width sizes varying between ¼ and ¾ in. Packing can be purchased in preformed rings, made to fit specific shaft/sleeve diameters and specific stuffing box bores. It can also be purchased in rolls and cut to the proper length as it is installed.

If packing is properly chosen to be compatible with the liquid pumped, is properly installed, and is allowed to be lubricated by the pumped liquid, it can seal relatively high pressures with only periodic replacement. It does leak, however, as indicated above, and thus is not acceptable for many corrosive or extremely high-pressure services.

If the liquid being pumped contains abrasives, flushing the packing with an external flush medium is a good idea. Otherwise, the abrasive particles can embed in the soft packing and act as a grinding wheel against the shaft or sleeve, damaging the shaft or sleeve and causing the packing to leak excessively. See further discussion of external flushing in Section III.E on *lantern rings*. If external flushing with a clean liquid is not assured for an abrasive application, the shaft or sleeve should be coated with a ceramic or other hard coating where it runs under the packing, to protect the shaft or sleeve against early failure.

Refer to Chapter 8, Section IV.A.2, for further guidelines on the proper installation of packing.

D. Packing Gland

The gland, sometimes called the *packing follower*, holds the packing rings in place in the stuffing box and provides a means of further compressing the packing as wear occurs on the packing, shaft, or sleeve. Most glands are split, so that they can be removed to allow more room for changing the packing without breaking apart the shaft coupling. The gland typically attaches to the stuffing box with studs, and tightening the stud nuts moves the gland toward the packing, tightening it further.

Some glands have a hollow cavity that allows the introduction of cooling liquid if the pumped liquid is hot. One important application for this *gland quench* feature is for boiler feed service, to prevent the water from flashing back to steam as it breaks down in pressure across the packing.

E. Lantern Ring

The lantern ring has the same general shape as a ring of packing, fitting into the stuffing box between packing rings. The lantern ring is typically made of bronze, stainless steel, or plastic, and is drilled with holes on its outside perimeter. The lantern ring is fitted into the stuffing box at the location of the drilled opening in the stuffing box. The lantern ring in this case has several possible functions. Located at roughly the middle of the set of packing rings, the lantern ring allows the introduction of liquid to lubricate the packing if the pumped liquid is unable to do this. For example, if the packing is used on a split case double suction pump (Figure 4.18) operating on a lift, the pressure just below the packing is below atmospheric. Therefore, the liquid does not tend to leak across the packing area as it would if the packing were sealing against a positive pressure. The usual arrangement is to run tubing from the pump discharge flange or from the high-pressure part of the casing around to the stuffing box, introducing high-pressure liquid into the lantern ring. This liquid serves the dual purpose of both lubricating the packing and sealing against air leaking into the suction of the pump.

The lantern ring can also be used to introduce an external lubricating medium if the liquid being pumped has poor lubricity (such as gasoline, for example).

If the pumped liquid contains abrasives, the packing can be lubricated by introducing an external clean flush liquid into the lantern ring. In some cases with abrasive liquids, the lantern ring and the drilled opening in the stuffing box are located at the bottom of the stuffing box rather than in the middle of the packing rings, to more completely eliminate the possibility of solids getting into the packing area.

Another use of a lantern ring is with high-pressure applications. For these applications, the lantern ring is often located at the bottom of the stuffing box, just above the stuffing box bushing, but again in line with a drilled hole in the stuffing box. The lantern ring allows high-pressure liquid to relieve back to suction, thus lowering the pressure against which the packing rings have to seal.

IV. MECHANICAL SEALS

A. Mechanical Seal Advantages

Mechanical seals have a number of advantages over stuffing box and packing arrangements. The most important advantages of mechanical seals are discussed below.

1. Lower Mechanical Losses

In Chapter 2, Section V, the mechanical losses are listed as one of the four factors causing a pump to be inefficient. One of the largest of these mechanical losses comes from the frictional drag of the shaft running against packing. A mechanical seal, on the other hand, has considerably less mechanical loss, thus improving pump efficiency.

2. Less Sleeve Wear

Section III.C above discusses the fact that the sleeve under the packing is subject to wear, especially if the pumped liquid has any abrasives in it or if the packing is tightened too much. With a mechanical seal, this wear of the shaft sleeve is eliminated.

3. Zero or Minimal Leakage

Whereas packing should leak around 60 drops per minute to be properly lubricated, the leakage from most mechanical seals is very nearly zero under normal operation. This is especially important when the pumped liquid is corrosive, volatile, toxic, or radioactive. Mechanical seal technology is continuously changing to ensure compliance with the emission requirements of the EPA and other regulatory agencies.

4. Reduced Maintenance

If (and these are big "ifs") the seal is properly selected for the application, and if the pump is properly aligned and balanced and not subjected to vibrations from other sources, a mechanical seal should require less periodic maintenance than a packed stuffing box. In general, pumps with mechanical seals are less forgiving of misalignment, imbalance, and vibration than packed pumps.

5. Seal Higher Pressures

Mechanical seals are able to seal against higher pressures and can be used with higher-speed pumps than packing.

B. How Mechanical Seals Work

There are many different types of mechanical seals. Understanding seal types is complicated by the fact that mechanical seal and pump suppliers do not all use the same terminology to describe the seal components. Seal types that this chapter describes include inside and outside seals, balanced and unbalanced seals, and single, double, and tandem seals.

Every mechanical seal has in common three sealing points, consisting of two static seals and one dynamic seal. The term "static" seal means that there is no relative motion between the two parts. Each mechanical seal has a static seal between the rotating assembly and the sleeve (or shaft if there is no sleeve), and a second static seal between the stationary part of the seal and the gland or seal housing. The dynamic seal is between the two seal faces, one rotating and the other stationary. The descriptions of the different sealing types to follow serve to clarify these definitions.

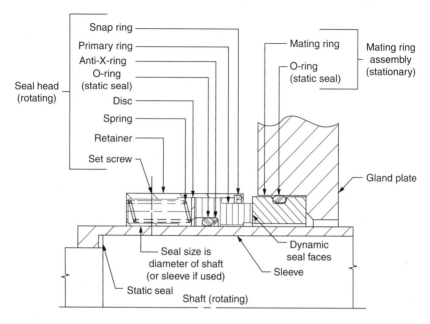

Figure 5.10 Basic single mechanical seal. (Courtesy of John Crane Inc., Morton Grove, IL.)

Figure 5.10 shows a basic single mechanical seal, with the liquid being sealed coming from the left side of the figure, and the shaft penetration to atmosphere being located on the right side. The rotating part of the seal (also called the *seal head*) includes a metal retainer, locked to the sleeve with set screws; a spring assembly; and the rotating dynamic seal face (also called the *primary ring, washer, seal ring face,* or *seal ring*). The stationary face of the dynamic seal (also called by some manufacturers the *mating ring,* or *seat*) is fixed to the seal housing or gland. In the case of the seal shown in Figure 5.10, the mating ring is held in place by an O-ring. Other designs have the mating ring clamped between the gland and the seal housing.

The springs in the rotating assembly keep the dynamic seal faces in uniform contact with each other even if the shaft moves axially due to thermal expansion or small amounts of misalignment. If the seal is working against a positive pressure, the pressure acting against the rotating seal face also keeps the two seal faces in contact. However, if the liquid being sealed against is below atmospheric pressure (as is the case for seals on a double suction pump operating on a suction lift), then the liquid being sealed does not push the two seal faces together. In that case, the springs must hold the two faces in contact. To maintain proper spring tension so that the two seal faces are always in contact, the location of the rotating part of the seal with respect to the sleeve (the distance to the edge of the sleeve) is an important dimension that the seal and/or pump manufacturer should provide.

On the seal shown in Figure 5.10, the static seal between the rotating assembly and the sleeve is an O-ring. This keeps liquid from leaking along the sleeve to atmosphere. The static seal between the stationary mating ring and the gland is also an O-ring, which holds and seals the stationary mating ring. Finally, the dynamic seal takes place at the contact area between the rotating and stationary seal faces.

The seal faces must be manufactured to extremely tight tolerances to make sure the seal faces are completely flat. The seal faces are polished, by a process called *lapping*, so precisely that their flatness is measured in light bands, and their surface finish in micro-inches. Care should be taken when handling mechanical seal faces. Even oily fingerprints can be enough to keep the seal faces from being in perfect contact with each other.

The spring assembly on the rotating part of the mechanical seal may consist of a number of small springs evenly spaced around the rotating part of the seal, or a single coil wrapped around the sleeve. The multiple spring arrangement provides a more uniform contact of the seal faces. However, the large single-coil spring is stronger and less likely to fail from corrosion or to fail to perform if the pumped liquid contains solids or viscous material. There are alternative designs that use various types of metallic or elastomeric bellows in the seal assembly.

Just as packing must be lubricated to keep from overheating and causing the pump to fail, so must the dynamic seal faces of a mechanical seal be wetted to work properly.

This is true even if one of the seal faces are made of carbon, a common seal face material that has good self-lubricating properties. Because of the very close contact of the mechanical seal faces, the liquid breaks down in pressure as it passes across the seal face, so that very little, if any, liquid actually escapes the seal.

In some applications, if the liquid being sealed against is close to its vapor pressure, there is the concern that the liquid might drop below the vapor pressure as it leaks across the seal faces. This could cause the dynamic seal faces not to be wetted across their entire surface, in turn causing them to run hot. The heat generated could cause the point along the dynamic seal faces at which the liquid flashes to vapor to move back progressively, eventually causing the seal faces to run completely dry and fail prematurely. This must be prevented from occurring, and is usually done by injecting liquid from the pump discharge into the seal chamber to increase the pressure of the liquid at the seal faces, so that it cannot drop below its vapor pressure prior to passing across the seal contact faces.

Mechanical seal materials must be carefully chosen to meet the requirements of the application. The metal parts of the seal must be able to handle the corrosive nature of the liquid. The elastomers used for the static seals (O-rings, wedges, gaskets) must be able to handle the temperature, pressure, and corrosiveness of the liquid. Elastomeric materials used for these components in mechanical seals include butyl, EPR, buna-N, neoprene, VITON® fluoroelastomer, and Teflon® PTFE. See Chapter 7, Section VI, for a discussion of O-ring matrials.

The rotating and stationary dynamic seal faces should be the best suited for the application. Various grades of carbon are often used for one of the dynamic sealing faces. Other seal face materials used for certain applications include ceramic, ni-resist, tungsten carbide, and silicon carbide.

To allow the seal and/or pump supplier to assist in making the optimum selection of seal type and material, it is important to provide as much information as possible about the conditions of service. This includes detailed information

Sealing Systems and Sealless Pumps

about the liquid being pumped (liquid type, pressure at the seal housing, temperature, pH, viscosity, vapor pressure, and solids concentration), both at normal operation and the expected range. Also, information about the type of operation (continuous vs. intermittent, pump speed, alignment and vibration tolerances) helps to select the best seal for the application. Finally, the dimensional data related to the pump sleeve outside diameter and depth of seal housing ensure that a correct seal size is chosen.

If a factory performance test is specified for a pump with a mechanical seal, the user may want to consider having the pump tested without the mechanical seal installed (substituting a lab seal or packing). Operation of the pump in the manufacturer's test facility may actually be more severe service than the pump can expect in the field, if the pump's service is a clean liquid with good lubricating properties. Consequently, the seal faces may not have been selected to work well in the environment of the test facility, and they may wear excessively during the test. As a compromise measure in this regard, the user's seal can be installed for the test, and then the faces re-lapped following the test.

C. Types of Mechanical Seals

1. Single, Inside Seals

Figures 5.10, 5.11, and 5.13 show single, inside mechanical seals. (In these figures, the sealed liquid is on the left side and atmosphere is on the right side.) They are called single seals because there is one dynamic seal face, and are called inside seals because the seal parts are located inside the seal housing, unobservable when the pump is running. The single, inside seal is the most common type of mechanical seal for pumps.

The seals shown in Figures 5.10, 5.11, and 5.13 have O-rings for the static seal between the primary ring and the sleeve, as well as for the static seal between the mating ring and the gland. The seal housing and gland provide virtually the same functions for mechanical seals as do, respectively,

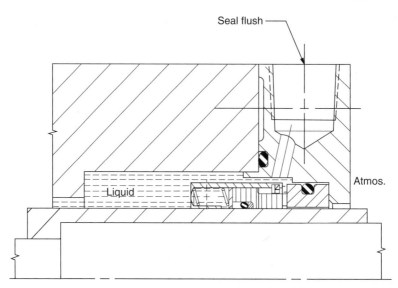

Figure 5.11 Single, inside, unbalanced seal. (Courtesy of John Crane Inc., Morton Grove, IL.)

the stuffing box and gland for packing — namely, to contain the seal and to hold it in place.

Note that the gland in Figure 5.11 contains a port for injecting liquid, either from the pump discharge or from an external source. This injection could be used to flush the dynamic seal faces. It could also be used to increase the pressure of the liquid at the seal faces, to keep it from dropping below the vapor pressure of the liquid as described in Section IV.B above. If the pumped liquid is used for injection, and if the pumped liquid contains abrasives or other solids, a separator or filter should be used to supply the seal with clean liquid.

Another important gland feature is a *vent and drain* connection. This feature is shown on the seal in Figure 5.13 and can be supplied as an option with most mechanical seals. (It may not be available in very small pumps due to space limitations in the seal area.) The function of the vent and drain connection is to carry away any liquid that leaks past the dynamic seal faces, to keep this liquid from moving along the shaft and then being slung off the shaft. Depending on the nature of the pumped liquid, failure to drain this liquid with a vent and drain connection could cause damage to the equipment or be a safety hazard, in addition to making a mess. Therefore, the vent and drain feature has a housekeeping function, serves to reduce maintenance expense, and may even have a safety function. Note that this is an optional feature for many pumps, and will not necessarily be supplied unless it is specified. It is a particularly worthwhile feature to include in any seal installation if it is available.

If the pumped liquid is volatile, there should be a nonsparking bushing located in the gland in the event the shaft comes in contact with the gland. The seal in Figure 5.13 shows such a bushing.

2. Single, Outside Seals

Figure 5.12 shows a single, outside mounted seal. Here, the entire rotating portion of the seal is outside the seal chamber. (In this figure, the sealed liquid is on the left side, atmosphere on the right side.) The static seal between the rotating assembly

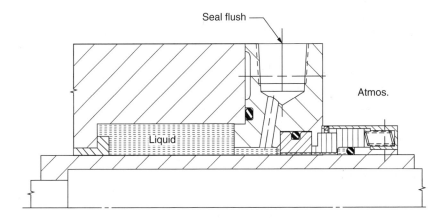

Figure 5.12 Single, outside, balanced seal. (Courtesy of John Crane Inc., Morton Grove, IL.)

and the sleeve is an O-ring, and the static seal between the mating ring and the gland is also an O-ring.

The main advantage of an outside seal is that the hardware items of the seal do not come in contact with the liquid, so this assembly is not subject to corrosive attack or other deteriorating influences from the pumped liquid. Also, the rotating assembly is observable during operation, and may be considered easier to get at for maintenance.

The primary disadvantage of the outside seal is that the pressure of the sealed liquid pushes the two seal faces apart, rather than forcing them together as is the case for an inside seal. The only thing holding the two seal faces together is the spring compression. Outside seals are limited to sealing pressures of about 25 to 30 psig unbalanced, and about 60 psig balanced. (Refer to the discussion in Section IV.C.3 below on balanced vs. unbalanced seals.) Another disadvantage of outside mounted seals is that they are subject to exposure to dust and other environmental contaminants.

The port shown on the seal housing in Figure 5.12 could be used to inject an external liquid to lubricate the dynamic seal faces, in the event that the sealed liquid pressure is below atmospheric, or if it contains abrasives. The restricting bushing shown at the bottom of the seal housing is used to limit the amount of externally injected liquid that is allowed to enter into the pumped liquid.

3. Single, Balanced Seals

A balanced seal is shown in Figure 5.13, with the pumped liquid coming from the left side and atmosphere being on the right side. The purpose of a balanced seal is to seal against higher pressures and speeds than possible in an unbalanced configuration. The balance is achieved by having a stepped or shortened sleeve, and by having a step in the primary ring. These reduce the hydraulic pressure at the dynamic seal faces, allowing the seal to handle higher pressure and reducing the power consumption of the seal. The seals shown in Figures 5.10 and 5.11 are unbalanced seals, while the one shown in Figure 5.12 is a balanced design.

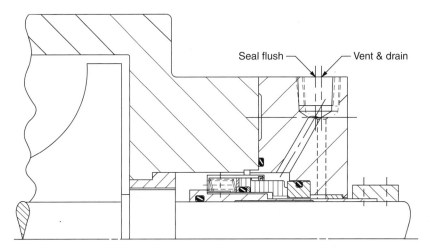

Figure 5.13 Single, inside, balanced seal. (Courtesy of John Crane Inc., Morton Grove, IL.)

The maximum pressure which can be handled by a mechanical seal in an unbalanced configuration is a function of the seal size (outside diameter of the sleeve), the pump speed, the seal material, and the liquid being pumped. Seal manufacturers provide curves that show the maximum pressure that can be handled with unbalanced seals based on these four variables. If the seal must handle higher pressures than this, then a balanced seal must be used.

4. Double Seals

Double seals are used for extremely tough services, where it is desirable to completely eliminate the possibility of leakage of pumped product into the environment, or where the dynamic seal faces must be isolated from the pumped liquid. Examples of liquids for which a double seal should be considered include extremely corrosive liquids, toxic liquids, and liquids containing abrasive solids. With the increasing amount of attention being given to controlling or eliminating the release of liquid and vapor from pump seals because of environmental considerations, double seals are considered by many as the best seal choice for this application.

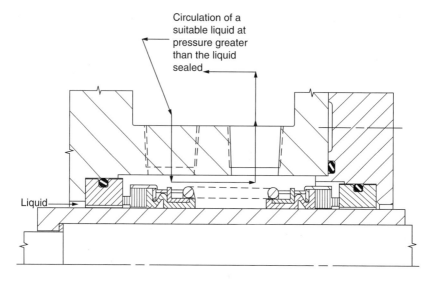

Figure 5.14 Double mechanical seal. (Courtesy of John Crane Inc., Morton Grove, IL.)

Figure 5.14 shows a double mechanical seal, with the liquid being pumped coming from the left side of the figure, and atmosphere being on the right side. A double mechanical seal is actually two seals, mounted back-to-back. The rotating parts of the two seals are in the middle of the seal, attached to the sleeve and rotating with it. There is a stationary mating ring in contact with the primary ring at each end of the rotating assembly. Each stationary mating ring has a static seal (shown as O-rings) between the mating ring and the housing or gland. There is also a static seal at each end of the rotating assembly (an elastomeric bellows), sealing against the sleeve.

A *buffer liquid* is injected into the cavity between the two seals described above, at a pressure higher than the pressure of the pumped liquid at the seal. If the outside dynamic seal of the double seal arrangement were to leak, the buffer liquid, rather than the pumped liquid, would leak out to the environment. If the inside dynamic seal were to leak, the buffer liquid would leak into the pumped liquid,

Sealing Systems and Sealless Pumps

rather than the other way around. Thus, the two dynamic seal faces are always kept completely free of the pumped liquid, and the pumped liquid cannot possibly leak into the environment.

The buffer liquid is chosen to be compatible with the application, so that it may leak to the environment or into the pumped product without severe negative consequences if either dynamic seal fails.

The buffer liquid is maintained at a pressure higher than pump pressure at the seal by an auxiliary pump, or by the use of a vessel with pressurized gas separated from the buffer liquid by a diaphragm. The loss of pressure on the gas side of this diaphragm or a drop in the liquid level in the barrier liquid vessel indicates a leak in one of the dynamic seal faces.

5. Tandem Seals

A tandem seal, Figure 5.15, is a special variation of the double seal, also for critical services. With a tandem seal, two seals are mounted back-to-front, or in tandem, rather than the back-to-back arrangement of the double seal shown in Figure 5.14. Some users refer to both tandem and double seals as *dual seals*.

The tandem seal has a different design consideration than a double seal. It is meant to keep the pumped liquid from leaking out to the environment, but it does this by providing a back-up seal, in the event of leakage of the primary seal. A tandem seal requires a buffer liquid, but the buffer liquid does not have to be at a pressure higher than the pumped liquid pressure, as is the case with the double seal.

The tandem seal shown in Figure 5.15 has the pumped liquid being sealed coming from the left side of the figure, and atmosphere on the right side. The seal on the left side of the figure is the primary seal, with the pumped liquid contained by the dynamic seal faces of this primary seal. The seal to the right of the primary seal is the secondary seal. Note that both the primary and secondary seals shown in Figure 5.15 are balanced seals. For these balanced seals, the rotating faces

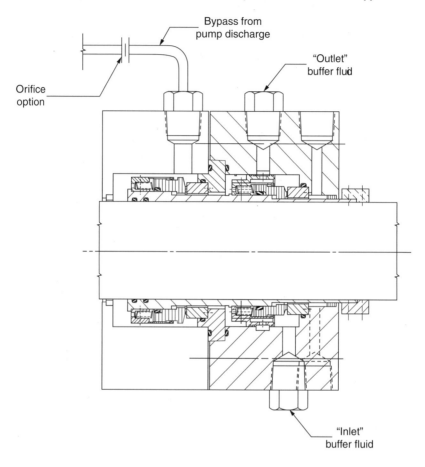

Figure 5.15 Tandem mechanical seal. (Courtesy of John Crane Inc., Morton Grove, IL.)

are tapered rather than stepped, as is the case for the previously examined balanced seals. The taper shown in Figure 5.15 is there to control the geometry of the rotating faces to reduce deflection, and to allow more liquid around the seal face outer diameter for cooling.

Ordinarily, no pumped liquid reaches the secondary seal, because the primary dynamic seal (along with its static seals) keeps the liquid at the primary seal. A buffer liquid must be introduced into the gland at the secondary seal to keep the

Sealing Systems and Sealless Pumps

dynamic seal faces of the secondary seal lubricated. This buffer liquid can be at a low pressure, and is sometimes simply supplied by a head tank mounted on the pump.

In the event that the primary seal fails, the secondary seal then takes over to keep the pumped liquid from escaping to the environment. Usually, the buffer liquid is monitored for pressure, pH, conductivity, or some other convenient variable, to signify a failure of the primary seal.

The pump would not have to be shut down immediately if the primary seal were to fail, however, because the secondary seal would then be sealing the pumped liquid. The idea of the tandem seal is that if the primary seal fails, the operator of the pump can plan an orderly shutdown of the pump to repair the primary seal, without any product escaping to the environment.

The ability of a tandem seal to perform its intended service of eliminating leakage to the environment depends on the promptness of the operator in repairing the primary seal if it fails. If the pump is shut down for a seal repair in a timely manner, the tandem seal works to eliminate any product leakage. If, however, the pump operates for an extended time after the primary seal has failed, it is, during that period of time, simply operating as a single seal. If the secondary seal were to begin leaking during that time, the pumped product could leak to the environment.

6. Gas Lubricated Non-Contacting Seals

A recent development in pump seal technology, gas lubricated seals provide a new approach to a double seal for controlling pump emissions. In addition, they provide reduction in power consumed at the seal. The seal illustrated in Figure 5.16 is a typical gas lubricated seal installation for a pumping application, with the liquid being sealed coming from the left side of the figure. Barrier gas from an external source (nitrogen, purified air, or other inert gas) is supplied between the two seals of this double seal, at a pressure of 25 to 30 psi higher than the process liquid. When the shaft begins to rotate, pressure is built up within spiral grooves which have been machined into the mating rings, separating the seal faces. This

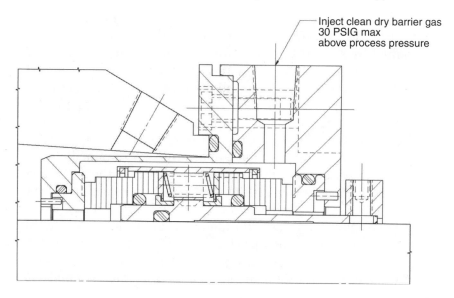

Figure 5.16 Gas lubricated noncontacting seal. (Courtesy of John Crane Inc., Morton Grove, IL.)

hydrodynamic effect results in noncontacting faces at both dynamic seal face positions.

The amount of barrier gas consumed is quite small. Horsepower losses at the faces are extremely small because frictional contact has been eliminated. This results in energy savings, as well as providing a seal that can achieve zero emissions both in operation and during standby. Applications for gas lubricated seals include abrasive, flashing, flammable, and toxic liquids.

V. SEALLESS PUMPS

A. General

There are some pump applications where it is desirable or necessary to have zero leakage of product, even less than the minimal leakage possible from a well-chosen and well-maintained mechanical seal. The applications that come to mind where even limited amounts of leakage would not usually be tolerated include:

Sealing Systems and Sealless Pumps 353

- Toxic liquids (e.g., phosgene, cyanide)
- Hot liquids (e.g., heat transfer liquids)
- Carcinogens or environmental hazards
- Noxious, malodorous liquids
- Highly corrosive liquids
- Radioactive liquids

With the Clean Air Act and other federal and state environmental regulations beginning to be enforced more vigorously, the list of liquids that must be pumped with zero leakage continues to grow.

Another incentive for eliminating shaft sealing completely is that packing and mechanical seal servicing is probably the most frequent type of regular maintenance that must be done on most pumps, and thus is the cause of a high percentage of overall pump maintenance expense. Included in the cost of double and tandem seal alternatives are the barrier liquids and the auxiliary systems necessary to maintain them. Plus, some processes cannot tolerate the possibility of contamination by barrier liquid leaking into the pumped product from a double seal.

Previous chapters in this book described several pump types that by their design do not require a mechanical seal or packing assembly. Chapter 1 discussed two types of positive displacement pumps that are sealless: the peristaltic pump (Section VI.C.4) and the diaphragm pump (Section VI.C.13). Both pump types, however, have limitations and shortcomings, chief among them being their hydraulic coverage and the fact that they produce large pressure pulsations. Still, if an application that requires sealless pumping falls within the hydraulic range these pumps can handle, and if the other characteristics of these pump types that Chapter 1 described can be tolerated, they may be a simple and relatively inexpensive sealless pumping choice.

Chapter 4, Section VIII, discussed the vertical column pump, with one of its advantages being that it, too, is a rather simple sealless pump. The pump does have a shaft penetration at the top; but because the shaft column is relieved back to the sump in which the pump is immersed, the shaft column is not pressurized. This pump type also has limitations, including hydraulic limitations; sleeve bearings often lubricated by the pumped liquid; and the fact that to be sealless, the pump must be mounted over a tank or sump, a configuration not always possible. Still, the vertical column pump may represent a reasonable alternative for sealless pumping.

The following two sub-sections describe *magnetic drive pumps* and *canned motor pumps*, two major types of sealless pumps that are generating a great deal of interest and consideration, both on the part of pump manufacturers and pump users.

B. Magnetic Drive Pumps

Magnetic drive pumps were invented more than 40 years ago and have been available commercially for more than 25; but until the late 1980s, their availability was largely limited to small nonmetallic pumps. There were several European pump suppliers that offered heavier-duty metal process pumps earlier. By the late 1980s and early 1990s, driven largely by the increased regulatory environment related to emissions from pumps, virtually all U.S. suppliers of centrifugal process pumps, and a number of makers of positive displacement process pumps, had introduced magnetic drive technology.

Figure 5.17 illustrates a magnetic drive (hereafter called "mag drive") process pump. The principle behind mag drive technology is that there is no shaft penetration of the pressurized part of the pump casing. Instead, the pump rotor is driven by magnetic flux. In Figure 5.17, the coupling end of the pump is attached to an outer cylinder made of a magnetic material. The outer magnet is supported by the bearings shown between the coupling shaft and the outer magnet, and rotates with the motor. Smaller versions are close-coupled, eliminating the ball or roller bearing system shown in Figure 5.17.

There is a second magnetic cylinder located inside the pressurized part of the pump casing and secured to the pump impeller shaft. The outer and inner magnetic cylinders are separated from each other by a containment can or shell, constructed of a corrosion-resistant alloy such as hastelloy, having

Sealing Systems and Sealless Pumps

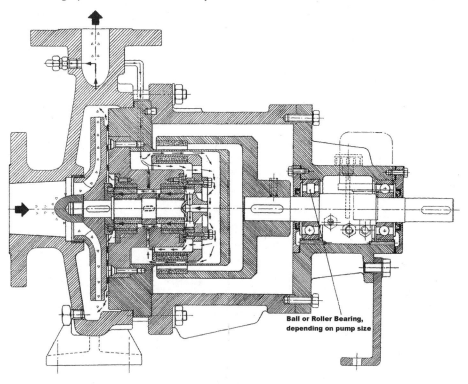

Figure 5.17 Magnetic drive process pump. (Courtesy of American Process Equipment, Inc./Caster Pumps, Bristol, WI.)

a thin wall to minimize loss of magnetic flux. The magnetic cylinder attached to the impeller shaft is able to rotate due to the magnetic flux between the inner and outer magnets. Thus, the pump impeller rotates, and liquid is pumped, without a shaft penetration from the pressurized pump casing.

For all practical purposes, magnetic drive couplings exhibit an "on" or "off" behavior; that is, both coupling halves are either running at the same speed or are disconnected (decoupled). The two halves are magnetically tied to each other through the sealed can until such time that the torque capability is exceeded. When the torque capability is exceeded, the pump stops quickly, while the motor and driving magnet continue to rotate at full speed. The two halves will reconnect when the motor is stopped or nearly at zero speed.

Generally, manufacturers rate magnet decouple as the maximum torque that the coupling will sustain as torque is raised gradually to the decouple point. Sudden pressure pulses or action from valves in the system may result in unexpected magnet decouple. High starting torque motors can exceed the decouple point upon start-up even though the pump has not reached rated pressure. These factors must be taken into consideration when sizing a magnet set. Also, the magnet decouple rating decreases with increasing temperature

Generally, it is not wise to depend on decouple to limit maximum pressure of the pump, for example, in place of a safety relief valve for a positive displacement magnetic drive pump. Decouple ratings are generally published as a minimum, and they can have a high tolerance range from magnet to magnet.

Advancements in magnet technology have helped move mag drive pumps into larger and larger pump applications, with current upper limits being in the range of 450 HP. The most common materials used for the magnets in mag drive pumps are ceramic, samarium cobalt, and neodymium iron boron. Each of these materials has different magnetic properties, cost, availability, and upper temperature limits.

Ceramic magnets (barium or strontium ferrite) are the most common magnet materials for smaller magnetic drive pumps. They are low cost, have excellent chemical resistance, moderate magnetic strength density, and high sustained temperature capability (around 572°F, or 300°C). Their downside is a considerable torque loss with increasing temperature (0.2 to 0.5% per degree C). They lose magnetism at a high rate as the temperature rises, but they tend to sustain higher temperatures than some of the rare earth magnets without permanent loss of magnetic strength. They are generally considered low-temperature magnets, but one would be wise to run the numbers before opting for the more expensive alternatives. If pressure requirements are low, a ceramic magnet can be a very good choice for high temperature.

Neodymium iron boron is a rare earth magnet material that is moderate in cost, especially in high production. These magnets have very poor chemical resistance (must be protected), very high magnetic strength density, and low torque

Sealing Systems and Sealless Pumps

loss with increasing temperature, until their upper temperature limit, which is about 250°F (121°C).

Samarium cobalt, another rare earth magnet, is relatively high in cost, has good chemical resistance, very high magnetic strength density, low torque loss with increasing temperature, and superior sustained high-temperature capability (above 482°F, or 250°C).

It is important to understand that mag drives are not a panacea. Despite the fact that they are taking the place of the high-maintenance mechanical seal, there are a number of technological limitations that currently exist with magnetic drives. The chief of these include the following:

1. Bearings in the Pumped Liquid

The pump impeller is still generating radial and thrust loads, just like any centrifugal pump. Unfortunately, these loads cannot be passed through the magnetic flux to bearings outside the pressurized pump casing. Mag drive pumps have sleeve-type radial bearings and plate or washer-type thrust bearings that are usually exposed to the liquid being pumped. The pump must have a circulation path to allow pumped liquid to lubricate these bearings (see Figure 5.17). Depending on the material of the bearings and on the nature of the liquid being pumped, the bearings are more or less subjected to abrasive wear and corrosion from the pumped product. The bearings are usually made of a hardened material such as silicon carbide or chrome oxide to resist abrasive wear and corrosion. Many mag drive pumps offer a screen built into the pump to attempt to remove some of the abrasives before the liquid from the pump is allowed to be introduced to the bearings. Most (but not all) manufacturers make their impellers for mag drive pumps in closed configuration, which, as discussed in Chapter 4, Section II.D, reduces axial thrust, compared with an open impeller. Other efforts are made to balance out radial and thrust loads as much as possible. Features such as external flush arrangements are available from some manufacturers.

Despite these design considerations, the sleeve and plate-type bearings used on a mag drive pump are no match for the ball and roller type bearings used to accommodate the radial and thrust loads of most centrifugal pumps. Also, the type of bearings used and the flow paths necessary to provide lubrication for these bearings make a mag drive pump more complex and include more wetted parts than most conventional centrifugal pumps.

2. Dry Running

In addition to providing lubrication to the pump bearings, the flush liquid circulating through the mag drive pump has a secondary function. A considerable amount of heat is generated due to the magnetic flux passing across the metal containment can, and this heat must be carried away by the liquid circulating through the pump. Most mag drive pumps do not tolerate running dry due to the heat given off from the magnetic flux, and a rapid failure occurs if flow into the pump is interrupted. Bearing temperature detectors or other instruments are sometimes incorporated into the pump to help indicate a problem in this regard, and shut down the pump if a loss of flow to the pump is indicated, before the pump fails.

3. Inefficiency

Because of the loss of magnetic flux as it passes across the metallic containment shell, mag drive pumps are considerably less energy efficient than their conventional sealed counterparts. This means that if a sealed pump in the field is being replaced with a mag drive pump, it is likely that a larger motor must be supplied, which may in turn require a larger base plate. An alternative design to significantly reduce these additional losses uses a plastic containment shell to avoid the flux losses that occur with the metallic containment shell. This design alternative has had only limited acceptance, however, due to misgivings on the part of industrial users about relying on a nonmetallic shell for pressure containment.

4. Temperature

As mentioned previously, magnet materials have upper temperature limits.

5. Viscosity

Because of the relatively small passageways for the recirculation of liquid through the pump to lubricate the bearings, most mag drive pumps have upper limits of viscosity that can be handled by the pump and still allow acceptable amounts of bearing flush circulation.

Despite the shortcomings discussed above, mag drive technology offers one solution to sealless pumping in process applications. The fact that so many makers of process pumps, including virtually all ANSI and many API pump makers, have committed so much effort and resources to developing mag drive offerings is a sign of the commitment of the pump industry to this technology. As magnet materials continue to improve, and as better ways to combat the limitations discussed above are developed, mag drive pumps will play a growing role in the process pump industry.

C. Canned Motor Pumps

Canned motor pump technology has been commercially available for more than 30 years, but is receiving a great deal of attention today as an alternative to mag drive technology for process pump applications in which sealless pumping is required. Figure 5.18 illustrates a canned motor pump. This pump technology relies on a specially designed motor, close-coupled to the pump, and hermetically sealed to prevent any leakage. The rotor portion of the motor is exposed to the pumped liquid for the same purpose as the circulating liquid in a mag drive pump, namely to lubricate the radial and thrust bearings. The stator of the motor is separated from the rotor by a metallic containment shell so that the motor stator remains dry.

Canned motor pumps have several limitations in common with mag drive pumps. The most important one is that the radial and thrust bearings of the canned motor pump must be wetted by the pumped liquid. Therefore, these bearings must be sleeve and plate type, hardened to keep from prematurely wearing from abrasive liquid. Manufacturers of canned motor

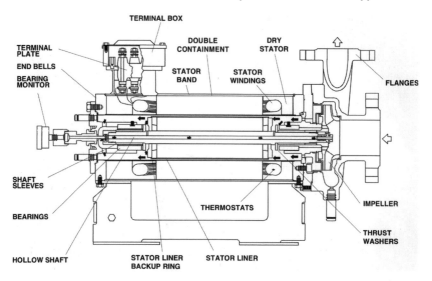

Figure 5.18 Canned motor pump. (Courtesy of Sundyne Corporation, Arvada, CO.)

pumps offer a variety of designs for circulation paths through the pump and motor to provide cooling and lubrication of these bearings. There are also upper limits of viscosity that are similar to those of mag drive pumps.

The proponents of canned motor pumps offer the following advantages compared to mag drive pumps.

1. Fewer Bearings

The mag drive pump must have a set of bearings in the pump motor, plus (except in the smallest sizes) a set of bearings supporting the rotating magnetic cylinder. The canned motor pump does not require these extra sets of bearings. While this argument is true, it must be pointed out that the extra bearings required by the mag drive pump are quite lightly loaded ball or roller bearings, not exposed to process liquids. Both pump types have a radial and thrust bearing system that is exposed to the process liquid.

2. More Compact

The close-coupled configuration of the canned motor pump takes up less space than the bedplate-mounted mag drive pump and motor. The proponents of mag drive technology would argue that this is a less structurally stable alternative than a bedplate mounted pump, because the bedplate on a mag drive pump can carry all loads down to the foundation.

3. Double Containment

The outer containment shell of a canned motor pump is constructed of pipe, creating, along with the inner containment shell between the motor rotor and stator, a double containment of the process liquid. Some, but not all, makers of mag drive pumps offer double containment. However, the outer containment shell is normally a cast part, with lower pressure-containing capability than the canned motor's outer containment pipe.

4. Lower First Cost

Based on their simplicity of construction, canned motor pumps should be less costly to manufacture than mag drive pumps in equal quantities. However, based on the commonality of some mag drive pump components with higher volume (such as ANSI pump) components, mag drive pumps may be less costly under some circumstances.

Offsetting the advantages of canned motor pumps is one fairly large concern. If a canned motor pump has a problem, the user must return to the manufacturer from whom the canned motor pump was purchased to get repairs or replacement components, because the motor is uniquely designed and manufactured. On the other hand, the motor for a mag drive pump is a standard motor, available locally to the user on very short notice from a number of manufacturers and suppliers. Thus, repair cost and time can be expected to be higher for a canned motor pump than for a mag drive pump. Recently, in areas of high pump density, factory-authorized repair facilities have been established by some manufacturers of canned motor pumps to address this shortcoming.

Another limitation of canned motor pumps is the upper horsepower limit of about 200 HP, vs. 450 HP for mag drive pumps.

6

Energy Conservation and Life-Cycle Costs

I. OVERVIEW

This chapter describes a number of methods for reducing the amount of energy consumed by centrifugal pumps. This includes methods of selecting the right pump in the first place so that efficiency can be optimized, as well as techniques for operating the pump to reduce energy consumption.

Maximizing pump efficiency by selecting the most efficient pump type and operating the pump with minimal energy consumption are important goals, particularly with larger horsepower pumps. For many pumps, small and large, the cost of the energy to operate them over their lifetimes is much more than their first cost or cumulative maintenance expense. The savings in energy that can be achieved by selecting the most efficient pump for a given service and operating it in the most efficient manner can represent a significant portion of total operating costs.

The importance of saving energy seems to ebb and flow in the minds of pump users, depending on the cost and availability of energy supplies. Many people have forgotten the oil crisis of the 1970s, when the cost of energy nearly quadrupled in a short time period. With inflation adjusted energy prices today being much lower than they were at their peak some years ago, some people may become complacent about energy, believing that relatively inexpensive supplies will always be available. This will most certainly not be the case.

This chapter includes a detailed discussion of variable-speed pumping, a technique being used more and more widely to reduce the energy consumed in pumping; and variable-speed drives, devices most commonly used to achieve variable-speed pumping.

In addition to lifetime energy costs, this chapter describes an analytical approach to the other life-cycle costs associated with pump ownership.

II. CHOOSING THE MOST EFFICIENT PUMP

It is possible to affect the efficiency, and thus the amount of energy consumed for a given flow and head, by the type of pump chosen for a particular application.

Figure 2.37 and its explanation in Chapter 2, Section XIII.C, indicated that it is possible to maximize the BEP of a pump chosen for a particular application by varying the number of stages or the speed of the pump chosen for the application (thus changing specific speed). However, choosing the most efficient pump does not necessarily make it the lowest first-cost choice, nor the choice with the lowest operating costs. There may be other considerations too, such as space constraints, ease of control, and noise. The following two examples illustrate the use of Figure 2.37 in examining how varying the number of stages or speed of the pump chosen can affect the efficiency. The examples show the type of analysis that might be performed to select the pump that minimizes total cost.

EXAMPLE 6.1: Selection of the number of stages to minimize total cost.

PROBLEM: A pump must be chosen to meet a particular rating. Several alternative pumps are analyzed, each with a different number of stages, to make the rating. Then the effect of the chosen pumps on energy, first cost, cost of maintenance, space, and ease of control are examined.

Energy Conservation and Life-Cycle Costs

Table 6.1 Expected Performance: One-, Two-, and Three-Stage Pumps

No. of Stages	N_s	Max Eff. (%)	BHP
1	655	68	111
2	1100	77	98
3	1495	80	95

Source: Figure 2.37.

GIVEN: Capacity = 500 gpm
Total head = 600 ft
Liquid = water at 60°F (s.g. = 1.0)
Pump speed = 3550 rpm

SOLUTION: Alternatives of single-stage, two-stage, and three-stage construction are considered. For each pump considered, the specific speed, N_s, is calculated, using Equation 2.22. For the single-stage pump, the value of H (head per stage) in the specific speed formula is 600, for the two-stage pump it is 300, and for the three-stage pump it is 200. Once specific speed is calculated, the expected efficiency at the BEP of the three alternatives can be found using Figure 2.37. Then, using the efficiency for each pump, the BHP for each pump can be computed, using Equation 2.15, with specific gravity of 1.0. Table 6.1 shows the results of these computations:

Note that the table shows three quite different pumps to make the same design rating point, all at the same speed of 3550, but with a varying number of stages.

The three-stage pump, being the most efficient, uses the least amount of energy to operate. Per Table 6.1, the three-stage pump requires 3 HP less than the two-stage pump, and 16 HP less than the single-stage pump. Using the technique described in Chapter 2, Section XII, with an assumed power cost of $0.10 per KW-HR and an assumed operating duty cycle of 6000 hours per year, the 16-HP difference between the three-stage and single-stage pumps is computed to be worth about $7160 per year.

Note that a four-stage pump has not been considered in this problem, because the amount of improved efficiency compared to a three-stage pump is negligible. That is because the specific speed for the three-stage pump places it near the flat part of the curve in Figure 2.37. Further increases in specific speed would not greatly increase the expected pump efficiency.

The problem statement suggests that other factors in addition to energy should be evaluated. These factors, and the likely best pump choice for each factor, are discussed below. It is clear from the discussion above and to follow that there is no best choice to optimize all the factors. Hopefully, however, this example provides some guidelines for making good engineering decisions as to the best choice for a given pump application.

- *First cost* — The single-stage pump would likely be the lowest first-cost alternative. Its configuration would be end suction, while the two-stage and three-stage alternatives would be radially or axially split multi-stage pumps, which would likely cost more. One factor that might lower the three-stage pump's first cost compared to the other two is the fact, per Table 6.1, that with a lower required BHP, it might be possible that the three-stage pump could use a 100-HP motor. (This assumes that the pump will not operate at any higher capacity than the design capacity, or that the motor service factor is used if the pump operates at higher capacities.) The single-stage pump, on the other hand, requires 111 HP to operate at the design capacity, so it would possibly need a larger motor than the three-stage pump. Still, this cost disadvantage of the single-stage pump is not likely to change its status of being the lowest first cost of the three alternatives.

- *Maintenance expense* — Experienced maintenance people would generally agree that the fewer the number of stages, the less the expected maintenance expense for a pump. Fewer impellers mean that it takes a shorter time to disassemble and reassemble the pump, that there are fewer required spare rotating parts (impellers, wear rings), and that there are fewer impellers to remachine or rebalance in the event work is done on them.

  An offsetting argument in favor of lower maintenance costs for the three-stage pump is the fact that because the three-stage pump makes the head with three impellers instead of one, each of the three impellers is quite smaller in diameter than the single-stage impeller. This means that the pump owner might be able to work on the smaller impellers in an in-house machine shop (e.g., trimming impellers, making wear rings, balancing impellers, etc.). The single-stage impeller, on the other hand, might have to be sent to an outside shop for the same operations, which could be more expensive. This argument, while having some merit, is not likely to offset the statement in the previous paragraph that the pump with the fewest number of stages is likely to have the lowest maintenance expense.

- *Space* — Space allowed for the pump may be an issue for a plant modification, for a skid-mounted assembly, in tight or confined spaces such as on ships or offshore platforms, and for machines or systems that include a pump as a part of the operation. If space is an issue, either the single-stage or the three-stage pump might be the best choice, depending on whether the space limitation is on the diameter or on the length. The single-stage pump would have the largest diameter but the shortest length, while the three-stage pump would be just the opposite, with the smallest diameter but the longest length.

- *Ease of Control* — The material in Chapter 2, Section VII, on specific speed indicates that the higher the pump specific speed, the steeper the resulting pump H–Q curve. And the discussion in Chapter 2, Section IX, on system head curves mentions the fact that a steeper H–Q curve is easier to control because it does not produce wide flow swings with variations of system head. Therefore, if a more constant flow is required without resorting to a flow control valve, then the higher specific speed pump (the three-stage pump) would be the best alternative.

EXAMPLE 6.2: Selection of pump speed to minimize total cost.

PROBLEM: A pump must be chosen to meet a particular rating. Two alternative single-stage pumps are analyzed, one at 1780 rpm and one at 3550 rpm, to make the rating. The likely effect of the chosen pumps on energy cost, first cost, maintenance expense, space, ease of control, and noise level are then discussed.

Energy Conservation and Life-Cycle Costs

Table 6.2 Expected Performance: 1780 and 3550 rpm Pumps

Speed	N_s	Max Eff. (%)	BHP
1780	825	72	31
3550	1650	80	28

Source: Figure 2.37

GIVEN: Capacity = 500 gpm
Total head = 175 ft
Liquid = water at 60°F (s.g. = 1.0)
Number of stages = 1

SOLUTION: For each pump considered, the specific speed, N_s, is calculated using Equation 2.22. Once the specific speed is calculated, the expected BEP efficiencies of the two alternatives can be found using Figure 2.37. Finally, using the efficiency for each pump, the BHP for each pump can be computed, using Equation 2.15, with specific gravity of 1.0. Table 6.2 lists the results of these computations:

The two alternative pumps considered in this example each make the desired rating point with one stage, but one of them does it while operating at 1780 rpm, the other while operating at 3550 rpm. They are two completely different pumps, each making the same rating point at a different running speed.

The two-pole (3550 rpm) pump results in the highest efficiency and the lowest energy cost. Does this mean that two-pole pumps should be used for all applications? Not necessarily, because per the problem statement, other factors in addition to energy costs must be considered. They are discussed below. Noise level concerns favor slower speed pumps, but noise is often not a concern for industrial pumps, especially in relatively small sizes such as this example. The following discussion shows that the only significant factor in favor of the slower-speed pump is maintenance expense. However, this factor is an extremely important one, and may in fact be the deciding one. If the pump owner has experience with two-pole pumps in this size, however, and is comfortable with the alignment and balance levels necessary for two-pole pumps, then the higher-speed pump is likely to be the choice representing the lowest total pump cost after factoring in all the variables.

- *First cost* — The two-pole pump would be the lowest first-cost machine because the pump required to make the rating at two-pole speed would be quite a bit smaller than the pump required to make the rating at four-pole speed. (This assumes that an end suction pump would be used in either case.) As an example, if ANSI pumps were used for this application, the two-pole pump selected would be a 3 × 4 - 8, while the four-pole selection would be a 3 × 4 - 13.

    The two-pole pump motor also would be less expensive because it too would have a physically smaller frame than the four-pole motor. For a given motor horsepower size, the higher the speed, the smaller the required frame size.

- *Maintenance Expense* — In general, the higher speed pump should cost more to maintain. There are several reasons for this. At higher speeds, pumps are less tolerant of misalignment and imbalance than they are at lower speeds. Also, at higher pump speeds, the seal's dynamic faces would see higher surface speeds, so the seal life on the higher-speed pump would likely be shorter. This is somewhat offset by the fact that the higher-speed pump would have a lower torque than the same horsepower pump at a lower speed, as per Equation 2.17. Therefore, the two-pole pump could get by with a smaller shaft diameter, and thus a smaller seal size. Another offsetting argument against the high-speed pump having the higher maintenance cost is the fact that the components of the high-speed pump are physically smaller than the same parts for the slower-speed pump. A replacement impeller or wear ring for the four-pole pump would cost more than the comparable part for the higher-speed pump.

Despite arguments to the contrary, most people experienced with pump operation and maintenance would agree that wear and tear on a two-pole pump is higher than that on a four-pole pump. Many users simply do not use two-pole pumps because of concerns about maintenance. On the other hand, users at other plants such as refineries and power plants are quite comfortable with pumps operating at this speed, and have good pump alignment and balancing techniques, so a pump at two-pole speed would not be a concern for them at all.

Note that the consideration of two-pole or four-pole speed assumes that the liquid being pumped is relatively clean. Applications involving paper stock or abrasive slurries are generally chosen at speeds slower than 1800 rpm in an effort to reduce excessive erosion in the high-velocity areas.

- *Space* — As covered in the discussion of first cost, the two-pole pump and motor would be smaller than the four-pole alternative.

- *Ease of control* — With the higher specific speed, the two-pole pump would have a steeper H–Q curve and thus would be easier to limit flow swings with system head variations.

- *Noise level* — Most of the noise from a centrifugal pump is caused by the motor. In general, for the same motor horsepower, the higher the speed, the higher the noise level. Thus, the two-pole pump would have a higher noise level than the four-pole pump. This may not be a problem for most industrial pumps, especially in the relatively small size of this example. It could, however, be a concern if the pumps were to be operated in an area near the general public or where noise levels need to be otherwise minimized.

The preceding examples focused on the tools that allow an engineer or pump user to select the most efficient pump for a given application, while being mindful that a selection based only on efficiency can also affect initial cost, maintenance, and operating parameters for the pump.

In the above examples, the pump head is a given value. It is the responsibility of the person who is sizing the pump and designing the system to determine the value of pump head. As demonstrated in Chapter 2, Section XII, oversizing the pump can lead to a considerable waste of energy.

Another point to consider in choosing equipment for minimal energy consumption is today's availability of premium-efficiency motors from many of the motor manufacturers. These motors are discussed in Chapter 4, Section XVI.I.

III. OPERATING WITH MINIMAL ENERGY

Once a particular pump has been chosen for an application, the best way to minimize energy consumption in the operation of the pump is to keep the pump operating as efficiently as possible. Make certain that the impeller settings for open

Energy Conservation and Life-Cycle Costs

impellers and the wear ring clearances for closed impellers are at their minimum recommended amounts considering the impeller size, materials of construction, any galling tendencies for the materials, temperature, and amount of abrasives present. (Recommended axial settings for open impellers and wear ring clearances for closed impellers are discussed in Chapter 4, Section II.A.) Impeller clearances should be reset as often as reasonable to achieve minimal leakage of liquid back to suction.

A centrifugal pump should not be operated at higher flow rates than required for the process. Remember, with most centrifugal pumps, the higher the flow rate, the higher the horsepower consumed.

Any throttling being done in the system (by valves or orifices) should be examined carefully, because this is a source of wasted energy. It may be that the pump has been oversized for the system requirements. An impeller trim or some other change can eliminate the unnecessary throttling in the system. (Refer to the discussion on the effects of oversizing pumps in Chapter 2, Section XII.)

If a pump must operate over a wide range of flow and head at different times, consideration should be given to multiple pumps operating in parallel or in series, as discussed in Chapter 2, Sections X and XI. Another possible solution if a multi-stage pump is being used might be to de-stage the pump during the time when head and flow requirements are lower. This is, for instance, commonly done with vertical turbine pumps in the mining industry. A third, more flexible alternative, variable-speed operation, is discussed in the following section.

IV. VARIABLE-SPEED PUMPING SYSTEMS

Variable-speed pumping has been around for many years, but its use has become more readily justified in recent years due to improvements in the technology for achieving variable-speed control of pumps and the reduction in the cost of such devices. Also, efforts by electric utilities to help their commercial and industrial customers reduce energy consumption through demand-side management have included incentives to incorporate variable speed into pump systems.

Variable speed is most easily justified when the pump must deliver a wide range of flow over time. Examples of pumping systems that demand a range of flow include process pumps in a variable-capacity process plant, municipal water and wastewater pumps, HVAC chilled water and cooling water pumps in commercial and institutional buildings, pipe line pumps, and power station pumps. In general, the wider the range of flow demand, the greater the likelihood that variable speed can be justified. However, an extremely wide range of flow is not necessarily required to justify the use of a variable-speed pumping system, since a 10% decrease in flow can reduce the power requirement by up to 21%.

In addition to applications that require variable flow, there are also a number of applications that require constant flow, but can still benefit from controlling the speed of the pump. The most common of these are applications that require a variable head pressure to deliver a constant flow. One example of this is a swimming pool filtration loop circulation pump. Another is a cooling tower filtration system that pumps water from the basin of a cooling tower, through a strainer, and back to the tower basin. If a constant-speed pump is used in either of these applications, the flow of filtered water will decrease as the filter loads. An adjustable-speed drive can increase the speed of the pump, making it possible to maintain a constant flow over a wide range of filter conditions. An additional benefit of this system is that variable-speed control can even indicate when the filter needs to be cleaned or replaced by signaling when the pump's speed is near maximum.

Primary hot or chilled water pumps and other pumps that normally provide constant flow against a constant head can also be candidates for variable-speed pumping. During the commissioning of the system, it is often necessary to reduce the capacity of the pump to achieve the proper temperature differential and flow. While this could be done by disassembling the pump, trimming its impeller, and reassembling the pump, this is not usually done. Instead, it is common to simply manually close a throttling valve to reduce the flow

Energy Conservation and Life-Cycle Costs

to the desired level. Doing this can introduce a significant, constant energy loss in the system. It can be economical to simply leave the valve fully open and manually adjust the speed of the pump to achieve the desired system performance. An additional advantage of using a variable-speed drive is that the flow can be easily readjusted in the future if the operating conditions of the system change. This would be more difficult to accomplish if flow were adjusted by trimming the impeller of the pump.

Variable-speed pumping saves energy by directly controlling the capacity of a pump by changing the pump's speed, rather than running the pump at full speed and using a valve to externally restrict or bypass excess flow. Using a variable-speed drive is much like driving a car down a highway and controlling its speed by adjusting the position of the gas pedal. By doing this, the output from the engine is directly controlled to meet the requirements of the system. In contrast, using a throttling value to control flow is much like driving a car down a highway by keeping the gas pedal floored and using the brakes to control the car's speed.

The energy savings that result from using a variable-speed drive are illustrated in Figures 6.1 through 6.6. When the speed of the pump is held constant, pump curve A in Figure 6.1 shows how this pump operates between flows 1 and 2. The higher flow rate is delivered by the pump when pump curve A intersects the system resistance curve X at point ax. The power associated with this flow is proportional to the flow times the pressure. A convenient way to visualize this is by observing the area of rectangle with a vertex at point ax.

To achieve the lower flow rate 1, a valve at the discharge of the pump can be closed. This increases the output pressure required by the pump and decreases the flow, as shown by system resistance curve Y and point ay in Figure 6.2. The power associated with this flow is illustrated by the area of the rectangle with a vertex at point ay. While this power is normally less than the power required for full flow (except for the case of an axial flow pump), there is a significant amount of waste. This wasted power is illustrated in Figure 6.3.

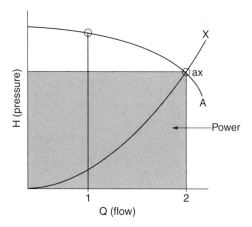

Figure 6.1 Power consumed in a pump is proportional to flow times pressure. (Courtesy of Danfoss Graham, Milwaukee, WI.)

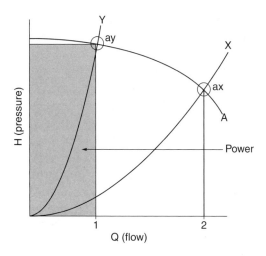

Figure 6.2 Using a throttling valve on a constant-speed pump effectively steepens the system resistance curve. (Courtesy of Danfoss Graham, Milwaukee, WI.)

The amount of power required to produce flow 1 on system resistance curve X is illustrated by the area of the bottom rectangle in Figure 6.3. The area of the top rectangle shows

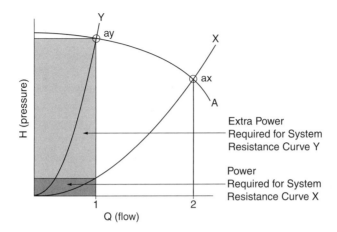

Figure 6.3 Illustration of power wasted by throttling. (Courtesy of Danfoss Graham, Milwaukee, WI.)

the additional power that is associated with creating the same flow while following system resistance curve Y. The additional power is required to overcome the pressure drop across the throttling valve that was added to the system.

Throttling the output of the pump can waste energy in two ways. First, if the pump was operating near its BEP at point *ax*, then at point *ay* it is operating at a reduced efficiency. Second, the pump is required to produce an *increased* head when it is producing the reduced flow. The system resistance curve shows that the system requires *reduced* pressure for reduced flow. The additional pressure drop is simply the pressure lost across the throttling valve. The power required to produce this pressure drop at this flow represents wasted energy.

Controlling the speed of the pump both helps to maintain high pump efficiency and to reduce energy consumption. When the speed of the pump is decreased, the flow is reduced by following the system resistance curve X to the lower flow in Figure 6.4. Instead of being forced to follow the full speed pump curve, adjusting the speed of the pump generates a new pump curve for each new speed. The system will now operate at point *bx* in Figure 6.4. Because no artificial pressure drops

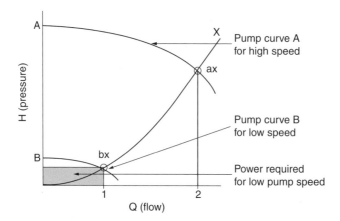

Figure 6.4 Using a slower pump speed to obtain a lower flow rate. (Courtesy of Danfoss Graham, Milwaukee, WI.)

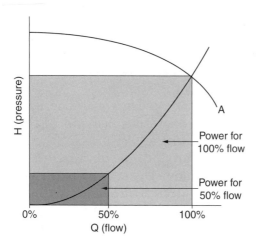

Figure 6.5 Illustration of power requirement for 100% and 50% flow in a system with no static head. (Courtesy of Danfoss Graham, Milwaukee, WI.)

are imposed on the system to reduce the flow, the energy loss is minimized. The area of the associated rectangle shows that the power associated with this flow is significantly less than the power that was required when a throttling valve was used to reduce flow.

While controlling the speed of the pump will always result in the optimum amount of energy saving, the actual energy savings that will result depends on the pressure requirements of the system, as indicated by the system resistance curve.

If no pressure is needed at zero flow (see Figure 6.5), the energy savings will be the greatest. In this case, the required pressure will proportional to the flow squared.

$$H \propto Q^2 \qquad (6.1)$$

As a result, because

$$HP \propto H \times Q \qquad (6.2)$$

Then,

$$HP \propto Q^3 \qquad (6.3)$$

This is the centrifugal pump affinity law, discussed in Chapter 2, which predicts that reducing the flow to 50% of maximum will reduce the power required to 12.5% of maximum. The analysis above only applies to systems whose pressure requirements approach zero when flow is reduced toward zero. In other situations, the system resistance curve looks like the one in Figure 6.6. Clearly, the power required to produce a given reduced flow with that system is greater than it would be for a system with no minimum pressure requirement. While controlling the speed of the pump will provide the greatest energy savings, the pressure requirement of the system will dictate the maximum energy savings potential.

Open pumping systems, such as lift station pumps and pressure booster pumps for potable water, often have a significant static pressure head requirement. Closed loop pumping systems, such as hot water and chilled water pumping loops in heating, ventilation, and air conditioning systems often have a minimum pressure set point requirement imposed by the control system to ensure proper operation under all flow requirements. This pressure set point acts like the static head in an open pumping system. To maximize energy savings, it is important to use the lowest system set point that provides acceptable system operation.

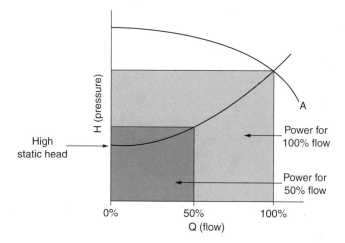

Figure 6.6 Illustration of power requirement for 100% and 50% flow in a system with some amount of static head. (Courtesy of Danfoss Graham, Milwaukee, WI.)

The reasons why the use of variable speed in a pumping system saves energy are further illustrated in Figure 6.7. The pump designated as curve A must operate between the flows marked as 1 and 2 in the figure. The higher flow rate delivered by the pump occurs where the pump H–Q curve intersects the system resistance curve X at point ax. Using a constant-speed pump, the lower flow rate is achieved by throttling the pump (i.e., closing a valve at the pump discharge) to steepen the system resistance curve until the new system resistance curve (curve Y) intersects the pump curve at the lower flow rate, at point ay in Figure 6.7.

Two things happen at the lower flow rate using a constant-speed pump. First, the pump is operating at a lower efficiency at this lower flow rate. (Refer to the efficiency curve A in Figure 6.7.) Second, the pump head at point ay (flow rate 1) is higher than the head at point ax (flow rate 2), because the constant-speed pump must operate on its characteristic head–capacity curve.

If instead of throttling the pump is slowed down, it would produce new head–capacity and efficiency curves, shown in

Energy Conservation and Life-Cycle Costs

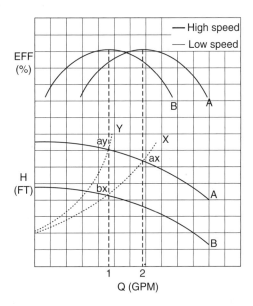

Figure 6.7 Pump H–Q and efficiency curves at two speeds, and system curve before and after throttling.

Figure 6.7 as curves B, which would follow the affinity laws. (Refer to discussion of the affinity laws in Chapter 2, Section VIII.) The intersection of this slower speed head–capacity curve (labeled curve B in Figure 6.7) with the original system head curve X, (point *bx* in Figure 6.7), represents the flow rate achievable by operating the pump at reduced speed without throttling. The difference in head between points *ay* and *bx* represents the head wasted across the control valve when the constant-speed pump is throttled instead of slowed down. Note that the efficiency curve for the pump at the lower speed shifts to the left. The pump still runs near its BEP when operating at the lower speed, provided the system curve is one similar to Figure 6.7. With a flatter system curve, the pump moves further to the left of BEP.

The use of variable speed instead of throttling a constant-speed pump saves energy in two ways in the above example. The pump operates near its best efficiency point, and no energy is wasted across the throttling valve. There are other benefits in addition to energy savings, as are detailed below.

Variable-speed pumping systems have evolved in their method of creating speed changes. Several pump driver types, such as steam turbines, hydraulic motors, air motors, and diesel engines, are inherently able to achieve a range of pump speeds. Even simpler mechanical devices such as gear boxes and belt drives can achieve incremental, although not variable, speed changes.

Magnetic slip has also been used to produce variable-speed pumping. Initially, this was done by altering the motor's slip by externally increasing the resistance of its rotor. While this allowed the speed of the motor's shaft to be adjusted, the energy lost through the external resistors and the maintenance of the brushes needed to connect these resistors to the motor's rotor were significant concerns. Eventually this system was replaced by external magnetic slip devices such as eddy current drives. Because these were coupled between the motor and the load, mechanical mounting was required. This required additional mounting space, and made it particularly difficult to retrofit a constant-speed system for variable speed. In addition, the absolute maximum theoretical efficiency of such a system is equal to the percentage of full output speed at which the pump is driven. So, when the pump was driven at 80% speed, over 20% of the energy delivered to the motor's shaft was wasted. This is a concern when energy conservation is the goal. The bearings required to support the shafts in the drive also require periodic maintenance.

The most efficient method of continuously adjusting the flow from a pump is to directly control the speed of the electric motor that drives the pump. Initially, it was only possible to efficiently control the speed of DC motors. This was relatively simple because the speed control only had to adjust the DC voltage that was applied to the motor while monitoring motor current. The drawbacks of such a system were the cost of the motor and the periodic maintenance that its brushes required.

When electronic power technology advanced sufficiently to allow the control of the speed of AC induction motors, the combination of an inexpensive, low-maintenance motor with a high-

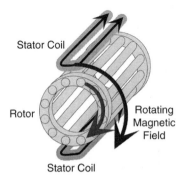

Figure 6.8 An AC induction motor. (Courtesy of Danfoss Graham, Milwaukee, WI.)

efficiency speed control produced an ideal package for controlling the flow from a pump. As a result, *AC adjustable frequency drives* or *variable frequency drives (VFDs)* have become the standard method for controlling the capacity of a pump.

Here is a brief overview of how a variable frequency drive operates. An AC induction motor's rotor (Figure 6.8) is driven by a rotating magnetic field that is produced by its stator coils. This is done by applying alternating current to the stator coils. As the alternating current in the stator coils changes, the induced magnetic field also changes. When three-phase electricity is applied to the coils, the resulting magnetic field rotates smoothly around the motor.

The frequency of the alternating current that is applied to the stator coils controls the rotational speed of this magnetic field, and thus the speed of the rotor. AC induction motors that are connected to the AC power line are single-speed devices because the frequency of the power mains is fixed. A variable-frequency drive allows the frequency of the power applied to the motor to be controlled. This allows the speed of the motor's shaft to be adjusted.

In addition to controlling the frequency of the power applied to the motor, a VFD must also control the voltage at the motor's terminals. This is because of the inductive reactance of the motor's coils. Inductive reactance, X_L, measures

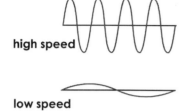

Figure 6.9 In a VFD, the voltage applied to the motor is controlled along with the frequency. (Courtesy of Danfoss Graham, Milwaukee, WI.)

the resistance that the coils offer to the flow of AC current. This varies with the frequency of alternating current applied to the coil according to the following equation:

$$X_L = 2\pi f L \tag{6.4}$$

From this formula it is clear that the inductive reactance, X_L, of a coil is low when the applied frequency, f, is low. As a result, if nothing else is changed, a motor would tend to draw more current as the frequency applied to it is reduced. To keep this from happening, voltage applied to the motor is controlled along with the frequency.

For full-speed operation, the drive applies rated frequency and voltage to the motor. For reduced-speed operation, the drive reduces both the frequency and the voltage of AC applied to the motor (Figure 6.9).

Some VFDs use a simple direct proportion to determine the motor's voltage. This is generally used when a constant torque is required at all operating speeds. This is often the case for positive displacement pump applications (Figure 6.10).

When a centrifugal pump is used, the torque that the motor must produce at slow speeds is generally quite low. Providing a constant ratio between motor voltage and frequency will produce a greater magnetic field in the motor than is needed to drive the pump at low speed. This extra magnetizing current produces extra heat in the motor and thus

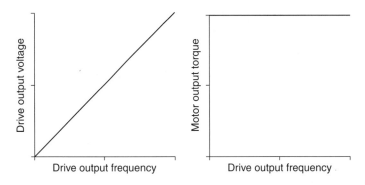

Figure 6.10 Drive output voltage–frequency pattern for constant motor torque. (Courtesy of Danfoss Graham, Milwaukee, WI.)

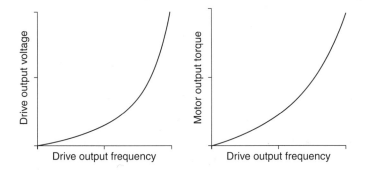

Figure 6.11 Drive output voltage–frequency pattern for reduced low-speed motor torque. (Courtesy of Danfoss Graham, Milwaukee, WI.)

reduces system efficiency. To maintain high efficiency throughout the speed range, some drives have a "variable torque" setting. This reduces the voltage applied to the motor to a greater extent at low speeds, minimizing the unnecessary motor current (Figure 6.11). Some advanced VFDs even offer an automatic energy optimizer feature. This automatically matches the drive's output voltage to the speed and torque requirements of the load.

While a number of methods are used to control the voltage and frequency produced by a VFD, there are some basic

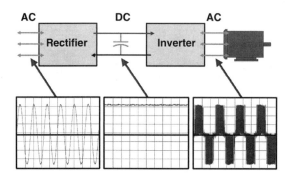

Figure 6.12 A modern adjustable frequency drive consists of three sections. (Courtesy of Danfoss Graham, Milwaukee, WI.)

similarities among all the common variable-frequency drive designs (see Figure 6.12).

First, the *rectifier* converts the incoming alternating current (AC) into direct current (DC). One purpose of this conversion step is to remove the line frequency from the power. In this way, the drive's output stage does not have to continually compensate for the variations in the incoming alternating current. The DC gives the output stage of the drive clean power to start from when it generates the output frequency that is needed for the desired output speed. It is interesting to note that the DC bus voltage that is produced is much closer to the peak voltage of the AC power line than it is to the average voltage that is generally used to describe the AC line voltage.

The middle section of a modern VFD is the *DC bus*. The main purpose of this is to filter out most of the residual "ripple" from the rectified AC line power, so that there is no significant interaction between the AC power line's frequency and the drive's output frequency.

This filtered DC is then fed to the drive's output section, which is called the *inverter*. Transistorized switches in the inverter direct the DC bus voltage to the appropriate motor lead. By alternately connecting each motor lead to the positive side of the DC bus and then to the negative side in the proper sequence, an appropriate AC output is applied to the motor's

Energy Conservation and Life-Cycle Costs   387

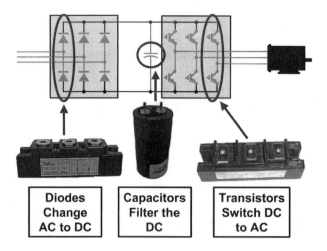

Figure 6.13 A pulse width modulation (PWM) variable-frequency drive. (Courtesy of Danfoss Graham, Milwaukee, WI.)

leads. The faster the polarity reversals take place, the higher the output frequency, and so the faster the motor's speed.

Electrically, a "low-voltage" drive is defined as one that operates on an AC line voltage of 600 volts rms or less. The vast majority of modern low-voltage VFDs are voltage source, *pulse-width modulation (PWM) drives*. The diagram in Figure 6.13 fills in some details for a basic PWM drive.

The rectifier section consists of a set of *diodes*. These simply act as "check valves," forcing the input current in the proper direction to produce DC. It is important to note that the DC bus voltage is not controlled; its value fundamentally depends on the incoming AC line voltage.

The DC bus consists of a bank of *capacitors*. The larger the size of the drive, the greater the number of capacitors. These drives are called "voltage source" because the capacitors act to filter the voltage of the DC bus, and attempt to maintain it at a constant level. This filtered DC bus is also used to provide control voltage to the rest of the drive. Because the DC bus capacitors tend to maintain their voltage throughout a short power loss, such drives have the ability to "ride through" brief line voltage sags with no significant interruption in operation.

The inverter section of a PWM variable-frequency drive consists of a set of *switching transistors.* Each motor lead is connected to at least two transistors. One transistor connects the positive side of the DC bus to the motor lead. The other transistor connects the negative side of the DC bus to the motor lead. A complete inverter consists of at least six transistors, one pair for each lead of the three-phase motor. By controlling the switching of the transistors, the frequency of polarity reversal of each motor lead can be set. This determines the frequency of the AC that is applied to the motor, and thus its speed.

In addition to controlling the polarity of the voltage that is applied to the motor, the inverter section of a PWM drive also controls the average voltage that is applied to the motor. This is done by sending pulses of voltage to the motor. When only a small amount of voltage is required by the motor, the pulse is turned on for a very short period of time. The average voltage of this narrow pulse is quite low. When a higher voltage must be applied to the motor, the pulse is turned on for a longer period of time. This wider pulse has a higher average voltage. Because this drive controls the average motor voltage by controlling the width of the pulses that are applied to the motor, such drives are called pulse width modulation, or PWM, drives. The frequency of the pulses applied to the drive is generally in the range of 2 to 20 kHz. This is called the drive's *switching frequency* or *carrier frequency.* The range of common switching frequencies is large because each carries its own advantages and disadvantages. The specific application generally dictates the ideal switching frequency.

In addition to controlling the voltage applied to the motor based on the output frequency of the drive, modern PWM drives also continually adjust the width of the pulses that are applied to the motor to simulate the smooth increase and decrease of voltage that a sine wave AC voltage would apply to the drive. Such "sine-coded PWM drives" simulate the AC voltage that a pure AC power source would apply to the drive, ensuring efficient motor operation. Some manufacturers of variable-frequency drives have developed their own proprietary voltage control algorithms to optimize motor performance and minimize motor heating.

It seems strange initially to apply pulses of voltage to the motor rather than use a transistor to "throttle" the voltage applied to the drive, producing a smooth, sine-wave output voltage. This is not done for two reasons.

First, it would be quite inefficient. Using a transistor to smoothly throttle back voltage while providing a significant amount of motor current would be at least as wasteful as using a throttling valve to control the flow in a pumping system. The only difference would be that the inefficiencies would be moved from the pumping system to the variable-frequency drive. Because the main purpose of using a VFD to control a pumping system is to maximize system efficiency, this would be quite unacceptable. By contrast, the energy loss associated with switching an inverter's transistors on and off at a high switching frequency is quite small.

Second, it is not necessary. The important concern for motor operation is the current that flows through its windings, because it is the current that produces the magnetic field that creates torque in the motor. The inductance of the motor's stator windings filters the motor's current, making it closely resemble a sine wave.

Insulated-gate bipolar transistors (IGBTs) are commonly used in the inverter section of modern PWM variable-frequency drives. These have become the modern standard for low-voltage drives because they are very reliable, have a high efficiency, and can produce the high output switching frequencies that are required.

Medium-voltage systems are defined as those with an AC line voltage greater than 600 V AC and less than 38 kV AC. It is difficult to obtain IGBTs that function well with the high voltages and currents associated with such applications. So, it is generally necessary to control these drives using *thyristors*, which are also known as *silicon controlled rectifiers (SCRs)*. These SCRs cannot be switched as quickly as IGBTs, so they are of little use in generating a PWM waveform to control voltage to the motor. Instead, a set of SCRs is used in the input rectifier to control the voltage and current that is applied to the DC bus. A large series inductor is then generally used to smooth the current in the DC bus. Because this inductor

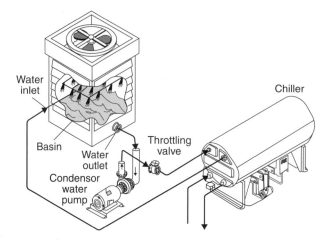

Figure 6.14 Traditional condenser pump system with throttle valve. (Courtesy of Danfoss A/S, Milwaukee, WI.)

acts to regulate the current passing through it, these drives are called *current-source* drives. Finally, a set of SCRs is used in the inverter section to switch the controlled DC bus to the motor. Such drives are more cumbersome than the PWM drives that are common for low-voltage drives. However, they are used in medium-voltage applications because appropriate IGBTs are not readily available.

The following example, courtesy of Danfoss A/S, Milwaukee, WI, illustrates the energy savings that can be achieved using a VFD in the condenser pump system shown in Figure 6.14 with a traditional throttling valve. For this system, a load profile, showing the percent of maximum flow required to satisfy the condenser loads during the various times of the day or days of the year, is prepared (see Figure 6.15). As an alternative to throttling the pump, a VFD controlled via a flowmeter is being considered (Figure 6.16). The system head curve and full-speed pump curve are shown in Figure 6.17. Three comparisons are presented for a 40HP/30KW condenser water pump with the system and load profile just described. The energy consumption during one year of operation is calculated for each. The comparisons are shown is Figure 6.18.

Energy Conservation and Life-Cycle Costs

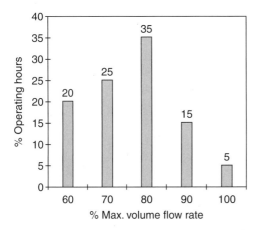

Figure 6.15 Load profile showing percent (%) operating hours and percent (%) of maximum flow rate. (Courtesy of Danfoss A/S, Milwaukee, WI.)

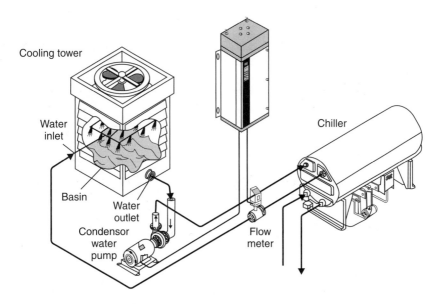

Figure 6.16 Condenser pump system with an adjustable frequency drive. (Courtesy of Danfoss A/S, Milwaukee, WI.)

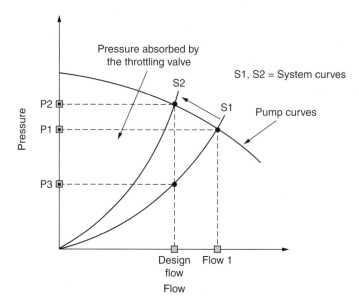

Figure 6.17 Throttling valve energy loss compared to variable-speed pumping. (Courtesy of Danfoss A/S, Milwaukee, WI.)

Configuration	% Flow	% Hours	Run hours	Power kW	Energy kWh
Discharge balancing valve, full speed pump	100	100	8,760	30	298,483
Adjustable frequency drive, reduced constant speed pump	100	100	8,760	25.22	264,514
Adjustable frequency drive at variable speed in closed loop control	60	20	1,752	5.45	12,613
	70	25	2,190	8.65	23,948
	80	35	3,066	12.91	48,542
	90	15	1,314	18.38	29,096
	100	5	438	25.22	1,326
Totals		100%	8,760		127,424

Figure 6.18 Energy savings example. (Courtesy of Danfoss A/S, Milwaukee, WI.)

In the first calculation, a 15% over-headed pump uses a discharge balancing valve to adjust the pump flow to the required system design flow. The pump operates at full speed,

100% of the time, at the P2 pressure and design flow, as shown in Figure 6.17.

In the second situation, the balancing valve is removed and the pump is operated by a variable-frequency drive at a constant reduced speed, adjusted manually (or by a flowmeter as in Figure 6.16) to achieve the required system 100% design flow all of the time. This results in pump operation at the intersection of P3 and the design flow, as shown in Figure 6.17. The result is an annual energy savings of 33,969 kW-hr with the adjustable frequency drive compared to the balancing valve, as shown in Figure 6.18.

In the last comparison, a variable-frequency drive operates in a closed loop control based on the system load, by controlling temperature in the cooling tower basin. The system load profile is shown in Figure 6.15. A minimum drive output to maintain at least 60% flow has been applied. The resulting variable-speed operation saves 171,059 kW-hr annually compared to the balancing valve. Energy savings is better than 50%, even maintaining a 60% minimum speed.

In addition to the energy savings associated with using variable-speed pumping, there are other important benefits related to the health of the pump. A pump operating at reduced speed to achieve a lower flow rate produces less head and operates closer to its BEP than the alternative of throttling a constant-speed pump. The slower speed pump produces lower radial and axial bearing loads, which should give longer bearing life and less chance of bearing or seal failure. Seal face wear should be less on the slower-speed pump. The pump operating closer to its BEP should produce less internal recirculation, which can cause erosive damage to the pump. Further, the pump is less likely to experience that instability sometimes associated with operating too far to the left of BEP.

Variable-speed pumping permits the continued ability to "tweak" a system so that the pump is performing optimally, despite changes in either the system or the pump. If head requirements change over time, for example, due to build-up of corrosion products in a pipe line, the pump can simply be speeded up to account for the higher head required. Or, as the pump impeller wears and clearances open up, the speed

Figure 6.19 Variable-frequency drive. (Courtesy of Danfoss Graham, Milwaukee, WI.)

can be adjusted upward to keep the pump from delivering less flow. The fudge factors often used in pump head calculations (and which may result in the oversizing of the pump as described in Chapter 2, Section XII) are essentially eliminated if variable-speed pumping is used.

The final advantage of variable speed pumping has to do with the possible reduction of electric power costs due to reduced-speed starting of the pump. For some high-energy pumps, the high amp draw when the motor is started at full speed can actually affect the overall rate that the electric utility charges the user. By starting the pump at a slower speed, the cost per unit of electric power may be reduced. This concept of using a VFD or other variable-speed device to *soft start* the pump is sometimes justification alone for a variable-speed system.

In summary, variable-speed pumping has been shown to be a highly effective way to reduce total pumping costs for systems that require a wide range of pump flow. It is used as an alternative to throttling of a single pump, or the use of multiple pumps in the system. Use of variable-frequency drives (Figure 6.19) is considered the most likely alternative to achieve meaningful savings. The advantages of using VFDs to achieve variable-speed pumping include:

Energy Conservation and Life-Cycle Costs

- Energy savings
- Ability to sometimes retrofit existing equipment without buying new motors
- Wide speed change achievable without serious energy loss consequences
- Ability to fine-tune speed in response to changes in the system or pump
- Lower bearing loads and overall better health of the pump
- Ability to start the pump at reduced speed

Surveys of pump users and engineers who design pumping systems using variable speed indicate that there are circumstances when applications for pumps as small as 10 to 20 HP can justify the addition of a variable-speed device.

The software introduced in Chapter 3, Section III, and demonstrated in the CD accompanying this book is an excellent tool for evaluating annual power savings achievable by using variable-speed pumping, and for comparing this option to alternatives of fixed speed with throttled flow or the use of multiple pumps in the system.

V. PUMP LIFE-CYCLE COSTS

This section is printed courtesy of, and with joint permission of the Hydraulic Institute, Parsippany, NJ; Europump, Brussels, Belgium; and the U.S. Department of Energy's Office of Industrial Technologies, Washington, D.C. The material is an executive summary of the 194-page book, *Pump Life Cycle Costs: A Guide to LCC Analysis for Pumping Systems*, published by the Hydraulic Institute and Europump. For more information on this publication, contact the Hydraulic Institute at 973-267-9700, or visit: www.pumps.org or Europump at 32 2 706 82 30, or visit: www.europump.org.

A. Improving Pump System Performance: An Overlooked Opportunity?

Pumping systems account for nearly 20% of the world's electrical energy demand and range from 25 to 50% of the energy usage in certain industrial plant operations (Figure 6.20).

Figure 6.20 In some industrial plant operations, pumping systems account for 25 to 50% of energy use. (Courtesy of, and with joint permission of the Hydraulic Institute, Parsippany, NJ; Europump, Brussels, Belgium; and the U.S. Department of Energy Office of Industrial Technologies, Washington, D.C.)

Pumping systems are widespread; they provide domestic services, commercial and agricultural services, municipal water/wastewater services, and industrial services for food processing, chemical, petrochemical, pharmaceutical, and mechanical industries. Although pumps are typically purchased as individual components, they provide a service only when operating as part of a system. The energy and materials used by a system depend on the design of the pump, the design of the installation, and the way the system is operated. These factors are interdependent. What's more, they must be carefully matched to each other, and remain so throughout their working lives to ensure the lowest energy and maintenance costs, equipment life, and other benefits. The initial purchase price is a small part of the life-cycle cost for high usage pumps.

Energy Conservation and Life-Cycle Costs

While operating requirements may sometimes override energy cost considerations, an optimum solution is still possible.

A greater understanding of all the components that make up the total cost of ownership will provide an opportunity to dramatically reduce energy, operational, and maintenance costs. Reducing energy consumption and waste also has important environmental benefits.

Life-Cycle Cost (LCC) analysis is a management tool that can help companies minimize waste and maximize energy efficiency for many types of systems, including pumping systems. This overview provides highlights from *Pump Life Cycle Costs: A Guide to LCC Analysis for Pumping Systems*, developed by the Hydraulic Institute and Europump to assist plant owners/operators in applying the LCC methodology to pumping systems. For information on obtaining a copy of the *Guide*, see Section V.J.

B. What Is Life-Cycle Cost?

The life cycle cost (LCC) of any piece of equipment is the total "lifetime" cost to purchase, install, operate, maintain, and dispose of that equipment. Determining LCC involves following a methodology to identify and quantify all of the components of the LCC equation.

When used as a comparison tool between possible design or overhaul alternatives, the LCC process will show the most cost-effective solution within the limits of the available data.

The components of a life cycle cost analysis typically include initial costs, installation and commissioning costs, energy costs, operation costs, maintenance and repair costs, downtime costs, environmental costs, and decommissioning and disposal costs. (See Figure 6.21.)

C. Why Should Organizations Care about Life-Cycle Cost?

Many organizations only consider the initial purchase and installation cost of a system. It is in the fundamental interest of the plant designer or manager to evaluate the LCC of different solutions before installing major new equipment or

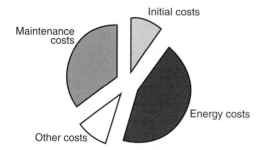

Figure 6.21 Typical life-cycle costs for a medium-sized industrial plant. (Courtesy of, and with joint permission of the Hydraulic institute, Parsippany, NJ; Europump, Brussels, Belgium; and the U.S. Department of Energy Office of Industrial Technologies, Washington, D.C.)

carrying out a major overhaul. This evaluation will identify the most financially attractive alternative. As national and global markets continue to become more competitive, organizations must continually seek cost savings that will improve the profitability of their operations. Plant equipment operations are receiving particular attention as a source of cost savings, especially minimizing energy consumption and plant downtime.

Existing systems provide a greater opportunity for savings through the use of LCC methods than do new systems for two reasons. First, there are at least 20 times as many pump systems in the installed base as are built each year; and second, many of the existing systems have pumps or controls that are not optimized since the pumping tasks change over time.

Some studies have shown that 30 to 50% of the energy consumed by pump systems could be saved through equipment or control system changes.

In addition to the economic reasons for using LCC, many organizations are becoming increasingly aware of the environmental impact of their businesses, and are considering energy efficiency as one way to reduce emissions and preserve natural resources.

D. Getting Started

LCC analysis, either for new facilities or renovations, requires the evaluation of alternative systems. For a majority of facilities, the lifetime energy and/or maintenance costs will dominate the life cycle costs. It is therefore important to accurately determine the current cost of energy, the expected annual energy price escalation for the estimated life, along with the expected maintenance labor and material costs. Other elements, such as the lifetime costs of downtime, decommissioning, and environmental protection, can often be estimated based on historical data for the facility. Depending upon the process, downtime costs can be more significant than the energy or maintenance elements of the equation. Careful consideration should therefore be given to productivity losses due to downtime.

This overview provides an introduction to the life cycle costing process. The complete *Guide* expands upon life cycle costing and provides substantial technical guidance on designing new pumping systems as well as assessing improvements to existing systems. The *Guide* also includes a sample chart, examples of manual calculation of LCC, and a software tool to assist in LCC calculation.

E. Life-Cycle Cost Analysis

In applying the evaluation process, or in selecting pumps and other equipment, the best information concerning the output and operation of the plant must be established. The process itself is mathematically sound, but if incorrect or imprecise information is used, then an incorrect or imprecise assessment will result. The LCC process is a way to predict the most cost-effective solution; it does not guarantee a particular result, but allows the plant designer or manager to make a reasonable comparison between alternate solutions within the limits of the available data.

Pumping systems often have a lifespan of 15 to 20 years. Some cost elements will be incurred at the outset, and others may be incurred at different times throughout the lives of the different solutions being evaluated. It is therefore practicable, and possibly essential, to calculate a present or discounted value of the LCC to accurately assess the different solutions.

This analysis is concerned with assessments where details of the system design are being reviewed. Here the comparison is between one pump type and another, or one control means and another. The exercise may be aimed at determining what scope could be justified for a monitoring or control scheme, or for different process control means to be provided. Whatever the specifics, the designs should be compared on a like-for-like basis. To make a fair comparison, the plant designer/manager might need to consider the measure used. For example, the same process output volume should be considered and, if the two items being examined cannot give the same output volume, it may be appropriate to express the figures in cost per unit of output (e.g., $/ton or Euro/kg). The analysis should consider all significant differences between the solutions being evaluated.

Finally, the plant designer or manager might need to consider maintenance or servicing costs, particularly where these are to be subcontracted, or spare parts are to be provided with the initial supply of the equipment for emergency stand-by provision. Whatever is considered must be on a strictly comparable basis. If the plant designer or manager decides to subcontract or carry strategic spares based entirely on the grounds of convenience, this criterion must be used for all systems being assessed. But, if it is the result of maintenance that can be carried out only by a specialist subcontractor, then its cost will correctly appear against the evaluation of that system.

Elements of the LCC equation are as follows:

$$LCC = C_{ic} + C_{in} + C_e + C_o + C_m + C_s + C_{env} + C_d \quad (6.5)$$

LCC = life cycle cost
C_{ic} = initial costs, purchase price (pump, system, pipe, auxiliary services)
C_{in} = installation and commissioning cost (including training)

Energy Conservation and Life-Cycle Costs

C_e = energy costs (predicted cost for system operation, including pump driver, controls, and any auxiliary services)

C_o = operation costs (labor cost of normal system supervision)

C_m = maintenance and repair costs (routine and predicted repairs)

C_s = downtime costs (loss of production)

C_{env} = environmental costs (contamination from pumped liquid and auxiliary equipment)

C_d = decommissioning/disposal costs (including restoration of the local environment and disposal of auxiliary services).

The following sections examine each element and offer suggestions on how a realistic value can be determined for use in computing the LCC. It should be noted that this calculation does not include the raw materials consumed by the plant in making a product.

1. C_{ic} — Initial Investment Costs

The pump plant designer or manager must decide the outline design of the pumping system. The smaller the pipe and fitting diameters, the lower the cost of acquiring and installing them. However, the smaller diameter installation requires a more powerful pump resulting in higher initial and operating costs. In addition, smaller pipe sizes on the inlet side of a pump will reduce the net positive suction head available (NPSH$_a$), thus requiring a larger and slower speed pump, which will typically be more expensive. Provisions must be made for the acceleration head needed for positive displacement pumps or the depth of submergence needed for a wet pit pump.

There will be other choices, which may be made during the design stage, that can affect initial investment costs. One important choice is the quality of the equipment being selected. There may be an option regarding materials having differing wear rates, heavier duty bearings or seals, or more extensive control packages, all increasing the working life of the pump. These and other choices may incur higher initial costs but reduce LCC costs.

The initial costs will also usually include the following items:

- Engineering (e.g., design and drawings, regulatory issues)
- Bid process
- Purchase order administration
- Testing and inspection
- Inventory of spare parts
- Training
- Auxiliary equipment for cooling and sealing water

2. C_{in} — Installation and Commissioning (Start-up) Costs

Installation and commissioning costs include the following:

- Foundations — design, preparation, concrete and reinforcing, etc.
- Setting and grouting of equipment on foundation
- Connection of process piping
- Connection of electrical wiring and instrumentation
- Connection of auxiliary systems and other utilities
- Provisions for flushing or "water runs"
- Performance evaluation at start-up

Installation can be accomplished by an equipment supplier, contractor, or by user personnel. This decision depends on several factors, including the skills, tools, and equipment required to complete the installation; contractual procurement requirements; work rules governing the installation site; and the availability of competent installation personnel. Plant or contractor personnel should coordinate site supervision with the supplier. Care should be taken to follow installation instructions carefully. A complete installation includes transfer of equipment operation and maintenance requirements via training of personnel responsible for system operation.

Commissioning requires close attention to the equipment manufacturer's instruction for initial start-up and operation. A checklist should be used to ensure that equipment and the system are operating within specified parameters. A final sign-off typically occurs after successful operation is demonstrated.

3. C_e — Energy Costs

Energy consumption is often one of the larger cost elements and may dominate the LCC, especially if pumps run more than 2000 hours per year. Energy consumption is calculated by gathering data on the pattern of the system output. If output is steady, or essentially so, the calculation is simple. If output varies over time, then a time-based usage pattern needs to be established.

The input power calculation formula is:

$$P = (Q \times H \times \text{s.g.})/(366 \times \eta_p \times \eta_m) \text{ [kW] (metric)} \quad (6.6)$$

$$P = (Q \times H \times \text{s.g.})/(3960 \times \eta_p \times \eta_m) \text{ [hp] (U.S. units)} (6.7)$$

P   = power
Q   = rate of flow, m³/hr (U.S. gpm)
H   = head, m (ft)
η_p = pump efficiency
η_m = motor efficiency
s.g. = specific gravity

The plant designer or manager needs to obtain separate data showing the performance of each pump/system being considered over the output range. Performance can be measured in terms of the overall efficiencies of the pump unit or of the energies used by the system at the different output levels. Driver selection and application will affect energy consumption. For example, much more electricity is required to drive a pump with an air motor than with an electric motor. In addition, some energy use may not be output dependent. For example, a control system sensing output changes may itself generate a constant energy load, whereas a variable-speed electric motor drive may consume different levels of energy at different operating settings. The use of a throttling valve, pressure relief, or flow by-pass for control will reduce the operating efficiency and increase the energy consumed.

The efficiency or levels of energy used should be plotted on the same time base as the usage values to show their relationship to the usage pattern. The area under the curve then represents the total energy absorbed by the system being reviewed over the selected operating cycle. The result will be in kWh (kilowatt-hours). If there are differential power costs at different levels of load, then the areas must be totaled within these levels.

Once the charge rates are determined for the energy supplied, they can be applied to the total kWh for each charge band (rate period). The total cost of the energy absorbed can then be found for each system under review and brought to a common time period.

Finally, the energy and material consumption costs of auxiliary services need to be included. These costs may come from cooling or heating circuits, from liquid flush lines, or liquid/gas barrier arrangements. For example, the cost of running a cooling circuit using water will need to include the following items: cost of the water, booster pump service, filtration, circulation, and heat extraction/dissipation.

4. C_o — Operation Costs

Operation costs are labor costs related to the operation of a pumping system. These vary widely, depending on the complexity and duty of the system. For example, a hazardous duty pump may require daily checks for hazardous emissions, operational reliability, and performance within accepted parameters. On the other hand, a fully automated nonhazardous system may require very limited supervision. Regular observation of how a pumping system is functioning can alert operators to potential losses in system performance. Performance indicators include changes in vibration, shock pulse signature, temperature, noise, power consumption, flow rates, and pressure.

5. C_m — Maintenance and Repair Costs

Obtaining optimum working life from a pump (Figure 6.22) requires regular and efficient servicing. The manufacturer

Figure 6.22 Maintenance and repair is a significant component of pumping system life-cycle costs, and an effective maintenance program can minimize these costs. (Courtesy of, and with joint permission of the Hydraulic Institute, Parsippany, NJ; Europump, Brussels, Belgium; and the U.S. Department of Energy Office of Industrial Technologies, Washington, D.C.)

will advise the user about the frequency and the extent of this routine maintenance. Its cost depends on the time and frequency of service and the cost of materials. The design can influence these costs through the materials of construction, components chosen, and the ease of access to the parts to be serviced.

The maintenance program can be comprised of less frequent but more major attention as well as the more frequent but simpler servicing. The major activities often require removing the pump to a workshop. During the time the unit is unavailable to the process plant, there can be loss of product or a cost from a temporary replacement. These costs can be minimized by programming major maintenance during annual shutdown or process change-over. Major service can be described as "pump unit not repairable in place," while the routine work is described as "pump unit repairable in place."

The total cost of routine maintenance is found by multiplying the costs per event by the number of events expected during the life cycle of the pump.

Although unexpected failures cannot be predicted precisely, they can be estimated statistically by calculating mean time between failures (MTBF). MTBF can be estimated for components and then combined to give a value for the complete machine.

It might be sufficient to simply consider best and worst case scenarios where the shortest likely life and the longest likely lifetimes are considered. In many cases, plant historical data is available.

The manufacturer can define and provide MTBF of the items whose failure will prevent the pump unit from operating or will reduce its life expectancy below the design target. These values can be derived from past experience or from theoretical analyses. The items can be expected to include seals, bearings, impeller/valve/port wear, coupling wear, motor features, and other special items that make up the complete system. The MTBF values can be compared with the design working life of the unit and the number of failure events calculated.

It must be recognized that process variations and user practices will almost certainly have a major impact upon the MTBF of a plant and the pumps incorporated in it. Whenever available, historical data is preferable to theoretical data from the equipment supplier. The cost of each event and the total costs of these unexpected failures can be estimated in the same way that routine maintenance costs are calculated.

6. C_s — Downtime and Loss of Production Costs

The cost of unexpected downtime and lost production is a very significant item in the total LCC and can rival the energy costs and replacement parts costs in its impact. Despite the design or target life of a pump and its components, there will be occasions when an unexpected failure occurs. In those cases where the cost of lost production is unacceptably high, a spare pump may be installed in parallel to reduce the risk. If a spare pump is used, the initial cost will be greater but the cost of unscheduled maintenance will include only the cost of the repair.

The cost of lost production is dependent on downtime and differs from case to case.

7. C_{env} — Environmental Costs, Including Disposal of Parts and Contamination from Pumped Liquid

The cost of contaminant disposal during the lifetime of the pumping system varies significantly, depending on the nature of the pumped product. Certain choices can significantly reduce the amount of contamination, but usually at an increased investment cost. Examples of environmental contamination can include cooling water and packing box leakage disposal, hazardous pumped product flare-off, used lubricant disposal, and contaminated used parts such as seals. Costs for environmental inspection should also be included.

8. C_d — Decommissioning/Disposal Costs, Including Restoration of the Local Environment

In the vast majority of cases, the cost of disposing of a pumping system will vary little with different designs. This is certainly true for nonhazardous liquids and, in most cases, for hazardous liquids also. Toxic, radioactive, or other hazardous liquids will have legally imposed protection requirements, which will be largely the same for all system designs. A difference may occur when one system has the disposal arrangements as part of its operating arrangements (for example, a hygienic pump designed for cleaning in place) while another does not (for example, a hygienic pump designed for removal before cleaning). Similar arguments can be applied to the costs of restoring the local environment. When disposal is very expensive, the LCC becomes much more sensitive to the useful life of the equipment.

F. Total Life-Cycle Costs

The costs estimated for the various elements making up the total life-cycle costs need to be aggregated to allow a comparison of the designs being considered. This is best done by means of a tabulation that identifies each item and asks for a value to be inserted. Where no value is entered, an explanatory comment should be added. The estimated costs can then be totaled to give the LCC values for comparison, and attention will also be drawn to nonqualitative evaluation factors.

There are also financial factors to take into consideration in developing the LCC. These include:

- Present energy prices
- Expected annual energy price increase (inflation) during the pumping system life time
- Discount rate
- Interest rate
- Expected equipment life (calculation period)

In addition, the user must decide which costs to include, such as maintenance, downtime, environmental, disposal, and other important costs.

For the calculation of the present worth of a single cost element, refer to Table 6.3 for present worth factor C_p/C_n, where C_p is the present cost of a single cost element and C_n is the cost paid after "n" years. For the calculation of the present worth of constant yearly expenditures, refer to Table 6.4 for the discount factor that applies for a particular real interest rate and number of years.

G. Pumping System Design

Proper pumping system design is the most important single element in minimizing the LCC. All pumping systems are comprised of a pump, a driver, pipe installation, and operating controls, and each of these elements is considered individually. Proper design considers the interaction between the pump and the rest of the system and the calculation of the operating

Table 6.3 Factor Cp/Cn for a Single Cost Element after n Years

No. of Years (n)	Real Discount Rate (interest rate minus inflation rate, in percent)												
	-2	-1	0	1	2	3	4	5	6	7	8	9	10
1	1.02	1.01	1	0.99	0.98	0.97	0.96	0.95	0.94	0.93	0.93	0.92	0.91
2	1.04	1.02	1	0.98	0.96	0.94	0.92	0.91	0.89	0.87	0.85	0.84	0.82
3	1.06	1.03	1	0.97	0.94	0.92	0.89	0.86	0.84	0.81	0.79	0.77	0.74
4	1.08	1.04	1	0.96	0.93	0.89	0.86	0.82	0.79	0.76	0.73	0.70	0.68
5	1.10	1.05	1	0.95	0.91	0.86	0.82	0.78	0.75	0.71	0.68	0.65	0.61
6	1.12	1.06	1	0.94	0.89	0.84	0.79	0.75	0.71	0.67	0.63	0.59	0.56
7	1.15	1.07	1	0.94	0.87	0.82	0.76	0.71	0.67	0.62	0.58	0.54	0.51
8	1.17	1.08	1	0.93	0.86	0.79	0.74	0.68	0.63	0.58	0.54	0.50	0.46
9	1.19	1.09	1	0.92	0.84	0.77	0.71	0.65	0.60	0.55	0.50	0.46	0.42
10	1.21	1.10	1	0.91	0.83	0.75	0.68	0.62	0.56	0.51	0.47	0.42	0.39
15	1.32	1.15	1	0.87	0.76	0.66	0.57	0.50	0.43	0.38	0.33	0.28	0.25
20	1.44	1.20	1	0.83	0.69	0.58	0.48	0.40	0.34	0.28	0.23	0.19	0.16
25	1.56	1.25	1	0.80	0.64	0.51	0.41	0.33	0.26	0.21	0.17	0.13	0.11
30	1.69	1.30	1	0.77	0.59	0.46	0.35	0.27	0.21	0.16	0.12	0.09	0.07

Courtesy of Hydraulic Institute, Parsippany, NJ, www.pumps.org; and Europump, Brussels, Belgium, www.europump.org.

Table 6.4 Discount Factor (df) for Constant Yearly Expenditures

No. of Years (n)	Real Discount Rate (interest rate minus inflation rate, in percent)												
	−2	−1	0	1	2	3	4	5	6	7	8	9	10
1	1.02	1.01	1.00	0.99	0.98	0.97	0.96	0.95	0.94	0.93	0.93	0.92	0.91
2	2.06	2.03	2.00	1.97	1.94	1.91	1.89	1.86	1.83	1.81	1.78	1.76	1.74
3	3.12	3.06	3.00	2.94	2.88	2.83	2.78	2.72	2.67	2.62	2.58	2.53	2.49
4	4.21	4.10	4.00	3.90	3.81	3.72	3.63	3.55	3.47	3.39	3.31	3.24	3.17
5	5.31	5.15	5.00	4.85	4.71	4.58	4.45	4.33	4.21	4.10	3.99	3.89	3.79
6	6.44	6.22	6.00	5.80	5.60	5.42	5.24	5.08	4.92	4.77	4.62	4.49	4.36
7	7.60	7.29	7.00	6.73	6.47	6.23	6.00	5.79	5.58	5.39	5.21	5.03	4.87
8	8.77	8.37	8.00	7.65	7.33	7.02	6.73	6.46	6.21	5.97	5.75	5.53	5.33
9	9.97	9.47	9.00	8.57	8.16	7.79	7.44	7.11	6.80	6.52	6.25	6.00	5.76
10	11.19	10.57	10.00	9.47	8.98	8.53	8.11	7.72	7.36	7.02	6.71	6.42	6.14
15	17.20	16.27	15.00	13.87	12.85	11.94	11.12	10.38	9.71	9.11	8.56	8.06	7.61
20	24.89	22.26	20.00	18.05	16.35	14.88	13.59	12.46	11.47	10.59	9.82	9.13	8.51
25	32.85	28.56	25.00	22.02	19.52	17.41	15.62	14.09	12.78	11.65	10.67	9.82	9.08
30	41.66	35.19	30.00	25.81	22.40	19.60	17.29	15.37	13.76	12.41	11.26	10.27	9.43

Courtesy of Hydraulic Institute, Parsippany, NJ, www.pumps.org; and Europump, Brussels, Belgium, www.europump.org.

duty point(s). The characteristics of the piping system must be calculated to determine required pump performance. This applies to both simple systems as well as to more complex (branched) systems.

Both procurement costs and operational costs make up the total cost of an installation during its lifetime. A number of installation and operational costs are directly dependent on the piping diameter and the components in the piping system.

A considerable amount of the pressure losses in the system is caused by valves, in particular control valves in throttle-regulated installations. In systems with several pumps, the pump workload is divided between the pumps, which together, and in conjunction with the piping system, deliver the required flow.

The piping diameter is selected based on the following factors:

- Economy of the whole installation (pumps and system)
- Required lowest flow velocity for the application (e.g., avoid sedimentation)
- Required minimum internal diameter for the application (e.g., solids handling)
- Maximum flow velocity to minimize erosion in piping and fittings
- Plant standard pipe diameters

Decreasing the pipeline diameter has the following effects:

- Piping and component procurement and installation costs will decrease.
- Pump installation procurement costs will increase as a result of increased flow losses with consequent requirements for higher head pumps and larger motors. Costs for electrical supply systems will therefore increase.
- Operating costs will increase as a result of higher energy usage due to increased friction losses.

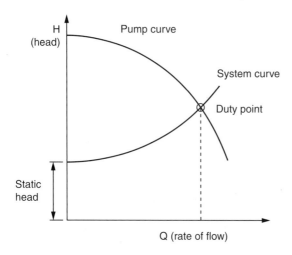

Figure 6.23 The duty point is the intersection between the pump and system curves. (Courtesy of, and with joint permission of the Hydraulic Institute, Parsippany, NJ; Europump, Brussels, Belgium; and the U.S. Department of Energy Office of Industrial Technologies, Washington, D.C.)

Some costs increase with increasing pipeline size and some decrease. Because of this, an optimum pipeline size may be found, based on minimizing costs over the life of the system.

The duty point of the pump is determined by the intersection of the system curve and the pump curve as shown in Figure 6.23.

A pump application might need to cover several duty points, of which the largest flow and/or head will determine the rated duty for the pump. The pump user must carefully consider the duration of operation at the individual duty points to properly select the number of pumps in the installation and to select output control. Many software packages are currently available that make it easier to determine friction losses and generate system curves (see Chapter 3, Section III). Most pump manufacturers can recommend software suitable for the intended duty. Different programs may use different methods of predicting friction losses and may give slightly different results. Very often, such software is also linked to pump-selection software from that particular manufacturer.

H. Methods for Analyzing Existing Pumping Systems

The following steps provide an overall guideline to improve an existing pumping system.

- Assemble a complete document inventory of the items in the pumping system.
- Determine the flow rates required for each load in the system.
- Balance the system to meet the required flow rates of each load.
- Minimize system losses needed to balance the flow rates.
- Effect changes to the pump to minimize excessive pump head in the balanced system.
- Identify pumps with high maintenance cost.

One of two methods can be used to analyze existing pumping systems. One consists of observing the operation of the actual piping system, and the second consists of performing detailed calculations using fluid analysis techniques. The first method relies on observations of the operating piping system (pressures, differential pressures, and flow rates); the second deals with creating an accurate mathematical model of the piping system and then calculating the pressures and flow rates within the model.

Observing the operating system allows one to view how the actual system is working, but system operational requirements limit the amount of experimentation that plant management will allow. By developing a model of the piping system, one can easily consider system alternatives, but the model must first be validated to ensure that it accurately represents the operating piping system it is trying to emulate. Regardless of the method used, the objective is to gain a clear picture of how the various parts of the system operate and to see where improvements can be made and the system optimized.

The following is a checklist of some useful means to reduce the life-cycle cost (LCC) of a pumping system.

- Consider all relevant costs to determine the LCC.
- Procure pumps and systems using LCC considerations.
- Optimize total cost by considering operational costs and procurement costs.
- Consider the duration of the individual pump duty points.
- Match the equipment to the system needs for maximum benefit.
- Match the pump type to the intended duty.
- Do not oversize the pump.
- Match the driver type to the intended duty.
- Specify motors to be high efficiency.
- Match the power transmission equipment to the intended duty.
- Evaluate system effectiveness.
- Monitor and sustain the pump and system to maximize benefit.
- Consider the energy wasted using control valves.
- Utilize auxiliary services wisely.
- Optimize preventative maintenance.
- Maintain the internal pump clearances.
- Follow available guidelines regarding the rewinding of motors.
- Analyze existing pump systems for improvement opportunities.
- Use the showcases in the *Guide* as a source for ideas.

I. Example: Pumping System with a Problem Control Valve

In this example, the Life-Cycle Cost analysis for the piping system is directed at a control valve. The system is a single pump circuit that transports a process fluid containing some solids from a storage tank to a pressurized tank (Figure 6.24). A heat exchanger heats the fluid, and a control valve regulates the rate of flow into the pressurized tank to 80 cubic meters per hour (m^3/hr) (350 gallons per minute [gpm]).

Energy Conservation and Life-Cycle Costs

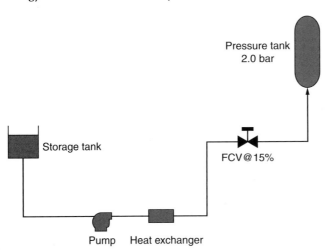

Figure 6.24 Sketch of pumping system in which the control valve fails. (Courtesy of, and with joint permission of the Hydraulic Institute, Parsippany, NJ; Europump, Brussels, Belgium; and the U.S. Department of Energy Office of Industrial Technologies, Washington, D.C.)

The plant engineer is experiencing problems with a flow control valve (FCV) that fails due to erosion caused by cavitation. The valve fails every 10 to 12 months at a cost of 4000 Euro or USD per repair. A change in the control valve is being considered to replace the existing valve with one that can resist cavitation. Before changing out the control valve again, the project engineer wanted to look at other options and perform a Life-Cycle Cost analysis on alternative solutions.

The first step is to determine how the system is currently operating and determine why the control valve fails, then to see what can be done to correct the problem.

The control valve currently operates between 15 to 20% open and with considerable cavitation noise from the valve. It appears the valve was not sized properly for the application. After reviewing the original design calculations, it was discovered that the pump was oversized; 110 m³/hr (485 gpm) instead of 80 m³/hr (350 gpm). This resulted in a larger pressure drop across the control valve than originally intended.

As a result of the large differential pressure at the operating rate of flow, and the fact that the valve is showing cavitation damage at regular intervals, it is determined that the control valve is not suitable for this process.

The following four options are suggested:

1. A new control valve can be installed to accommodate the high pressure differential.
2. The pump impeller can be trimmed so that the pump does not develop as much head, resulting in a lower pressure drop across the current valve.
3. A variable frequency drive (VFD) can be installed, and the flow control valve removed. The VFD can vary the pump speed and thus achieve the desired process flow.
4. The system can be left as it is, with a yearly repair of the flow control valve to be expected.

The cost of a new control valve that is properly sized is 5000 Euro or USD. The cost of modifying the pump performance by reduction of the impeller diameter is 2250 Euro or USD. The process operates at 80 m^3/hr for 6000 hr/year. The energy cost is 0.08 Euro or USD per kWh and the motor efficiency is 90%.

The cost comparison of the pump system modification options is contained in Table 6.5. Figure 6.25 shows the pump and system curves showing the operation of the original system and the modified impeller.

By trimming the impeller to 375 mm (Option 2), the pump's total head is reduced to 42.0 m (138 ft) at 80 m^3/hr. This drop in pressure reduces the differential pressure across the control valve to less than 10 m (33 ft), which better matches the valve's original design point. The resulting annual energy cost with the smaller impeller is 6720 Euro or USD per year. It costs 2250 Euro or USD to trim the impeller. This includes the machining cost as well as the cost to disassemble and reassemble the pump.

A 30-kW VFD (Option 3) costs 20,000 Euro or USD, and an additional 1500 Euro or USD to install. The VFD will cost

Table 6.5 Cost Comparison for Options 1 through 4 in the System with a Failing Control Valve System

Cost	Option 1 Change Control Valve	Option 2 Trim Impeller	Option 3 VFD and Remove Control Valve	Option 4 Repair Control Valve
Pump Cost Data				
Impeller diameter	430 mm	375 mm	430 mm	430 mm
Pump head	71.7 m (235 ft)	42.0 m (138 ft)	34.5 m (113 ft)	71.7 m (235 ft)
Pump efficiency	75.1%	72.1%	77%	75.1%
Rate of flow	80 m³/hr (350 USgpm)	80 m³/hr (350 USgpm)	80 m³/hr (350 USgpm)	80 m³/hr (350 USgpm)
Power consumed	23.1 kW	14.0 kW	11.6 kW	23.1 kW
Energy cost/year	11,088 Euro or USD	6,720 Euro or USD	5,568 Euro or USD	11,088 Euro or USD
New valve	5,000 Euro or USD	0	0	0
Modify impeller	0	2,250 Euro or USD	0	0
VFD	0	0	20,000 Euro or USD	0
Installation of VFD	0	0	1,500 Euro or USD	0
Valve repair/year	0	0	0	4,000 Euro or USD

Courtesy of, and with joint permission of the Hydraulic Institute, Parsippany, NJ; Europump, Brussels, Belgium; and the U.S. Department of Energy Office of Industrial Technologies, Washington, D.C.

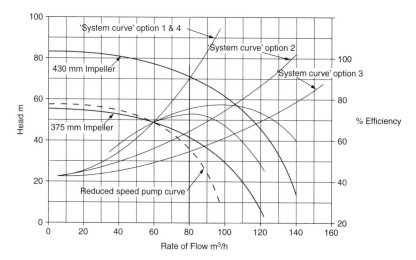

Figure 6.25 Pump and system curves showing the operation of the original system and the modified pump impeller. (Courtesy of, and with joint permission of the Hydraulic Institute, Parsippany, NJ; Europump, Brussels, Belgium; and the U.S. Department of Energy Office of Industrial Technologies, Washington, D.C.)

500 Euro or USD to maintain each year. It is assumed that it will not need any repairs over the project's eight-year life.

The option to leave the system unchanged (Option 4) will result in a yearly cost of 4000 Euro or USD for repairs to the cavitating flow control valve.

LCC costs and assumptions are as follows:

- The current energy price is 0.08 Euro or USD per kWh.
- The process is operated for 6000 hours/year.
- The company has an annual cost for routine maintenance for pumps of this size at 500 Euro or USD per year, with a repair cost of 2500 Euro or USD every second year.
- There is no decommissioning cost or environmental disposal cost associated with this project.
- This project has an eight-year life.

- The interest rate for new capital projects is 8%, and an inflation rate of 4% is expected.

The life-cycle cost calculations for each of the four options are summarized in Table 6.6. Option 2, trimming the impeller, has the lowest life-cycle cost and is the preferred option for this example.

J. For More Information

To order *Pump Life Cycle Costs: A Guide to LCC Analysis for Pumping Systems,* contact the Hydraulic Institute or Europump.

1. About the Hydraulic Institute

The Hydraulic Institute (HI), established in 1917, is the largest association of pump producers and leading suppliers in North America. HI serves member companies and pump users by providing product standards and forums for the exchange of industry information. HI has been developing pump standards for over 80 years. For information on membership, organization structure, member and user services, and energy and life-cycle cost issues, visit www.pumps.org.

Hydraulic Institute
9 Sylvan Way
Parsippany, NJ 07054
973-267-9700 (phone)
973-267-9055 (fax)

2. About Europump

Europump, established in 1960, acts as spokesman for 15 national pump manufacturing associations in Europe, and represents more than 400 manufacturers. Europump serves and promotes the European pump industry. For information regarding Europump work in the field of life-cycle cost issues, please e-mail secretariat@europump.org. For information on Europump, visit www.europump.org.

Table 6.6 LCC Comparison for the Problem Control Valve System

	Option 1 Change Control Valve	Option 2 Trim Impeller	Option 3 VFD and Remove Control Valve	Option 4 Repair Control Valve
Input				
Initial investment cost:	5,000	2,250	21,500	0
Energy price (present) per kWh:	0.080	0.080	0.080	0.080
Weighted average power of equipment in kW:	23.1	14.0	11.6	23.1
Average operating hours/year:	6,000	6,000	6,000	6,000
Energy cost/year (calculated) Energy price × Weighted average power × Average operating hours/year:	11,088	6,720	5,568	11,088
Maintenance cost (routine maintenance)/year:	500	500	1,000	500
Repair every 2nd year:	2,500	2,500	2,500	2,500
Other yearly costs:	0	0	0	4,000
Downtime cost/year:	0	0	0	0
Environmental cost:	0	0	0	0
Decommissioning/disposal (salvage) cost:	0	0	0	0
Lifetime in years:	8	8	8	8
Interest rate (%):	8.0	8.0	8.0	8.0
Inflation rate (%):	4.0	4.0	4.0	4.0
Output				
Present LCC value:	91,827	59,481	74,313	113,930

Courtesy of, and with joint permission of the Hydraulic Institute, Parsippany, NJ; Europump, Brussels, Belgium; and the U.S. Department of Energy Office of Industrial Technologies, Washington, D.C.

Europump
Diamant Building, 5th Floor
Blvd. A Reyers 80, B1030
Brussels, Belgium
+32 2 706 82 30 (phone)
+32 2 706 82 50 (fax)

3. About the U.S. Department of Energy's Office of Industrial Technologies

DoE's Office of Industrial Technologies (OIT), through partnerships with industrial companies and trade groups, develops and delivers advanced energy efficiency, renewable energy, and pollution prevention technologies for industrial applications. The OIT encourages industry-wide efforts to boost resource productivity through a strategy called Industries of the Future (IOF). IOF focuses on nine energy- and resource-intensive industries: agriculture, aluminum, chemicals, forest products, glass, metal casting, mining, petroleum, and steel. Visit www.oit.doe.gov to learn more about these programs and services.

U.S. Department of Energy
Office of Industrial Technologies
1000 Independence Avenue, SW
Washington, D.C. 20585
clearinghouse@ee.doe.gov
1-800-862-2086

7

Special Pump-Related Topics

I. OVERVIEW

The fundamentals of centrifugal pump design have remained largely unchanged over the past 50 years. The design basics of pump impellers, volutes, and diffusers; the shapes of pump performance curves; and the characteristics of pumps operating in systems are little different from the way they appeared in pumps and systems half a century ago. However, the consequences of misapplying pumps (i.e., choosing inappropriate configurations, poor material selections, incorrect sizing of pumps, or operating pumps too far from BEP on their H–Q curves) are far more significant than they were 50 years ago. The reasons for this include the fact that energy is much more expensive, maintenance of pumps is far more costly, and downtime in a production facility is a far more expensive thing to consider than was the case 50 years ago.

The above paragraph is not meant to imply that nothing new has happened in pump technology in the past 50 years. A better understanding of suction specific speed and hydraulic shaft loads has contributed significantly to pump reliability. There have been a great many achievements in manufacturing techniques resulting in higher quality and more durable pumps. Improvements in these areas include better metal casting techniques such as investment casting, producing cast parts of greater integrity and requiring less machining and repair; and more precise machine tools and techniques, producing more accurate fits and closer tolerances.

Many improvements in pump reliability and the ability to handle highly corrosive and abrasive liquids have been achieved through the development of superior materials, including both metal alloys and nonmetallic materials.

A number of new pump configurations have come to prominence in the past 20 years, replacing earlier pump configurations because of higher reliability, lower cost, and/or other benefits. Examples include submersible pumps supplanting vertical lineshaft or column type pumps; wet rotor technology being used in place of coupled pumps for residential hot water circulators; and air-operated diaphragm pumps replacing other positive displacement or centrifugal alternatives.

There have also been numerous advances in specific aspects of pump mechanical design. Some examples include the development of better packing and mechanical seal types and materials, a wider variety of bearing materials and lubrication systems, and material options to reduce abrasive wear in areas subject to such damage.

This chapter describes a few special topics related to pumps that are considered to have contributed significantly to the broadening of the application range of pumps and to the improvement in pump reliability. The topics discussed here are not meant to be an all-inclusive listing of all of the technological advances. Some of these topics have been covered in other chapters of this book as well.

II. VARIABLE-SPEED SYSTEMS

Variable-speed pumping systems are covered in Chapter 6, Section IV. The technology has evolved through a number of alternative techniques for achieving variable-speed operation of pumps. The use of variable frequency drives (VFDs) to achieve this end is considered by many to be the best alternative for the broadest range of pump applications. The use of variable-speed technology for those pumping applications

Special Pump-Related Topics

where it is appropriate has resulted in significant energy savings, improved pump performance, and reduced maintenance costs, compared with constant-speed alternatives.

The growth in applications for VFDs has resulted in a dramatic increase in the number of suppliers of VFD equipment; improvements in the quality, reliability, and durability of the products; and a reduction in the cost of the equipment. There are presently more than 50 suppliers of variable-frequency drives in the United States, including virtually all of the major makers of electric motors and electronic control devices.

III. SEALLESS PUMPS

Sealless pumps are covered in Chapter 5, Section V. Several types of positive displacement pumps (diaphragm and peristaltic pumps) are discussed as sealless alternatives in that chapter, as is the rather mundane vertical column centrifugal pump.

The two primary alternatives for sealless centrifugal pumping — magnetic drive and canned motor pumps — are compared and contrasted in Chapter 5. Both technologies have been around for many years, but both are continuing to expand to higher horsepower pumps, more aggressive liquids, and higher operating temperatures.

The need for sealless pumping is continuing to grow as more and more liquids are put into the category for which zero leakage is accepted, and as the search for greater pump reliability attempts to eliminate one of the leading contributors to pump downtime, the shaft sealing system.

As Chapter 5 points out, neither the magnetic drive nor the canned motor design type is a panacea, with the major weakness for both design types being the fact that radial and thrust loads generated by the pump must be accommodated by sleeve bearings and thrust plates that are often exposed to the liquid being pumped. Other limitations include upper limits of viscosity, concerns for inadvertent dry running (especially for mag drive pumps), special motors required for canned motor pumps, complication of design compared to sealed pumps, and the resulting high level of maintenance and reduced reliability that these factors suggest.

Despite these shortcomings, manufacturers are working hard to improve sealless pump designs, and there is no doubt that sealless technology will play an important role in the pump industry in the 21st century.

IV. CORROSION

Corrosion attack occurs on many components in a plant in addition to pumps, so many of the comments in this section are relevant to other equipment as well as pumps. The discussion in this section is restricted to corrosive attack on metals. The chemistry and nature of corrosion-like attack on nonmetallic materials is quite different from that for metals. Section V to follow discusses properties of nonmetallic components used in pumps, while Section VI covers the properties of elastomers used for O-rings in pumps.

When discussing the corrosion of metals, the concept of a local cathode and anode on a metallic surface provides a general description of how most corrosive attacks occur. No matter what the type of corrosion in metals, the basic nature is always the same: a flow of electricity between two areas of a different electrical potential, through a solution capable of conducting an electric current. For corrosion to occur in metals, three separate conditions must occur:

1. There must be a difference in electrical potential (i.e., there must be an anode and a cathode).
2. The areas that are different in electrical potential must be in an electrolyte.
3. There must be a metallic connection between the areas of potential difference.

The area of the lower electrical potential is called the *anode* or the negative pole and is the area where the corrosion attack will occur. The area of higher electrical potential is the *cathode* or the positive pole and is normally not subject to corrosive attack. A difference of electrical potential will result from two different metals or alloys, or it may occur between two points on the same material, due to local defects in chemical composition, a variation of mechanical properties from stress or machining, or variations in the environment, such as a partially submerged component.

An *electrolyte* is a liquid that conducts electricity. Most liquids will conduct electricity and therefore are good electrolytes. Seawater is an excellent electrolyte. On the other hand, pure hydrocarbons are nonpolar, that is, they will not conduct electricity. Therefore, metals immersed in hydrocarbons generally are not subject to corrosion.

The metallic connection between the anode and the cathode provides a path for the flow of electrons from the anode to the cathode, and allows the current to flow from the cathode to the anode. It can be a separate metallic connection, or merely exist by the fact that the cathode and anode are in contact with each other.

When corrosion occurs on a metal component that contains iron (Fe), the result is the formation of $Fe(OH)_3$ (ferric hydroxide), otherwise known as rust. Meanwhile, on the cathode, a layer of H_2 gas at the cathode surface restricts corrosion. This corrosive-resistant layer of gas can be removed by velocity or abrasion.

Although all corrosion on metals is electrochemical in nature as just described, it is possible nevertheless to classify corrosion by type from the visual appearance and the environment in which it takes place. These distinct types are caused by specific influences and specific environmental factors. The following list is generally accepted by most corrosion engineers to include the common forms of corrosion in metals.

- Galvanic, or two-metal corrosion
- Uniform, or general corrosion
- Pitting corrosion
- Intergranular corrosion
- Erosion corrosion
- Stress corrosion
- Crevice corrosion
- Graphitization or dezincification corrosion

These corrosion types are discussed in more detail below.

Table 7.1 Galvanic Series of Common Metals and Alloys in Seawater

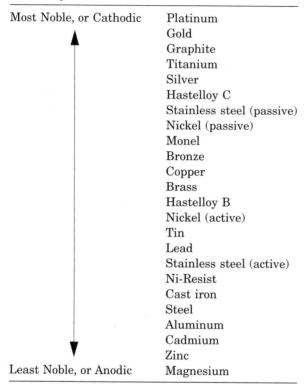

Most Noble, or Cathodic	Platinum
	Gold
	Graphite
	Titanium
	Silver
	Hastelloy C
	Stainless steel (passive)
	Nickel (passive)
	Monel
	Bronze
	Copper
	Brass
	Hastelloy B
	Nickel (active)
	Tin
	Lead
	Stainless steel (active)
	Ni-Resist
	Cast iron
	Steel
	Aluminum
	Cadmium
	Zinc
Least Noble, or Anodic	Magnesium

1. Galvanic, or Two-Metal Corrosion

When two metals separated by a spread in electrical potential are connected in an electrolyte, the one with the lower potential becomes the anode and will corrode. On the other hand, the higher potential part, the cathode, will not corrode. Table 7.1 shows the electrochemical order (also called the *galvanic series*) of a number of common metals in a seawater environment. The higher potential part is also called the more *noble* of the two materials. Note that some materials, such as stainless steel and nickel, come in two forms, an active and passive form, with widely varying electrical potential.

When there is a potential for galvanic corrosion in pump parts, the following practices should be observed to help restrict or reduce the amount of corrosion.

Special Pump-Related Topics

1. If two different metals will be wetted and in contact with one another in a pump, select a combination of metals as close together as possible in the galvanic series. The corrosion rate is greater the farther apart the two metals are in the galvanic series.
2. Avoid the combination of a small anode in contact with a large cathode. The higher the ratio of cathode size to anode size, the more accelerated the corrosion rate of the anode.
3. Insulate the two materials, if possible, to eliminate the metallic connection. The insulation can be in the form of plastic washers, gaskets, or sleeves.
4. Avoid wet threaded areas, or other crevice areas with two dissimilar materials.
5. Be careful about using metal coatings that are high in the galvanic series. A very small pinhole in the high galvanic metal coating will result in a small anode/large cathode relationship. Materials lower on the galvanic series, such as zinc, often make superior corrosion-resistant coatings.
6. Install a third (sacrificial) metal that is anodic to both working metals. This form of cathode protection is commonly used to protect large metal tanks and other metal structures. Aluminum, zinc, and magnesium are commonly used as the anode material.

2. Uniform, or General Corrosion

This is a type of corrosion that occurs uniformly over the entire exposed metal surface. The metal becomes thinner as the corrosion works its way through the part, and eventually the part fails mechanically. A plate immersed in an acid will dissolve at a uniform but rapid rate. A steel plate exposed to the atmosphere will rust at a uniform, although much slower rate. Uniform corrosion is by far the most common corrosion attack in metals. Because the metal loss is uniform based on the environment, it is the most predictable type for common applications. Corrosion handbooks provide data on suitability of different metals for different corrosive environments, along with corrosion loss in mils per year for different metals exposed to different corrosive liquids. The corrosion loss (in mils per year) is multiplied by the design life (in years) to obtain the *corrosion allowance*. The resulting corrosion allowance is then added to the thickness of the material that is required for structural or functional purposes, and the result is the minimum thickness that the part can be.

3. Pitting Corrosion

Pitting corrosion is a local spot where the surface protection is attacked at an isolated location. It is the prevalent form of attack with passive metals, but it can occur on any metal under the right conditions. Pitting corrosion involves the localized breakdown of the passive film that otherwise protects the metal. This area is adjacent to a large area of high potential or cathode, with the result that an accelerated corrosive attack occurs in the area of the pit. Pitting generally occurs in an environment where the particular alloy ordinarily exhibits only a moderate corrosion rate. Frequently, the breakdown in a passive film is caused by a minor trace element, such as fluorine, chlorine, bromine, or iodine, with chlorine being the most common.

4. Intergranular Corrosion

Intergranular corrosion is a localized attack, occurring at or near grain boundaries, with very little corrosion of the base material. The austenitic (300 series) stainless steels are particularly susceptible to this type of corrosive attack, especially if they are not properly heat treated. If the material is allowed to remain too long in the temperature range between 950 and 1450°F (the *heat sensitization range*), a chromium carbide precipitate forms along the grain boundaries. At the same time, this causes areas adjacent to the grain boundaries to become depleted of chromium. The chromium-depleted area next to the grain boundary becomes an anode, adjacent to a large chromium-rich cathode, and the anode area is subject to rapid localized corrosion in environments in which stainless steel would not normally be expected to corrode.

Special Pump-Related Topics

Intergranular corrosion can be prevented by proper heat-treating (solution annealing) of the material. The heat treatment consists of heating the part to between 1950 and 2050°F, followed by a quick quench to ensure transition through the heat sensitization range in less than 3 minutes. In cases where the part is too big or not practical to heat treat, another preventative method is to lower the carbon content to 0.03% maximum (such as the case with 304L or 316L stainless steel). Still another way to prevent this corrosion attack is by adding stabilizing elements, such as columbium (347 stainless) or titanium (321 stainless).

5. Erosion Corrosion

Erosion corrosion is the wearing away, by high velocity or abrasion, of the protective surface film that is resistant to corrosion. Thus, this attack takes place with a combination of corrosion and erosion. Metal loss associated with cavitation is a form of erosion corrosion. With an iron or steel impeller subject to cavitation, a corrosive film of $Fe(OH)_3$ (rust) forms on the surface of the impeller vane. Along comes a bubble and, when it collapses, the collapsing bubble causes destruction of the protective film and removal of the unprotected material. Then a new rust film forms, and the process keeps repeating until ultimately the damaged area gets large enough to be critical.

Another example of erosion corrosion is the effect of velocity on the corrosion rate of a particular part exposed to moving liquid. For some materials, there is a breakaway velocity above which the corrosion rate rapidly accelerates. Other materials (e.g., titanium) show almost no change in corrosion rate even when exposed to high fluid velocities. Still other materials actually exhibit a higher acceleration rate at lower velocities than at higher velocities (e.g., stainless steel in slow-moving or stagnant seawater will corrode faster than in higher-velocity seawater).

6. Stress Corrosion

Stress corrosion occurs when a metal is subjected to a combination of mechanical stress and a corrosive medium. The condition is accelerated at higher temperatures. Chemical environments that are likely to produce stress corrosion cracking include chlorides, caustics, and ammonia. The stress in the metal component can be induced by a combination of pressure, heat treatment, machining, and/or forming.

7. Crevice Corrosion

Crevice corrosion occurs at a crevice in a pump part (typically occurring at a place where two surfaces that are the same material are in loose contact, or are separated by a gasket that has allowed wetting of the gasket surface). The crevice typically has an area of depleted oxygen, while the liquid being pumped past the crevice is richer in oxygen. Therefore, there is localized corrosion in the portion of the crevice that is depleted of oxygen.

8. Graphitization or Dezincification Corrosion

Graphitization corrosion occurs when one element of a material is selectively removed, leaving a part that has more or less the same appearance as the original part, but is porous and has considerably reduced physical properties. This corrosion type occurs when cast iron is exposed to certain acid water environments. In this environment, iron is selectively removed, leaving a part that has a higher percentage of carbon (graphite) than it previously did, although the part may not look any different. Because of the higher graphite composition, the part may fail mechanically because it is not as strong.

Dezincification, a similar corrosion process, can occur in the high zinc bronzes, where selective loss of zinc can occur.

Material selection on pumps is often performed by an expert on metallurgy and corrosion. In the absence of such expertise, Table 7.2 outlines some generalized recommended materials for pumps handling liquid of known pH value.

V. NONMETALLIC PUMPS

Industrial equipment made of plastic used to be thought of as a cheap but weaker alternative, resulting in shorter service

Table 7.2 Recommended Pump Materials for Different pH Liquids

pH Value	Materials of Construction
10–14	Corrosion-resistant alloys or nonmetals
8–10	All iron
6–8	Bronze fitted or standard fitted
4–6	All bronze
0–4	Corrosion-resistant alloys or nonmetals

life. As a result, nonmetallic components were seldom specified by industrial pump users. The past 15 to 20 years have witnessed the introduction of many new plastic materials that are uniquely suited for nonmetallic pump components and pumps made entirely of plastic. With the proper material selection, the use of nonmetallic components can result in several benefits, which include superior corrosion and abrasion resistance, extended service life, elimination of contamination in ultrapure applications, lower weight, and reduced cost.

The selection of the correct plastic for a particular application requires careful attention to all of the application parameters, as is the case with the selection of materials in a metal pump. The most important factors to consider when selecting plastic pumps or components are resistance against corrosion and abrasion of the particular plastic in the specific liquid, the liquid temperature and pressure range to which the material will be exposed, the liquid velocity, and the variation of stresses to which the components will be exposed.

Plastics are broadly categorized as either *thermoplastics* or *thermosets*, depending on the nature of their molecular structure. Thermoplastics offer greater resistance to corrosion and abrasion, and can be used in ultrapure applications (such as ultrapure water used for computer chip manufacturing, or food-grade applications) in their virgin form. Thermosets, while having lower resistance to abrasion and often to corrosion, have higher mechanical tensile strength.

Table 7.3 summarizes the most popular plastics currently available for use in many pump types, including some

Table 7.3 Application Highlights for Plastic Pump Components

Material	Chemical Resistance	Upper Temp. Limit (°F)	Other Important Characteristics
PVC (Polyvinyl chloride)	Resists many acids, alkalies, and other chemicals	140	Relatively low cost Not useful against many solvents
CPVC (Chlorinated polyvinyl chloride)	Similar to PVC	212	Superior in abrasion resistance and stronger than PVC
PP (Polypropylene)	Useful in many acid, alkali, and solvent services	185	Relatively low cost Poor with strong oxidizing acid or chlorinated hydrocarbons Lightest of the thermoplastics
PE (Polyethylene)	Similar to PP	200	Similar in mechanical properties to PP, but not as light in weight
PVDF (Polyvinylidene fluoride)	Useful for most acids, alkalies, solvents, and many halogens	275	Strong and abrasion resistant Excellent in virgin state for ultrapure services
ECTFE (Ethylene chlorotrifluoro-ethylene)	Similar to but better than PVDF	300	Beats PVDF in abrasion resistance and for ultrapure applications
PTFE (Polytetrafluoro-ethylene), commonly called Teflon®	Broadest chemical resistance commonly available for pumps	400	Highest cost of nonmetallic choices
FRP/GRP (Fiberglass or glass-reinforced polyester)	Useful in many corrosive services	230	Not very resistant to abrasives

comments on their major application strengths, weaknesses, and limitations.

In addition to the plastics cited in Table 7.3, other nonmetallic materials can be used in pumps to extend service life. For example, ceramics such as chrome oxide can be applied at shaft journals under sleeve bearings (such as in vertical turbine pumps as described in Chapter 4, Section XI), or onto shaft sleeves of any pump fitted with packing, to achieve superior abrasion resistance compared to most metals. The ceramic material has very weak mechanical properties but is subject to almost no mechanical loading when used as a coating on a shaft or sleeve. The proper preparation of the surface to be coated with the ceramic and the application of this coating is essential to its good service life. Ceramic coatings can handle highly corrosive environments and can operate in temperatures above 500°F.

VI. MATERIALS USED FOR O-RINGS IN PUMPS

A. General

Before discussing materials most commonly used for O-rings in pumps, a few terms should be defined. The following material on O-rings and O-ring materials is printed with permission of the Parker Hannifin Corporation, Seal Group.

Polymer: A polymer is the result of a chemical linking of molecules into a long chain-like structure. Both plastics and elastomers are classified as polymers. In this book, polymer generally refers to a basic class of elastomer, members of which have similar chemical and physical properties. O-rings are made from many polymers, but a few polymers account for the majority of O-rings produced, namely Nitrile (Buna N), EPDM, and Neoprene.

Rubber: Rubber-like materials first produced from sources *other* than rubber trees were referred to as "synthetic rubber." This distinguished them from natural gum rubber. Since then, usage in the industry has broadened the meaning of the term "rubber" to include both natural as well as synthetic materials having rubber-like qualities. This book uses the broader meaning of the word "rubber."

Elastomer: Though *elastomer* is synonymous with *rubber*, it is formally defined as a "high molecular weight polymer that can be, or has been modified, to a state exhibiting little plastic flow and rapid, and nearly complete recovery from an extending or compressing force." In most instances we call such material before modification "uncured" or "unprocessed" rubber or polymer.

When the basic high molecular weight polymer, without the addition of plasticizers or other diluents, is converted by appropriate means to an essentially nonplastic state and tested at room temperature, it usually meets the following requirements to be called an elastomer. The American Society for Testing and Materials (ASTM) uses these criteria to define the term "elastomer."

- It must not break when stretched approximately 100%.
- After being held for 5 minutes at 100% stretch, it must retract to within 10% of its original length within 5 minutes of release. (Note: Extremely high hardness/modulus materials generally do not exhibit these properties even though they are still considered elastomers.)

Compound: A compound is a mixture of base polymer and other chemicals that form a finished rubber material. More precisely, a compound refers to a specific blend of chemical ingredients tailored for particular required characteristics to optimize performance in some specific service.

The basis of compound development is the selection of the polymer type. There may be a dozen or more different ones from which to choose. The rubber compounder may then add various reinforcing agents such as carbon black, curing or vulcanizing agents such as sulfur or peroxide, activators, platicizers, accelerators, antioxidants, or antiozonants to the elastomer mixture to tailor it into a seal compound with its own distinct physical properties. Because compounders have thousands of compounding ingredients at their disposal, it seems reasonable to visualize two, three, or even one hundred-plus compounds having the same base elastomer, yet exhibiting marked performance differences in the O-ring seal.

Special Pump-Related Topics   437

The terms *compound* and *elastomer* are often used interchangeably in a more general sense. This usage usually references a particular type or class of materials such as *nitrile compounds* or *butyl elastomers*. Please remember that when one specific compound is under discussion in this book, it is a blend of various compounding ingredients (including one or more base elastomers), with its own individual characteristics and identification in the form of a unique compound number.

B. Eight Basic O-Ring Elastomers

The following are brief descriptions of eight of the most commonly used O-ring elastomers. There are, of course, many other specialized polymers. The ones listed below, however, account for well over 95% of all O-rings.

1. Nitrile (Buna N)

Due to its excellent resistance to petroleum products, and its ability to be compounded for service over a temperature range of –65 to +275°F (–54 to +135°C), nitrile is the most widely used elastomer in the seal industry today. Most military rubber specifications for fuel- and oil-resistant MS and AN O-rings require nitrile-based compounds. It should be mentioned, however, that to obtain good resistance to low temperature in a nitrile material, it is almost always necessary to sacrifice some high-temperature fuel and oil resistance. Nitrile compounds are superior to most elastomers with regard to compression set, cold flow, tear, and abrasion resistance. Inherently, nitrile-based compounds do not possess good resistance to ozone, sunlight, or weather. However, this specific weakness has been substantially improved through compounding efforts.

Nitrile is recommended for general purpose sealing, petroleum oils and fluids, water, silicone greases and oils, di-ester based lubricants (MIL-L-7808), and ethylene glycol based fluids (hydrolubes).

2. Neoprene

Neoprenes can be compounded for service at temperatures from –65 to +250°F (–54 to +121°C). Most elastomers are either resistant to deterioration from exposure to petroleum lubricants or oxygen. Neoprene is unusual in having limited resistance to both. This characteristic, combined with a broad temperature range and moderate cost, accounts for its use in many sealing applications.

Neoprene is recommended for refrigerants (freons, ammonia), high aniline point petroleum oils, mild acid resistance, and silicate ester lubricants.

3. Ethylene Propylene

Ethylene propylene has won broad acceptance in the sealing world because of its excellent resistance to Skydrol and other phosphate ester type hydraulic fluids. Ethylene propylene has a temperature range from −65 to +300°F (−54 to 150°C) for most applications.

Ethylene propylene is recommended for phosphate ester based hydraulic fluids (Skydrol, Fyrquel, Pydraul), steam to +400°F (+204°C), water, silicone oils and greases, dilute acids, dilute alkalies, ketones (MEK, acetone), alcohols, and automatic brake fluids.

4. Fluorocarbon (Viton)

Fluorocarbon elastomers were first introduced in the mid-1950s. Since that time, they have grown to major importance in the seal industry. Due to their wide-spectrum chemical compatibility and broad temperature range, fluorocarbon elastomers represent one of the most significant elastomer developments in recent history.

The working temperature range of fluorocarbons is considered to be from −20 to +400°F (−29 to +204°C), but some formulations have been known to seal at −65°F (−54°C) in some static low-temperature applications.

Recent developments in material formulation have further improved the characteristics of this very useful seal material. Fluorocarbon materials should be considered for use in aircraft, automotive, and other devices requiring maximum resistance to deterioration by environment and fluids.

Special Pump-Related Topics

Fluorocarbon is recommended for petroleum oils, di-ester based lubricants (MIL-L-7808, MIL-L-6085), silicate ester based lubricants, silicone fluids and greases, halogenated hydrocarbons, selected phosphate ester fluids, and acids.

5. Butyl

Prior to the introduction of ethylene propylene, butyl was the only elastomer that was satisfactory for Skydrol 500 service over a temperature range from −65 to +225°F (−54 to +107°C).

In addition, butyl exhibits excellent resistance to gas permeation, which makes it particularly useful for vacuum applications.

Butyl is recommended for phosphate ester type hydraulic fluids (Skydrol, Fryquel, Pydraul), ketones (MEK, acetone), silicone fluids and greases, and for vacuum service.

6. Polyacrylate

This material has outstanding resistance to petroleum-based fuels and oils. In addition, polyacrylate has good resistance to oxidation, ozone, and sunlight, combined with an excellent ability to resist flex cracking. Compounds of polyacrylate have been developed that are suitable for continuous service in hot oil over a temperature range from 0 to 300°F (−18 to +150°C).

Resistance to hot air is slightly superior to nitrile polymers, but tear strength, compression set, and water resistance are inferior to many other polymers. There are several polyacrylate types available commercially, but all are essentially polymerization products of acrylic acid esters.

The greatest use of polyacrylate elastomers is by the automotive industry in automatic transmission and power steering devices using Type A transmission fluids.

Polyacrylate is recommended for petroleum oils, automatic transmission fluids, and power steering fluids.

7. Silicone

The silicones are a group of elastomeric materials made from silicone, oxygen, hydrogen, and carbon. As a group, the silicones have rather poor tensile strength, tear, and abrasion resistance. Special silicone compounds have been developed that exhibit exceptional heat and compression set resistance. High-strength materials have also been developed but their strength does not compare to conventional elastomers. Silicones have excellent resistance to temperature extremes. Flexibility below −175°F (−114°C) has been demonstrated, and there are compounds that will resist temperatures up to +700°F (+371°C) for short periods. The maximum temperature for which silicones are recommended for continuous service in dry air is +450°F (+232°C). The ability of silicone to retain its original physical properties at these high temperatures is superior to most other elastomer materials.

Silicone compounds are not normally recommended for dynamic sealing applications due to silicone's rather low abrasion resistance.

Silicones are recommended for high-analine point oils, dry heat, and chlorinated diphenyls.

8. Fluorosilicone

Fluorosilicone elastomers combine the good high and low temperature properties of silicone with basic fuel and oil resistance. The primary uses of fluorosilicone are in fuel systems at temperatures up to +350°F (+177°C) and in applications where the dry heat resistance of silicone is required but the material may be exposed to petroleum oils and/or hydrocarbon fuels. The high temperature limit for fluorosilicone is limited because temperatures approaching +350°F (+177°C) may degrade certain fluids, producing acids that attack the fluorosilicone elastomer. High strength fluorosilicone materials are available and certain ones exhibit much improved resistance to compression set.

Fluorosilicone is recommended for petroleum oils and fuels.

Table 7.4 summarizes and compares the properties of commonly used elastomers found in O-rings, including those described above. Figure 7.1 summarizes the temperature ranges for common elastomeric materials.

Special Pump-Related Topics

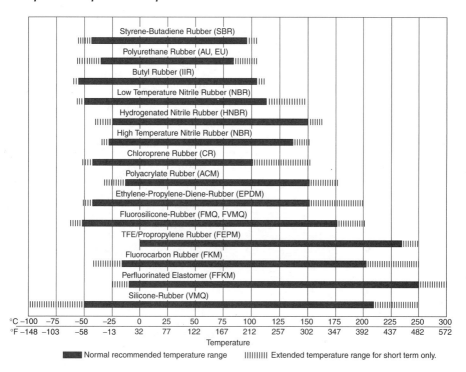

Figure 7.1 Temperature range for common elastomeric materials. (Courtesy of Parker Hannifin Corporation, Seal Group, Lexington, KY.)

VII. HIGH-SPEED PUMPS

The high-speed centrifugal pump has roots in the aerospace industry and in military applications. The pump (Figure 7.2) uses an integral gearbox drive to achieve speeds up to 25,000 rpm. This allows the pump to achieve heads to about 6300 ft with a single stage, with flows up to about 400 gpm. Flow rates up to 1000 gpm are achievable at lower heads with a single stage. Two pumps in series driven from a common gearbox allow heads to about 15,000 ft at 400 gpm, or to about 6300 ft at 1000 gpm.

Applications for high-speed, high-head pumps are principally in the refining and petrochemical industries. These pumps are also used in paper processing, power generation, steel production, and other heavy-duty industrial applications.

Table 7.4 Comparison of Properties of Commonly Used Elastomers

Elastomer Type (Polymer)	Parker Compound Prefix Letter	Abrasion Resistance	Acid Resistance	Chemical Resistance	Cold Resistance	Dynamic Properties	Electrical Properties	Flame Resistance	Heat Resistance	Impermeability	Oil Resistance	Ozone Resistance	Set Resistance	Tear Resistance	Tensile Strength	Water/Steam Resistance	Weather Resistance
AFLAS (TFE/Prop)	V	GE	E	E	P	G	E	E	E	G	E	E	G	PF	FG	GE	E
Butadiene	D	E	FG	FG	G	F	G	P	F	F	P	P	G	GE	E	FG	F
Butyl	B	FG	G	E	G	F	G	P	G	E	P	GE	FG	G	G	G	F
Chlorinated Polyethylene	K	G	F	FG	PF	G	G	GE	G	G	FG	E	F	FG	G	F	E
Chlorosulfonated Polyethylene	H	G	G	E	FG	F	F	G	G	G	F	E	F	G	F	F	E

Special Pump-Related Topics

Elastomer	Code																
Epichlorohydrin	Y	G	FG	G	GE	G	F	FG	FG	GE	E	E	PF	G	G	F	E
Ethylene Acrylic	A	F	F	FG	G	F	F	F	FG	E	F	F	E	G	F	G	PF	E
Ethylene Propylene	E	GE	G	E	GE	GE	G	P	E	E	G	P	E	GE	GE	E	E	E
Fluorocarbon	V	G	E	E	PF	F	E	E	G	E	G	E	E	GE	F	GE	FG	E
Fluorosilicone	L	P	FG	E	GE	P	F	G	E	G	P	E	E	GE	P	F	F	E
Isoprene	I	E	FG	FG	G	F	G	E	F	E	F	G	E	P	G	E	FG	F
Natural Rubber	R	FG	FG	G	E	G	P	F	F	F	P	P	GE	G	E	E	FG	F
Neoprene	C	G	FG	FG	FG	F	F	G	P	F	G	FG	GE	F	FG	G	F	E
HNBR	N	E	FG	G	GE	F	F	P	P	E	G	E	G	GE	FG	E	F	G
Nitrile or Buna N	N	G	F	FG	G	GE	F	E	P	G	G	E	P	GE	FG	GE	FG	F
Perfluorinated Fluoroelastomer	V	P	E	E	PF	F	E	E	E	G	E	E	E	G	PF	FG	GE	E
Polyacrylate	A	G	P	P	P	F	P	E	P	G	E	E	F	FG	F	F	P	E
Polysulfide	T	P	P	G	G	F	P	F	P	P	P	E	E	P	FG	P	F	E
Polyurethane	P	E	P	FG	G	E	F	P	F	F	G	P	E	F	GE	E	P	E
SBR or Buna S	G	G	F	FG	G	G	G	G	FG	FG	F	P	P	G	FG	GE	FG	F
Silicone	S	P	FG	GE	E	P	E	F	E	F	P	FG	E	GE	P	P	F	E

P = Poor; F = Fair; G = Good; E = Excellent

Source: From Parker Hannifin Corporation, Seal Group, Lexington, KY. With permission.

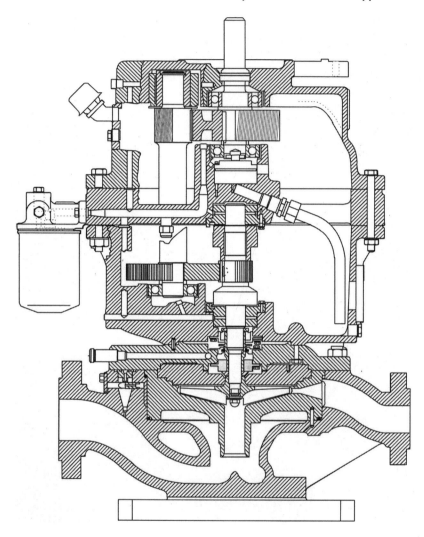

Figure 7.2 High-speed centrifugal pump. (Courtesy of Sundyne Corporation, Arvada, CO.)

Many high-speed pump designs use what is called a *partial emission* design (Figure 7.3). This design uses impellers whose vanes are oriented in a straight radial direction, called *Barske impellers*, rather than being curved backward like most impellers. Also, the partial emission design has the

Special Pump-Related Topics

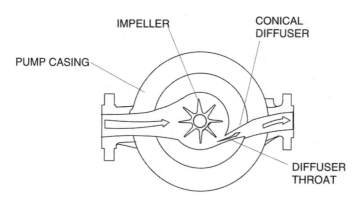

Figure 7.3 Partial emission pumps can achieve higher efficiencies at low specific speeds. (Courtesy of Sundyne Corporation, Arvada, CO.)

impeller concentrically located in the casing, and has a diffuser that only allows flow from a small portion of the impeller to be delivered to the pump discharge at any one time. Partial emission designs are able to produce higher efficiency pumps at low specific speeds. Beyond the range of partial emission impellers, *full emission* impellers using traditional impeller designs with double volutes or diffusers are employed.

High-speed centrifugal pumps have a number of advantages compared to the alternatives of single-stage centrifugal pumps with larger impeller diameters, multi-stage centrifugals, multiple pumps operating in series, or reciprocating positive displacement pumps. The advantages include higher efficiency, reduced sensitivity to dry running due to the larger running clearances, and reduced size and weight. The smaller size and weight can mean lower first cost than other alternatives, particularly in alloy construction.

There are several areas that deserve special attention with this unique pump design. These include the fact that the gearboxes required to produce the high speed necessitate additional cost and maintenance for seals, bearings, and gear sets. Some VFD designs are available to achieve higher speeds without the use of gearboxes. Also, high-speed designs may require inducers to achieve a reasonable $NPSH_r$. Dynamic balance of the rotor, and alignment between the pump, gearbox, and motor, are much more critical at the high speeds at which these pumps operate. Finally, selection of the proper mechanical seal is more critical and service is more severe because the seals are running at higher surface speeds. This last area of concern can be eliminated as sealless options become more available for high-speed pumps.

VIII. BEARINGS AND BEARING LUBRICATION

A number of advances have been made in the field of bearing design and bearing lubrication systems through the years. More precise manufacturing techniques have permitted the life of traditional ball and roller type bearings to be significantly increased. Plate-type thrust bearings can be used in pump types that generate higher thrust loads than are capable of being handled with traditional ball and roller type bearings. Precision alignment techniques, discussed in Section IX below, have added considerably to the expected life of many bearing systems.

The newest development in the area of bearing design is in the area of *magnetic bearings*. The use of a magnetic field to center a pump shaft and to carry the radial and thrust loads of a pump offers several major benefits. The lack of mechanical contact in the bearing system increases pump efficiency, extends bearing life, reduces maintenance, and eliminates the need for bearing lubricants. For sealless centrifugal pumps (magnetic drive and canned motor pumps), magnetic bearings may also allow dry running, pumping liquid containing abrasives, as well as operation over a wider flow range, because the bearings do not depend on the pumped liquid for lubrication.

The technology of magnetic bearings is still in its infancy, with few working prototypes in the field. The prospects are exciting, however, and may be the breakthrough that is necessary to allow magnetic drive and canned motor technologies to reach their full potential.

In the area of bearing lubrication, *oil mist lubrication* of bearings for a wide range of equipment in large plants is being used with more and more regularity. This lubrication system, often based on a central system distributing an oil mist to a variety of pieces of rotating equipment in the plant, offers a number of potential benefits compared to traditional grease or oil lubrication of individual pump bearings. These benefits can include significantly longer bearing life, a reduction in the amount of bearing oil consumed, and a reduction in oily waste pollution.

IX. PRECISION ALIGNMENT TECHNIQUES

The importance of coupling alignment for centrifugal pumps is discussed in Chapter 8. Lack of proper alignment may very well be the single most important cause of premature failure of frame-mounted pumps. Like many other activities, the degree of pump alignment is not a specific point, but rather a spectrum, with more precise alignment generally being reserved for larger, higher-speed, more-expensive, and more-critical equipment. To achieve the full range of possible alignment accuracies, many approaches can be used, ranging from very simple and rudimentary alignment using straight edges and feeler gauges, to the use of laser alignment equipment at the other end of the spectrum. In between these two extremes of the spectrum are other alignment techniques, such as single dial indicator (*rim and face*) alignment, *reverse indicator* alignment, and alignment using electronic gauges. The details of these alignment methods and instruments are beyond the scope of this book, but the interested reader is referred to Ref. [9] for more information on alignment techniques and instrumentation.

Two of the terms used to express the degree of misalignment are *offset* and *angle*. *Offset* indicates the difference between the two shaft centerlines at the coupling center, or the amount of *parallel misalignment*, expressed in mils (1 mil = 0.001 in.). *Angle* is the change in gap between the coupling faces, divided by the distance across the faces, or the *angular misalignment*, expressed in mils per inch. Angular and parallel misalignment are illustrated in Chapter 8, Figure 8.1 and Figure 8.2.

In recent years there has been a significant increase in the use of optical and laser alignment instruments. These devices have become more widely available, more affordable, and easier to use. Added to that is the growing understanding of the extension of the mean time between failure that can be achieved by more precise alignment efforts. The benefits of these precision alignment techniques include longer bearing life, longer seal life, improved pump reliability, reduced maintenance costs, and lower noise levels.

The most precise and sophisticated alignment method is not always the correct choice of alignment technique for every pump application. The decision of the degree of alignment accuracy is a trade-off between alignment precision (and presumably pump life), alignment time, and the cost of the instrumentation used for alignment. Using the most sophisticated alignment techniques, misalignment can be measured far more accurately than is practical to correct. Also, some of the most sophisticated alignment methods require that the pump equipment be brought into a certain level of alignment, using simpler equipment and techniques, before the more precise instrumentation can even be used.

Pump speed is also an indicator of the degree of alignment precision required. For example, a pump running at 900 rpm requires much less accurate alignment than one running at 3600 rpm. If no other guidelines are available from the pump manufacturer or from company policies, the alignment tolerances shown in Table 7.5 can be used.

Table 7.5 Pump Alignment Tolerances

rpm	Offset (mils)	Angle (mils/in.)
600	5.0	1.0
900	3.0	0.7
1200	2.5	0.5
1800	2.0	0.3
3600	1.0	0.2
7200	0.5	0.1

Special Pump-Related Topics

X. SOFTWARE TO SIZE PUMPS AND SYSTEMS

Chapter 3, Section III, describes computer technology that allows the modeling and optimization of complex piping networks, so that pump head and the reaction of the pump and system to various operating modes can be accurately determined. These computer software programs, one of which is demonstrated on the CD accompanying this book, make it much more feasible to accurately analyze new piping systems during the design phase, or to model existing pumping systems for modification or improvement. System designers and engineers will find the use of software particularly helpful and cost effective if the system is complex, has multiple flow branches, employs multiple pumps, or contains liquids that are appreciably different from water in terms of their effect on pipe friction losses and pump performance.

8

Installation, Operation, and Maintenance

I. OVERVIEW

This chapter provides a step-by-step procedure for installing and starting up a centrifugal pump in the field. The most important consideration for the successful installation and start-up of many pumps is careful attention to alignment. After the pump is started, the benchmarks used to monitor performance throughout the life of the pump are established. These benchmarks include hydraulic performance, temperature, and vibration.

This chapter discusses the criteria that should be considered in determining minimum continuous flow rate of a pump. The criteria that might affect the determination of minimum flow include temperature rise, radial and axial thrust bearing loads, prerotation, recirculation, separation, settling of solids, noise, vibration, and power consumption.

This chapter also includes a discussion of ten ways to prevent low flow damage in pumps.

Regular and preventive maintenance techniques for pumps are discussed, along with the benefits of establishing benchmarks of performance for new or recently overhauled pumps.

If a problem does occur with an operating pump, this chapter outlines some of the things that can be done to find the cause of the problem. A troubleshooting chart is provided as a guide for such an investigation.

Finally, if repair of the pump is necessary, this chapter provides some guidelines for successful pump repair.

II. INSTALLATION, ALIGNMENT, AND START-UP

A. General

This section describes a step-by-step checklist for installing and starting up a frame-mounted centrifugal pump. The order of the steps and the specific start-up activities vary for each type of pump, and the reader is urged to consult the manufacturer's instruction manual for the specific piece of equipment. For light-duty pumps below 25 HP, it may not be cost effective to carry out the multiple alignment checks and some of the other procedures discussed in this section. Within the bounds of reason, however, attention to the spirit of this checklist, if not to the letter, is recommended for all industrial pumping equipment.

For a pump supplied with a coupling, alignment is undoubtedly the most important aspect of the start-up procedure, and it is repeated a number of times during the start-up procedure, with each successive alignment check being more precise than the preceding one. Further discussion on the subject of alignment can be found in Chapter 7, Section IX. Some users have the mistaken impression that they can specify that a frame-mounted pump be aligned at the factory, thus eliminating the need for careful alignment in the field. This is not generally true because the vibration to which the equipment is exposed during its shipment to the job site most likely causes the pump to be knocked out of alignment. Also, the pump alignment can be further compromised during activities such as the completion of the piping. There is no substitute for careful field alignment.

Another aspect related to alignment that causes some confusion to certain pump owners is the so-called "flexible" couplings with which some pumps are equipped. Some users mistakenly believe that flexible pump couplings are designed to withstand large amounts of misalignment, but this is not true. Flexible couplings are meant to withstand only very minor amounts of misalignment (refer to Chapter 4, Section XV), and they should by no means be considered a substitute for good coupling alignment.

B. Installation Checklist

1. Tag and Lock Out

This important safety function should always be the first step performed before beginning work on any pump. The equipment should be tagged and locked out at the starter or switch gear, so that it cannot possibly be started, either remotely or locally, while the pump is being serviced.

2. Check Impeller Setting

The impeller may have been set at the factory, but it is a good idea to recheck the setting prior to installing the pump. As pointed out in Chapter 4, Section II.A, the axial setting of the impeller is most critical for open impellers, as this affects pump efficiency. A common axial setting for open impellers is 0.012 to 0.015 in. Also, Chapter 4, Section II.D, discusses the trade-off between efficiency on the one hand, and thrust load and stuffing box pressure on the other, in setting of open impellers.

For closed impellers, the axial setting is much less critical, with the setting being merely to ensure that the impeller is not rubbing against the front or back side of the casing. Many closed impellers have no method of adjusting impeller setting. If there is a method for setting the impeller, it can be set ⅛ to ⅜ in. (depending on pump size) from the front side of the casing.

3. Install Packing or Seal

The mechanical seal can be installed at the factory, but this runs the risk of damaging the faces due to the vibration the pump undergoes in shipment. This is especially true for carbon seal faces (a very common seal face material), which are subject to cracking if they receive a shock in transit.

The packing should be changed as part of the pump start-up procedure, especially if the pump has been stored for an extended period of time before start-up.

4. Mount Bedplate, Pump, and Motor

The pump bedplate should be mounted reasonably level on its foundation, using shims if required. Some bedplates are equipped with leveling screws. The pump and motor are then mounted on the bedplate if this has not already been done.

Note that for pumps smaller than about 40 HP, a *stilt-mounted base* can be considered. This type of base requires no grouting, is above the floor so that corrosion damage is minimal, and is designed to move as a unit with pipe load, so that the pump, base, and motor maintain original alignment. This last benefit has been questioned by some, whose experience has been that the stilt-mounted base causes more pipe strain and deflection than a fully grouted bedplate and ultimately shortens the life of mechanical seals.

5. Check Rough Alignment

This first alignment check is done to ensure that the pump and motor can be brought into more precise alignment later on with more sophisticated alignment methods. (See discussion on alignment techniques in Chapter 7, Section IX.) Accordingly, this first rough alignment check can be performed with feeler gauges between coupling halves to check angular alignment (Figure 8.1) and a straight edge along coupling edges to check parallel alignment (Figure 8.2).

6. Place Grout in Bedplate

Most bedplates, whether they be cast or fabricated steel, have hollow cavities on their undersides. These cavities should be filled with grout to keep the bedplate from flexing while the

Installation, Operation, and Maintenance 455

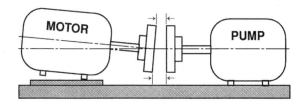

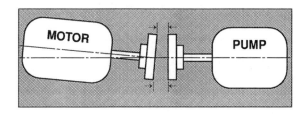

Figure 8.1 Rough angular alignment check using feeler gauges.

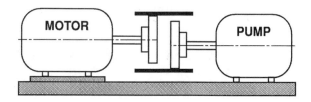

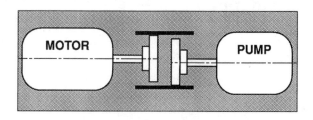

Figure 8.2 Rough parallel alignment check using straight edge.

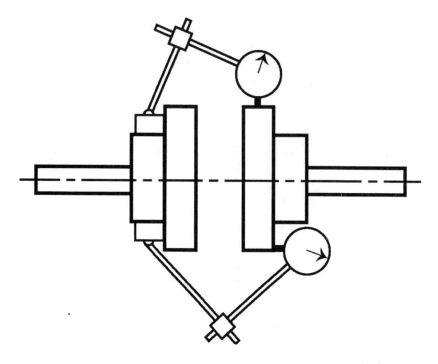

Figure 8.3 Rim and face alignment technique using dial indicator.

pump is operating, and to ensure that all loads are evenly transmitted to the foundation below the pump.

Grout should be allowed to fully dry, which generally takes 24 to 48 hr, before proceeding to the next step, because the bedplate may shift or flex while the grout is drying.

Stilt-mounted baseplates require no grouting, as discussed in Step 4.

7. Check Alignment

This alignment check should generally be done with a dial indicator, using the rim and face method shown in Figure 8.3 as a minimum. (More precise alignment techniques are discussed in Chapter 7, Section IX.) If the pump can operate at high temperature (or if the driver is a steam turbine which

Installation, Operation, and Maintenance

operates at a high temperature), the expected growth of the pump or turbine due to thermal expansion must be estimated or obtained from the supplier and taken into account in the parallel alignment procedure.

8. Flush System Piping

The system piping has not yet been connected to the pump. Before connecting the piping, it should be flushed to remove dirt, cutting chips, weld debris, or other foreign material from the piping. Temporary strainers can be installed in the piping to collect this debris.

9. Connect Piping to Pump

If the system piping is brought up to the pump, there is a high probability that it will not be exactly in line with the pump suction and discharge flanges (Figure 8.4). Applying

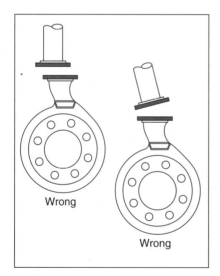
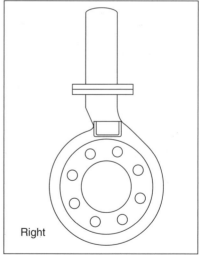

Figure 8.4 If piping is not in alignment with pump flanges, this can lead to excessive deflection of the pump stuffing box. (Courtesy of Goulds Pumps, Inc., ITT Industries, Seneca Falls, NY.)

leverage on the pipe to bring it in line with the flanges is liable to cause undue strain on the pump flanges, which could cause the stuffing box to deflect, eventually causing premature seal failure.

This can be avoided by bringing the system piping only up to the closest pipe restraint rather than all the way up to the pump. Then, the piping is run from the pump out past the closest pipe restraint. In this way, any strain that has to be put on the pipe to get the last two flanges to connect is carried by the pipe restraint, rather than by the pump flange.

When bolting the pump flange to the connecting piping, care should be taken if the pump flange is a cast material like ductile iron or bronze. Because the mating steel pipe often has a male register, this results in a gap between the mating pipe and pump flanges. If the flange bolts are over-torqued, this gap can cause the pump flange to break or to bend. Bolts should be tightened snugly, but not overly, and should be tightened uniformly, a little at a time on each bolt, rather than fully tightening a single bolt.

One way to avoid the possibility of breaking or bending the pump flange is to machine off the register on the pipe flange that attaches to the pump flange, thereby eliminating the aforementioned gap. If the register on the pipe is machined off, a full face gasket should be used between the two flanges.

The pressure-containing capability of a gasket is the bolt force used, divided by the gasket area. For the same bolt force, increasing the gasket area (which would be done if the register is machined off the pipe flange and a full face gasket is used) lowers the maximum pressure that the gasket can contain. This should be checked before the pipe flange register is machined off to ensure that the resulting gasket pressure capability is adequate.

Note that if the connecting pipe has a register on it, the gasket size is the same whether or not the pump has a machined register. This means that the pressure containing capability of the gasket is the same whether or not the pump has a register. Specifying a raised face flange on a cast pump that normally comes with a flat face flange does absolutely nothing in the way of increasing the pressure carrying capability of the gasket,

Installation, Operation, and Maintenance 459

while probably increasing the cost and delivery time for the pump.

10. Check Alignment

This alignment check is made to verify that no excessive strain has been applied to the pump by the piping connection. If the alignment has changed from the previous check (Step 7), the piping should be re-run.

11. Turn Pump by Hand

It should be possible to turn a new or newly overhauled pump by hand (with perhaps the aid of a strap wrench for larger pumps). Turning the pump by hand, the technician makes note of any unusual sounds or rubbing. This could be an indication of foreign matter such as dirt, weld spatter, or cutting chips located in the pump; a bent shaft; or a pump wear ring that is not machined concentrically or that has been installed improperly. Any unusual observation should be thoroughly checked prior to proceeding.

12. Wire and Jog Motor

Before the coupling halves between the pump and motor are connected, the motor wiring is completed, and the motor jogged, or bump started to confirm that the motor has the correct rotation. This is especially important for pumps such as ANSI pumps that typically have the impeller threaded on to the end of the shaft, tightening with rotation. If this type of pump is allowed to rotate in the opposite direction from that which is intended, it simply backs the impeller off until it hits the casing with the front side of the impeller, possibly damaging both the impeller and casing. Many other pump designs can be damaged if operated in reverse rotation, so this rotation check is important for all pumps.

13. Connect Coupling

Coupling halves are connected, completing the assembly of the pump unit.

14. Check Shaft Runout

Checking the shaft runout with a dial indicator after the alignment has been finalized and the coupling halves connected is a final assurance against improper assembly, a bent shaft, or imprecise machining of any of the mating components in the pump bearing housing, shaft, sleeve, or coupling. The dial indicator should be mounted on the pump base and the runout reading taken with the pin of the indicator located on the shaft at the pump seal, and the shaft rotated by hand from the coupling end. A well-made pump should be able to achieve 0.002 in. maximum total indicated runout in one revolution of the shaft. A runout in excess of this may not be permissible (for example, with pumps built to the API 610 specification described in Chapter 4, Section XIV.C). A higher runout can be tolerated unless it is otherwise specified, but 0.002 in. is a worthy goal.

15. Check Valve and Vent Positions

Valves in the suction line should be fully open, and drain lines on the pump should be closed. Vent connections on the pump should be open if the suction pressure is above atmospheric pressure; if not, provisions must be made to prime the pump (Step 17). Pump discharge valve should be closed on most radial and mixed flow pumps. (Higher horsepower units often require the discharge valve to be opened about 10%. Consult the instruction manual or manufacturer's recommendation for the specific pump.) Finally, all valves associated with the pump lubrication and cooling systems should be opened.

16. Check Lubrication/Cooling Systems

Lubrication systems include lubrication of the pump and driver. Refer to further discussion of lubrication systems in Section IV.A.1 to follow.

17. Prime Pump if Necessary

A centrifugal pump mounted above the source of suction requires priming prior to start-up. Priming fills the suction piping and pump casing, and vents the air out of the pump.

Installation, Operation, and Maintenance 461

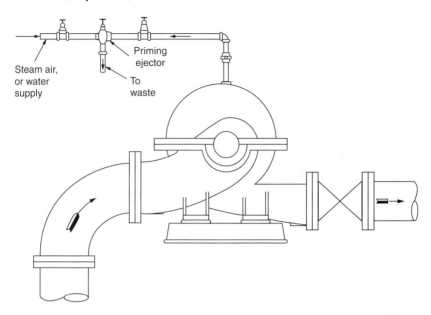

Figure 8.5 Use an ejector (shown) or vacuum pump for priming. (From *Pump Handbook*, I.J. Karassik et al., 1986. Reproduced with permission of McGraw-Hill, Inc., New York.)

If the pump has a foot valve in the suction piping, priming consists of filling the suction line from an external source while venting the casing. If there is no foot valve, the suction line must be filled by pulling a vacuum at the top of the pump casing, using either an ejector or a vacuum pump (Figure 8.5). The vacuum lifts the liquid up the suction line, priming the pump.

18. Check Alignment

This alignment check is done to ensure that filling the pump during the priming process has not caused the alignment to shift.

19. Check System Components Downstream

Make certain that all system piping, components, and the discharge vessel are ready to receive flow and pressure.

20. Start and Run Pump

Run for at least 1 hr, at design operating temperature if possible.

21. Stop Pump and Check Alignment

This final alignment check should be done hot (i.e., while still at operating temperature) if possible, to make certain that the thermal expansion allowance was correctly made. As pointed out in Chapter 4, Section XIV.C on API 610, a centerline-mounted case design minimizes movement of the casing due to thermal growth.

22. Drill and Dowel Pump to Base

If the pump operation is satisfactory, and if the pump is within specification on alignment and vibration, secure the pump and motor to the bedplate with tapered dowel pins to help maintain the alignment.

23. Run Benchmark Tests

Now that the pump has been successfully started, and while the pump is optimally aligned and balanced, this is the best time to run the *benchmark tests*. Benchmark tests are designed to create a signature of the performance characteristics of the pump while all components are new (or newly rebuilt). These performance characteristics can then be periodically remeasured and compared with the benchmark characteristics as a way to monitor any deterioration of the pump.

The most commonly measured benchmarks are the pump's hydraulic performance, temperature, and vibration. These are discussed in more detail in Section IV.C below.

III. OPERATION

A. General

Probably the best advice that can be given relative to good pump operation is that the operator should try to keep the pump operating at a healthy point on the head–capacity

Installation, Operation, and Maintenance

curve. If the pump is allowed to run too far out on the curve, it can experience problems with cavitation, cause excessive power load on the motor, and cause the pump radial bearings to be excessively loaded. Refer to Section III.B below for a discussion of the detrimental effects of operating a pump for an extended period of time below the minimum recommended flow for that pump.

Aside from running in a healthy operating zone, other operating practices should focus on minimizing the energy consumed. Keep the pump near its best efficiency point if possible. Variable-speed pumps should not be run at a higher speed than necessary to meet system requirements. Excessive throttling should be avoided if possible, as this wastes energy and will likely increase maintenance problems for valves and pumps. Pumps should never be operated with a closed suction valve. Valves on the discharge side should usually not remain closed for more than a minute, although this varies with pump size. High-energy pumps larger than several hundred horsepower are generally not recommended to ever run with the discharge valve completely closed, but rather should be cracked open about 10% as a minimum. Refer also to the discussion to follow on minimum continuous operating flow.

For shutdown of the pump, the pump should be de-energized prior to closing of any block valves. If the pump is pumping against a static head, and if there is no check valve in the pump discharge (as there should be), it may be necessary to close the discharge block valve immediately prior to shutting off the pump. There should be no delay between these two actions, however. This means that if the controls are not local to the pump, this operation may require two people.

Finally, operators of pumps should ensure that the equipment is well maintained, which is the subject of Section IV to follow.

B. Minimum Flow

A determination of the minimum acceptable continuous operating flow of a pump is necessary to set limits on control equipment and instrumentation, establish operational procedures, and determine the need for and size of minimum flow bypass systems. Unfortunately, there is no single answer to this question for every pump and every circumstance. Rather, each pump must be carefully examined in the context of its application to arrive at the minimum flow. For example, many small pumps handling ambient temperature water can safely operate at flows as low as 25% of best efficiency flow for long periods of time. However, a much larger pump, or one handling a liquid near its boiling point, must be operated at a much higher percent of best efficiency flow, perhaps as high as 80% of BEP.

The factors that should be considered in arriving at the minimum continuous flow for a particular pump are discussed in the paragraphs that follow. Not all of these considerations are applicable to every case. The determination of which of the factors discussed below dominates a particular pump application is not easy, and good engineering judgment must be used. In general, a conservative approach is to supply pump systems with a system to prevent the pump from operating below its recommended minimum flow.

1. Temperature Rise

As liquid passes through a centrifugal pump, its temperature increases due to several effects, including friction and the work of compression. Assuming that all the heat generated remains in the liquid, the temperature rise ΔT in °F is:

$$\Delta T = \frac{H}{778 C_p \eta} \quad (8.1)$$

where:
H = total head, in feet
C_p = specific heat of the liquid, in Btu/lb°F
η = pump efficiency as a decimal

At very low flow rates, with pump efficiency being very low, temperature rise is the highest. Temperature rise limits are determined by the difference between $NPSH_a$ at the pump suction and the NPSH required by the pump. An arbitrary but usually conservative practice calls for limiting temperature rise

Installation, Operation, and Maintenance

in centrifugal pumps to about 15°F, although for pumps handling cold liquids, a rise of 50°F or more may be quite acceptable. For light hydrocarbons with low specific heat and high vapor pressure, and where $NPSH_a$ is close to $NPSH_r$, it would be wise to double-check to ensure that the temperature rise does not elevate the liquid temperature to the boiling point.

2. Radial Bearing Loads

As discussed in Chapter 1, Section V, radial bearing loads can affect both shaft deflection and bearing life. As Figure 1.7 shows, the radial loads are at a minimum at the BEP and may be considerably higher at reduced flow rates. This can limit minimum flow if the excessive radial bearing load causes excessive shaft deflection and premature seal or bearing failure. Of course, whether or not this is a factor depends on how conservatively the pump bearing system was originally designed. As an example, many manufacturers of ANSI process pumps only offer two or three different bearing frames for over 20 pump sizes. There may be six or seven pump sizes using the same bearing frame. Obviously, the smallest pump size in a given group has a greater design safety factor on its bearing system than does the largest pump in the same group.

3. Axial Thrust

Axial thrust is discussed in Chapter 4, Section II.D. Axial thrust at shutoff is higher than the axial thrust at BEP, in the same proportion as the ratio of pump head at shutoff to the head at BEP. As per the discussion above about radial bearings, the design safety factor of the thrust bearing system must be considered to determine if this factor might restrict minimum flow.

4. Prerotation

This is the change in the inlet velocity triangle as a pump operates off the BEP, causing the liquid to spiral in the suction ahead of the impeller. The net effect of pre-rotation is to lower the pump's suction head and efficiency. The effect is more pronounced the farther to the left of BEP the pump operates, and is more pronounced on larger pumps.

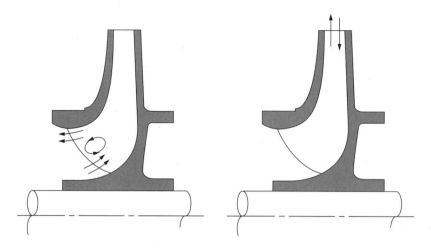

Figure 8.6 Recirculation occurs at impeller inlet and/or outlet at capacities below BEP. (From *Pump Handbook*, I.J. Karassik et al., 1986. Reproduced with permission of McGraw-Hill, Inc., New York.)

5. Recirculation and Separation

Recirculation is a flow reversal at the suction and/or discharge tips of impeller vanes (Figure 8.6). Suction recirculation causes noise and cavitation-like damage at the impeller inlet, while discharge recirculation can cause similar damage on the pressure side of the outlet edge of the vane. The flow also separates from the vanes at these points. Other symptoms of recirculation include cavitation-like damage to the shrouds near the vane outlets, and unexpected shaft breakage or thrust bearing failure.

It is important to note that this cavitation-like damage occurs on the pressure side of the impeller vane. When looking at the inlet of an impeller, the pressure side of the vane is the side that is usually not able to be seen directly with the eye, but which requires a mirror to see behind the vane. This is the opposite side of the impeller vane from that where conventional cavitation damage occurs.

Suction and discharge recirculation damage can occur as the pump moves away from its BEP. While the onset of recirculation alone is not necessarily a cause for concern, if the pump is allowed to operate too long in a period where there is too much of this recirculation, it can cause damage to the pump. Thus, this is a criterion used for determining minimum flow on some pump types. The pump types most likely to exhibit problems with recirculation are high-energy pumps (i.e., more than 100 HP per stage), and pumps with high suction specific speed.

The percent of BEP capacity at which suction recirculation begins is a function of the pump specific speed N_s; the suction specific speed S; and whether the pump is single suction, double suction, or multi-stage (Ref. [1]). At higher values of both N_s and S, the minimum acceptable flow (as a percentage of BEP) to prevent recirculation increases. Also, the flow at which recirculation begins (as a percentage of BEP) is higher for double suction pumps than for single suction pumps, and higher still for multi-stage pumps. For example, for N_s = 1000 and S = 7500, the percent of BEP at which recirculation begins is about 50%, 55%, and 65%, respectively, for single suction, double suction, and multi-stage pumps. For the same value of N_s but S = 10,000, the percent of BEP at which suction recirculation begins is about 63%, 73%, and 83% of BEP for the same three pump types. For higher S values, these percentages are even higher. This effect is one reason why large-eye impellers for reduced $NPSH_r$ (higher S value) must be used with caution, as this can restrict the safe operating range of the pump.

Note that there is neither complete agreement within the pump industry as to exactly when the onset of recirculation begins for a given pump type, nor is there complete agreement as to how the onset of recirculation should relate to the recommended minimum flow for that particular pump. Some manufacturers are more conservative in their approach than others, and some industrial pump end users such as refinery owners tend to be more conservative than some pump manufacturers in setting minimum flows. Good communication between the pump end user and manufacturer is essential in establishing the recommended minimum flow to avoid recirculation for a given pump, especially a high-energy pump.

6. Settling of Solids

For liquids containing solids, the minimum flow must be high enough to prevent the solids in the liquid from settling in the pump or in the piping.

7. Noise and Vibration

At reduced flow rates, pump operation may be noisy due to a combination of the effects of prerotation, recirculation, and cavitation. Vibration may be caused by the same factors, plus the higher bearing loads and greater shaft deflection that occur at low flows.

8. Power Savings, Motor Load

As discussed in Chapter 2, Section VII, the BHP curve for higher specific speed pumps may be flat, or may even slope upward at reduced capacities. Operating a pump with high specific speed to the left of BEP may actually cause the motor load to be higher than at BEP.

C. Ten Ways to Prevent Low Flow Damage in Pumps

Centrifugal pumps have a minimum operating flow rate, below which the pump should not be run for long periods without sacrificing reliability. Naturally, it is best, from the standpoint of long-term reliability and operating efficiency, to operate pumps close to their BEP; but there may be periods where reduced flow demand or system changes cause the pump to run at reduced flow rates. Minimum flow is usually expressed as a percentage of the flow at the best efficiency point (BEP) of the pump, for a given impeller diameter. Extended operation below recommended minimum flow can lead to excessive vibration, impeller damage, and premature bearing and seal failures. With most sealless (magnetic drive and canned motor) pumps, the allowable runtime below the minimum flow rate may be only a matter of minutes before significant damage to the pump occurs.

Installation, Operation, and Maintenance

The recommended minimum flow rate varies considerably from one size and type of pump to another, ranging from 10 to 80% of the BEP flow. The major characteristics of the pump that influence the determination of minimum flow were discussed in the previous section, and include the energy level (horsepower per stage) of the pump, the specific hydraulic design of the impeller inlet, the mechanical design of the pump shaft and bearing system, and the cost and criticality of the pump. For sealless pumps, the amount of heat generated in the canned motor or across the magnets is also a consideration, as is the specific heat of the pumped fluid.

There is no accepted industry standard for minimum flow that applies to all pump types, and even different manufacturers of the same pump type may have a range of acceptable minimum flows for a given application. That being said, however, the pump manufacturer is still the best place to start for the recommended minimum flow for a particular pump installation. Some manufacturers show this information on the pump performance curve.

What to do, if anything, to protect the pump from the consequences of low flow damage is an economic decision made by the user. This analysis considers the cost of the pump, minimum flow protection system, and downtime/lost production, in addition to energy and maintenance costs. Other potential factors may include health, safety, and environmental risks.

A significant majority (estimated at over 80%) of centrifugal pumps have no minimum flow protection whatsoever. The vast majority of the pumps installed annually are fairly low horsepower pumps used for transfer or cooling, which are not normally expected to operate over a wide range of flow. These pumps are unlikely to operate below the minimum flow point, except for the inadvertent closure of the main discharge valve or other inadvertent blockage of the system. Furthermore, many of these relatively low-cost pumps are not deemed worth the capital expenditure for minimum flow protection. This is especially true for noncritical pumps used in residential, commercial, and light-duty industrial services.

For the 20% or so of pump applications that do require minimum flow protection, there are a number of choices available to the user or system designer. The determination of which choice to use considers accuracy, reliability, cost, and criticality, and is very specific to the application. One other factor in the selection process is whether the pump needs to be protected for minimum flow in a modulating fashion (i.e., keeping the pump operating but with a certain amount of flow bypassed), or whether it is sufficient to simply alarm or trip off the pump in the event that the flow rate drops below the recommended minimum flow. Finally, additional protection obtained from the same instrument should be considered (e.g., a power monitor can protect against both high and low flow damage to a pump, while a relief valve will only protect against low flow).

For the 20% or so of pump applications that do require minimum flow protection, the following are ten different methods that can be considered for protection of the pump. All of these are used in industry, and some systems use a combination of these methods for protection against low flow excursions.

1. Continuous Bypass

This may be the lowest capital cost method of protecting a pump, whereby a bypass line with an orifice allows a fixed amount of flow to be pumped continuously back to the suction source. This always ensures that the pump delivers its recommended minimum flow, even if the main line is shut off completely. The biggest negative aspect of this system is that the pump must be oversized in the first place to allow for the continuously bypassed flow. Second, and sometimes more importantly, is the fact that energy is wasted due to the extra horsepower required to accommodate the bypassed flow. There may also be a potential for product damage when being forced through an orifice. Still, this alternative is chosen by many industrial users for pumps in the range of 50 HP and below.

2. Multi-Component Control Valve System

This type of system relies on a continuous flow measurement in the system. When the flow drops below the recommended minimum flow, a signal is sent to a valve in the bypass line that either opens it completely or modulates the valve so that it gradually opens. This valve may be a solenoid type if it is strictly on/off, which is generally the least costly method, or may be a pneumatically or electro-mechanically actuated control valve. This method of bypass eliminates the energy waste of continuous bypass, but relies on considerably more complexity than a continuous bypass system. The system includes multiple components, each of which could fail. It requires a power supply and, if pneumatically actuated, an air supply. Maintenance costs are typically higher than other alternatives. As such, it is one of the more costly methods of minimum flow protection. However, it is deemed by many users to be the best approach, especially if the system already includes a reliable method of flow measurement.

The operating cost of a pump system with a bypass control valve low flow protection system can be much less than that of a system with a continuous bypass system. In the system shown in Figure 8.7, a 50-HP ANSI process pump

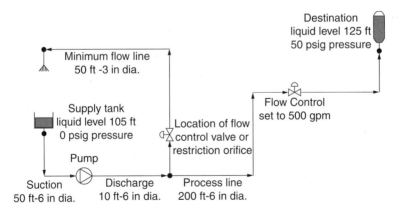

Figure 8.7 Simple piping system with a minimum flow recirculation line. (Courtesy of Engineered Software, www.eng-software.com, Lacey, WA.)

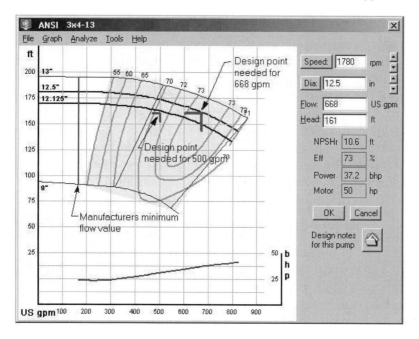

Figure 8.8 The design point on this ANSI pump is selected at 668 gpm to deliver 500 gpm through the system and ensure a 168 gpm continuous bypass. (Courtesy of Engineered Software, www.eng-software.com, Lacey, WA.)

(Figure 8.8) must deliver 500 gpm of water for 6000 hr annually, and must operate at the specified minimum flow of 168 gpm for 100 hr per year. Power costs are assumed at $0.10USD per kWh. Total annual pumping costs are calculated with two minimum flow protection systems: (1) a continuous bypass and (2) a bypass control valve, which only opens when required. An analysis of annual pumping costs using these two methods shows that the continuous bypass method costs $3320 more per year to operate. For details on how this calculation was made, visit the following Web address: http://www.eng-software.com/kb/item.aspx?article=1310.

3. Variable Frequency Drive

Variable frequency drives (VFDs), described in detail in Chapter 6, Section IV, change the frequency of the electric motor

Installation, Operation, and Maintenance

on the pump to slow the pump down when the demand for lower flow is called for by the process. For most systems, this keeps the pump operating near its BEP at all times, and prevents the moving of the pump to a lower percentage of BEP that causes the damage to pumps. VFDs are being used more and more in process applications and have eliminated the need for other minimum flow protection when they are being used. (Note that with canned motor and magnetic drive pumps, there will still be a minimum flow required to carry away the heat caused by the motor or magnetic flux, and to lubricate the bearings.) VFDs are relatively expensive, although their cost has reduced dramatically in recent years with rapid improvements in technology. Other benefits of VFDs include lighter loading of pump seals and bearings, and the ability to "soft start" equipment at slower speeds, thus reducing the strain and high current caused by across-the-line starts.

4. Automatic Recirculation Valve

An automatic recirculation valve (Figure 8.9) combines the features of a check valve, bypass valve, and a flow sensing element, as well as providing pressure breakdown, and has a number of advantages compared to other approaches. Compared to the multi-component flow control valve system, it has fewer components, requires lower installation and operating costs, has less environmental effect (no dynamic seals), and does not require air or electricity. Compared to systems that just shut down the pump, it keeps the pump and the system operating (does not shut down the process). Disadvantages include its relatively high cost, and the fact that these valves are not normally available in alloys higher than stainless steel, thus eliminating many chemical services. Also, automatic recirculation valves are generally unsuitable for fluids containing solids.

5. Relief Valve

This system simply relies on a pressure relief valve in the pump discharge piping being set to relieve back to suction

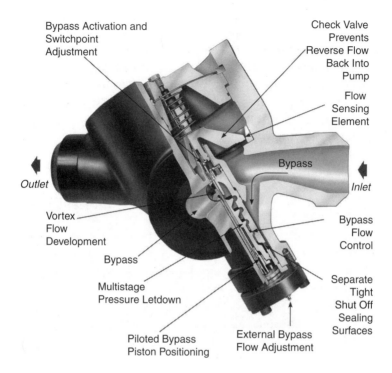

Figure 8.9 Automatic recirculation valve. (Courtesy of Tyco Valves and Controls, Blue Bell, PA.)

when the pressure put out by the pump reaches a certain set point pressure. The characteristic performance curve of all centrifugal pumps is such that as the pump delivers a lower capacity (flow), the pressure (head) that the pump produces gets higher. Some pumps exhibit a steeper capacity vs. pressure curve than others. A pressure relief valve is particularly appropriate for pumps with fairly steep capacity vs. pressure curves. Thus, this is the normally chosen minimum flow protection system for regenerative turbine style pumps. It is also the accepted method for protection of many fire pump systems. (Note: With liquid returning to the suction, the internal temperature of the pump may rise, so this may limit the use of the relief valve for an extended period of time.) Also, relief valves are subject to maintenance and testing/calibration regimens.

Installation, Operation, and Maintenance

Finally, it is import to know that most relief valves are not designed to operate continuously (i.e., for a long period of time), so the pump should not be sized such that the relief valve is continuously operating.

6. Pressure Sensor

This device relies on the fact that as the flow decreases with a centrifugal pump, the amount of pressure produced by the pump increases. This high-pressure signal is then used to either open a bypass valve at a high pressure (low flow) indication, or to simply trip the pump. For pumps with relatively steep head vs. capacity curves, this method can be economical and reliable. For pumps with flatter head–capacity curves in the low flow range, it is considered to be less reliable than other approaches.

7. Ammeter

The amp draw of the electric motor varies across the range of flow produced by a pump. For many pumps, the amp draw of the pump is lower at lower flow rates and increases with increasing flow. Thus, it is possible with many pump types to monitor amp draw, and to alarm or trip the pump when the amps drop below a certain set point level. While this is a relatively inexpensive way to protect the pump against low flow damage, it has some potential drawbacks. It may be subject to unacceptable inaccuracy due to current fluctuations in the system and the fact that the amp draw curve can be fairly flat at lower flow rates. In general, the lower the nominal speed of the driver, the less practical amp monitoring becomes, due to the flatter curve and resulting smaller amp range. The device may also need to be disabled during start-up of the pump due to the high current draw that occurs then. On the plus side, an ammeter can also be used to protect a pump from damage due to excess flow.

8. Power Monitor

Power monitors measure motor horsepower. Because most pump curves have a horsepower curve that rises with increasing flow, it is possible to set the motor to shut off if the power drops below a minimum setpoint, so this is a reliable protection against low flow problems. Power monitors are typically more reliable that ammeters because they are not subject to fluctuating results with variations in line current. For pumps with relatively flat head–capacity curves, where pressure measurements are not reliable, the power monitor may be the best choice for low flow protection. Power monitors can be programmed to protect against excessive flow (high power) as well as minimum flow (low power). They can also be programmed to ignore momentary power spikes, where an ammeter might trip the motor. They are adjustable to allow altering setpoints should the process requirements change. They are not appropriate for many mixed-flow pumps, which may have a nearly flat horsepower curve as a function of pump flow. If the power monitor measures motor input power rather than motor output power, it may not be as accurate, because the efficiencies of small motors at low power can be quite low.

9. Vibration Sensor

Some pump systems have vibration monitors to alarm or trip the pump if the pump begins to vibrate excessively. One of the things that occurs at lower flow rates is that the pump may indeed vibrate significantly higher than normal. Note that high vibration levels may also be an indication of other problems with the pump, such as misalignment, imbalance of the impeller, or cavitation. This device, while relatively expensive, is part of the low flow protection system on many critical process pumps. If vibration is associated with pump wear or other factors, such as bearing degradation, it is also possible to project the time of failure and plan preventive maintenance.

10. Temperature Sensor

At very low flow rates, the temperature of the pumped liquid increases due to friction, the work of compression, and the recirculation of the liquid within the pump. Thus, if the pump discharge is shut off by a closed main valve, the temperature of the liquid inside the pump will begin to rise. The rate of temperature rise depends on several factors, including the pump efficiency, the specific heat of the liquid, pump head, and the volume of liquid in the pump. One method of protecting the pump against this occurrence is by monitoring the temperature in the pump casing (or containment shell in the case of a magnetic drive pump), and tripping off the pump when the temperature rises above a certain setpoint value. This may be relatively inexpensive but not necessarily too reliable because, by the time it shuts off the pump, damage may have already occurred in the pump.

IV. MAINTENANCE

This section divides maintenance activities into two categories: (1) *regular maintenance* is performed according to a fixed schedule, and is done to keep the pump running optimally, and (2) *preventive maintenance* has as its specific goal the prevention of an unplanned emergency shutdown of the pump, which often results in a much more expensive pump repair when repair becomes necessary, and also often involves costly disruptions in production activities.

A. Regular Maintenance

1. Lubrication

Grease lubricated pumps should be regreased about every 2000 hr (or about every 3 months for continuous-duty pumps). If the grease cavity is not vented properly, it may be possible to apply too much grease, causing the pump to run hot.

Oil-lubricated pump bearing housings typically have the oil level in the bearing housing set by the manufacturer. Either a too high or too low oil level can cause the bearings to not be ideally lubricated, reducing their life.

A common system for maintaining a constant oil level in the bearing housing is shown in Chapter 4, Figure 4.12. A leveling bar below the sight glass is set by the manufacturer at the prescribed oil level. If the level drops below the setpoint, it creates an air path that allows oil to flow from the sight glass into the bearing housing until the prescribed oil level is reached again. Thus, the sight glass serves as an inventory of spare oil to make certain the oil level stays constant, and as a visual indicator to alert the operator that the oil seals are leaking and need replacement, or that additional oil needs to be added.

Oil mist lubrication of rotating equipment through a centralized plant oil mist system is becoming more popular. This type of lubrication system, although requiring capital investment to implement, can result in the benefits of longer bearing life and reduced oil consumption.

Note that some coupling types (e.g., gear and grid spring types) require periodic lubrication.

2. Packing

Cutting and installing packing is an art, and proper installation techniques can do a lot to extend the service life of the packing. When making packing rings from a roll, a dummy shaft of the same size as the pump shaft or sleeve (if the pump has a sleeve) can be wrapped with the packing to ensure that the rings are cut to the right length. Cutting the rings on a diagonal rather than straight across helps minimize leakage through the split (the point where the two ends of a ring come together when they are wrapped around the shaft or sleeve).

When the rings are installed, each ring should be compressed as it is installed, with a bushing that can slide over the shaft. If this is not done, the lower rings of packing are not compressed when the gland is tightened down, and only the final outer rings of packing do any sealing.

Another good idea is to stagger the splits of the individual rings 90° apart from each other as the packing rings are installed, to prevent the possibility of liquid leaking out through the ring splits.

Before installing new packing, it is a good idea to examine the old packing and the stuffing box, especially if the packing has prematurely failed. This can sometimes provide a clue as to why the packing failed, and can prevent a recurrence of the problem. For example, if the outside surfaces of the packing rings (next to the stuffing box bore) are worn

Installation, Operation, and Maintenance

instead of the inside surfaces (next to the shaft or sleeve), the packing was likely not installed properly, or else the packing size used was too small for the stuffing box bore. If the outer rings of packing (the ones nearest the gland) are worn but the inner ones are not worn, chances are the inner rings were not compressed with a bushing by the technician when the packing was installed, so the outer rings of packing were the only ones doing any work.

Another thing to consider at the time that packing is being replaced is the possibility of hardening the shaft or sleeve that runs under the packing with a metallic, ceramic, or other type of hard coating. This minor modification of the pump can substantially increase the operating runtime before the next repacking. Excessive leakage around the shaft is often the first thing to go wrong with a pump, especially if the sleeve is not hardened or if the liquid contains abrasives.

3. Seals

When doing maintenance on a mechanical seal, care should be taken in handling the seal's dynamic running faces. Very small impurities, even oil residue on the technician's hands, can reduce the seal's performance or life.

Lubricant used to mount O-rings or other elastomeric parts of seals should be compatible with the elastomeric material. Hydrocarbon-based oils should not be used for lubricant because these oils can cause some elastomeric materials to swell. A vegetable-based oil such as corn oil or a synthetic lubricant is generally better for this function because these will not cause swelling of most seal elastomers.

Set screws, O-rings, and, if possible, springs should be replaced if a seal is being overhauled.

B. Preventive Maintenance

A preventive maintenance program relies on the following maintenance operations being periodically performed on pumps to lengthen the runtime between repairs, limit the damage done to pumps, and prevent unscheduled outages of the equipment. Each preventive maintenance program is unique and depends on the size, cost, location, and importance of the pumps; the size and experience level of the pump owner's maintenance organization; and the severity of the pump services.

1. Regular Lubrication

As discussed in Section IV.A.1, regular lubrication can increase the life of the pump and driver bearings, and the coupling if it requires lubrication.

2. Rechecking Alignment

A quick periodic alignment check when the pump is shut down ensures that the equipment maintains good alignment and does not move out of alignment due to vibration or structural shifts.

3. Rebalance Rotating Element

This should be done during a pump overhaul if the impellers or coupling are machined or otherwise modified, or if they have suffered any wear, erosion, or cavitation damage.

4. Monitoring Benchmarks

The three most important benchmarks to monitor as part of a preventive maintenance program are the (1) pump hydraulic performance, (2) temperature, and (3) vibration. These three benchmarks are discussed in more detail below.

C. Benchmarks

1. Hydraulic Performance

Chapter 3, Section VI, discusses procedures and instrumentation for measuring pump capacity, total head, and power in an installed system. This data should be recorded when the pump is first put into service to the best extent possible with the testing instrumentation that is available at the installation.

Installation, Operation, and Maintenance

If no flow measuring instrumentation is available at the site, pump total head (or discharge pressure if the liquid density will not change) can be measured with the discharge valve in its normally open position and fully closed. (Check with the manufacturer, however, before testing with a fully closed discharge valve, as some pumps cannot be operated, even briefly, with the discharge valve fully closed.)

When the hydraulic performance parameters are periodically rechecked, this provides an indication of the deterioration of pump performance caused by the wear in the leakage joint, other corrosive and/or erosive wear on the pump impeller or casing, and cavitation.

2. Temperature

A pump begins to run hotter if its bearing is about to fail, if it has inadequate lubrication or cooling, or if the pump is misaligned. These are excellent reasons to monitor the temperature of the pump at the bearing housing and to investigate any increase in bearing temperature. Many engineers and maintenance technicians make a habit of feeling the bearing housing with their hands to see if the bearings are running hot. Unfortunately, the human hand does not make a very good thermometer, as anything over about 130°F feels "too hot" to the hand, while pump bearings can run much hotter than this. The upper temperature limit is when the oil or grease lubricant begins to carburize, which is in the range of 180 to 200°F for most lubricants. A much better indication of the pump bearing temperature can be obtained using a thermocouple or thermowell.

If the pump bearings appear to be running hot, consideration should be given to simple reasons why the pump might be running hot before pushing the panic button. The oil level in the bearing housing should be checked to make certain that the bearings have adequate lubrication. Perhaps the process liquid is hotter than usual, which would conduct heat along the shaft and cause the pump bearings to run hotter. Maybe the pump has lost cooling water to the bearing housing or seal, or perhaps the cooling water temperature is higher than normal. Consider the ambient temperature, because a pump tends to run hotter when the ambient temperature is hotter. If none of these considerations apply, a vibration check (discussed below) should be made prior to shutting down the pump. After the pump is shut down, the cause of the high temperature can be further investigated. Alignment should be checked first; and following that, other possible causes can be considered using the troubleshooting techniques discussed in Section V below.

Note that a new or newly overhauled pump should be run for an hour or so (at operating temperature) before establishing the temperature benchmark. It can actually run a little hotter when the pump is first started, until the oil seals run in a bit.

3. Vibration

Figure 8.10 shows the recommended maximum acceptable field vibration of horizontal end suction frame-mounted pumps handling clear liquids. The figure shows maximum vibration velocity in inches per second RMS (and above that, in millimeters per second RMS), unfiltered, as a function of input power. This chart shows, for example, that a pump with an input power of 10 HP (7.5 kW) has a maximum allowable field vibration of about 0.14 in./sec (3.5 mm/sec), while one with an input power of 100 HP (75 kW) has a maximum vibration allowance of about 0.21 in./sec (5.3 mm/sec). Vibration velocity is measured on the bearing housing of horizontal pumps and at the top motor bearing of vertical pumps. Measurements should be taken in all three planes (H, V, and A in Figure 8.10), with Figure 8.10 showing the maximum allowable vibration measured in any plane.

Some pump users are reluctant to check vibration readings on pumps. Their argument is that because they do not have the sophisticated multi-spectrum vibration analysis equipment that measures frequency as well as displacement or velocity of vibration, they are not able to determine the cause of the vibration. It is true that to perform a thorough analysis to determine if vibration is being caused by imbalance, misalignment, bad bearing, etc., one must have instrumentation to

Installation, Operation, and Maintenance

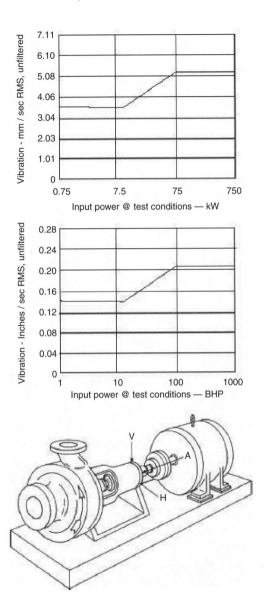

Figure 8.10 Acceptable field vibration limits for centrifugal pumps handling clean liquids. (Courtesy of the Hydraulic Institute, Parsippany, NJ; www.pumps.org and www.pumplearning.org.)

determine the frequency as well as the amplitude of vibration. However, a simple measurement of vibration displacement or velocity (comparing it to the benchmark reading taken when the pump was first put into service) is enough to call attention to the fact that a problem is developing. The user can then bring in more sophisticated analysis equipment or do other checks to help determine the root cause of the problem, hopefully before long term damage is done to the pump. A simple hand-held probe to measure vibration displacement or velocity is adequate to serve as a preventive maintenance tool for setting a benchmark level and comparing periodic readings against this value.

If more sophisticated vibration equipment is available (along with the proper training on its use), vibration readings can be used to do a much more in-depth troubleshooting analysis of a pump. Table 8.1 is a vibration troubleshooting summary chart that can serve as a guide to vibration root cause.

Table 8.1 can serve as a guide to vibration root causes, and how to interpret vibration symptoms. It is not meant to include all possibilities, and is in the order of the frequency value observed, not in order of likelihood or importance to reliability.

In studying Table 8.1, be mindful of the following definitions:

- *Runout:* false vibrations picked up by a proximity probe, actually reflecting shaft scratches, gouges, undulations, etc., or static shaft bends or misalignment ("mechanical runout"), or eddy current sensitivity variations along the shaft surface ("electrical runout"). Runout can be quantified by observing the apparent shaft orbit with a dial indicator when the shaft is slowly turned.
- *Inboard (IB):* the coupling side of the pump or driver.
- *Outboard (OB):* the end of the pump or driver opposite to the coupling.
- *Narrowband:* vibration response that is at a single frequency.
- *Broadband:* vibration response that covers a band of frequencies on a spectrum plot.

Table 8.1 Vibration Troubleshooting Summary Chart

Multiples of Running Speed	Other Symptoms	Probable Causes
0.05 X to 0.35 X	Vibration response is broad in frequency	Diffuser or return channel stall (flow vortices around the front of the diffuser)
Exactly ½ X, ⅓ X, ¼ X, etc.	Vibration may increase dramatically shortly after this appears	Light rub, combined with low shaft or bearing support natural frequency
0.41 X to 0.49 X	Near shaft natural frequency and orbit "pulses," forming an inside loop (a smaller orbit inside of a larger orbit) Vibration onset is sudden at a speed roughly twice the excited natural frequency, and "locks" onto the natural frequency despite speed increase	Stable fluid whirl if whirl frequency is below critical speed, or rotordynamic instability due to fluid whirl in close clearances if whirl frequency becomes equal to shaft critical speed (e.g., bearing "oil whip")
0.6 X to 0.93 X	Smaller peaks at (1–f) X, and at ± (1–f) X "sidebands" of the first several multiples of running speed (where f = 0.6 to 0.93) This is often accompanied by rumbling noise and "beating" felt in the foundation It occurs at part-load capacities, but disappears at very low capacity; and independently depends on speed and flow	Internal flow recirculation probably at suction, particularly if accompanied by the crackling noise of cavitation at flow rates below BEP
Less than 1 X	Increased broadband vibration and noise level below running speed as NPSH decreases, especially at high flows Often accompanied by cracking noise	Classical cavitation (i.e., cavitation without recirculation)

Table 8.1 Vibration Troubleshooting Summary Chart (continued)

Multiples of Running Speed	Other Symptoms	Probable Causes
1 X	Stronger on shaft than on housing. Hydraulic performance and/or suction pressure normal. Axial vibrations within normal limits and vibrations above runout increase with roughly speed squared	Imbalance in rotating assembly
	Vibration highest on driver/driven IB bearing housings	Pump coupling imbalance
	Vibration highest on driver IB or OB bearing housing	Driver rotor imbalance
	Vibration high on pump IB or OB housing, low on driver	Pump rotor imbalance
	Natural frequency near 1 X, as determined by bump test	Resonance
	Axial vibrations are over ½ of H or V vibrations, or vibrations increase much slower than the square of the speed. Also, bearing oil temperature is high, and/or temperature readings show one side of the bearing housing is hotter than the other by more than 5°F	Pump/driver misalignment at the coupling
	Discharge pressure pulsations are strong at 1 X but not at impeller vane pass in a single volute pump	Clogged or damaged impeller passage
	Discharge pressure pulsations are strong, particularly at vane pass, especially at flows far above or below the design point	Volute tongue designed too close to impeller OD, or excessive impeller/volute eccentricity

Installation, Operation, and Maintenance

Frequency	Symptoms	Probable Causes
2 X	Axial vibrations are low	Clogged or damaged impeller passage
	Both shaft and housing vibrations are strong, and discharge pressure pulsations are strong, but impeller vane pass vibrations are low in a twin volute pump	Twin volute vanes designed too close to impeller OD, or clogged or damaged volute, or impeller/volute eccentricity
	Same as above, but with high vane pass vibrations and discharge pressure pulsations	Looseness in bearing support or cracked shaft
	Shaft vibrations much stronger than housing vibrations, and approach or exceed bearing clearance. Decrease in shaft first bending natural frequency. Other multiples of running speed may be stronger than usual, but especially 2 X	
	Shaft vibrations stronger than housing, and torque pulses	Torsional excitation
	In motor driven pumps, with speed equal to electric line frequency, and highest vibration at motor IB housing, with vibration exactly at 2 X line current frequency of 50 or 60 Hz	Electrical problem with motor
	Axial vibrations are over ½ of horizontal vibrations, 1 X vibrations are also high, and bearing oil temperature is high	Pump/driver misalignment at the coupling

Table 8.1 Vibration Troubleshooting Summary Chart (continued)

Multiples of Running Speed	Other Symptoms	Probable Causes
Number of impeller vanes X running speed	Discharge pressure pulsations reasonably low and both shaft and housing vibrations high	Volute too close to impeller OD due to design or excessive rotor eccentricity
	Discharge pulsations relatively low, but housing vibrations much higher than shaft vibrations	Piping mechanical resonance at vane pass
	Discharge pressure pulsations high at vane pass frequency but suction pressure pulsations reasonably low	Acoustic resonance in discharge pipe
	Suction pressure pulsations high	Acoustic resonance in suction pipe
	Pump rotor or casing natural frequency close to vane pass	Resonance
Several multiples of running speed, including 1 X, 2 X, 3 X, 4 X and possibly higher	Orbit shows sharp angles or shows evidence of "ringing", and/or spectrum shows evidence of exactly ½ or ⅓ X response; grinding noises and speed changes may be evident	Internal rub or poorly lubricated gear coupling
	Orbit is "fuzzy" but does not pulse or "ring"; spectrum may also exhibit ⅓ or ½ running speed	Jammed, clogged, or damaged seal
	Orbit pulses, usually in one direction much more than the other and shaft vibrates more than housing; harmonics of exactly ½ X may also be present	Shaft support looseness, especially of bearing to bearing housing or bearing housing to casing
	Housing and casing vibrate much more than shaft, combined with vibration response over a broad range of frequencies below running speed	Looseness in pump casing, pedestal, or foundation

Courtesy of William D. Marscher, Mechanical Solutions, Inc., Parsippany NJ, www.mechsol.com.

Installation, Operation, and Maintenance

V. TROUBLESHOOTING

A typical on-the-job problem faced by readers might be an operating pump making a lot of noise, or with bearings running too hot. These symptoms could be due to any of a number of causes. They might be caused by vibration due to imbalance, vibration due to misalignment, cavitation, air entrainment, or a number of other possible causes. Troubleshooting such problems is like detective work. The problem is solved by making a list of suspects and trying to eliminate the suspects one at a time, starting with the ones that are most easily eliminated. The ones most easily eliminated are usually those that can be checked out without shutting down the pump. The next ones to be eliminated are those that can be analyzed by shutting down the pump, but without having to disassemble the pump. Finally, the last group of suspects are those that can only be checked by disassembling the pump.

Table 8.2 is a troubleshooting chart that lists eight of the most commonly observed symptoms of centrifugal pump problems, along with the possible causes for these symptoms.

VI. REPAIR

A. General

A thorough discussion of pump repair would fill another entire book and is therefore beyond the scope of this section. Some very light-duty pumps such as residential pumps have very little repair capability and must simply be replaced when they wear out. Commercial and industrial pumps are generally repairable to some degree or another, with the degree of complexity and therefore the amount of repair that can be economically done varying considerably. Some sophisticated pump users maintain elaborate repair shops to do their own pump repairs, while other users send out their equipment to repair centers that are especially suited to the repair of rotating equipment.

If it is available, the manufacturer's instruction manual should be consulted prior to beginning any repair operations.

Table 8.2 Pump Troubleshooting Chart

Symptom	Possible Causes
Insufficient flow	1-2-3-4-6-7-10-11-12-13-20
Pressure too low	1-2-3-4-6-9-10-11-12-14-20
High amp reading	5-6-7-8-9-10-13-15-17-18-22-23-26
Packing leaks too much	15-17-21-22-23-24-25-27-28
Seal/packing fails early	1-7-11-15-17-19-21-22-23-24-25-26-27-28
High vibration/noise	1-2-3-7-8-11-13-15-16-17-18-19-20-24-25
Bearings short lived	1-6-7-9-11-15-17-24-25-28-29-30
Bearings run too hot	1-7-11-15-18-19-24-25-29-30-31-32

1. Air entrainment
2. Suction obstructed
3. Poor sump design
4. Speed too low
5. Speed too high
6. Incorrect rotation
7. Pump flow too low
8. Pump flow too high
9. Change in density
10. Change in viscosity
11. Cavitation
12. Leakage joint excessive
13. Foreign matter in impeller
14. Loose impeller
15. Misalignment
16. Foundation not rigid
17. Shaft bent
18. Impeller or wear ring rubbing
19. Worn bearings
20. Damaged impeller
21. Shaft scored at packing/seal
22. Packing installed improperly
23. Incorrect type of packing
24. Excessive shaft runout
25. Impeller or coupling unbalanced
26. Gland too tight
27. Stuff box bushing clearance high
28. Dirt or grit in pumped liquid
29. Bearing cooling water failure
30. Inadequate bearing lubrication
31. High ambient temperature
32. High process liquid temperature

Whether the pump is to be repaired in the owner's own shop or in an outside repair center, it is good to remember that not all pump repairs are equal. A very minimal repair job might not include any checking of the fits and running clearances in the pump. It might ignore any analysis of what went wrong to cause the pump to require repair. It might merely consist of changing a few very standard components (such as bearings and seals) with factory parts.

At the other extreme, a much more thorough repair procedure might include a careful measurement and recording of all fits and clearances (and a repair of any component which is out of specification in this regard); a check of shafts for

Installation, Operation, and Maintenance 491

straightness; a rebalancing of all impellers after any welding, grinding, or machining operations have been completed; a concentricity check of all bearing housings, stuffing boxes, and similar components; etc.

Obviously, the more thorough repair job takes longer and is more expensive than the "quick-and-dirty" one. Hopefully, however, the more thorough repair job results in a longer runtime before the pump must be overhauled again. With the high cost of pulling the pump and placing it back in service, and the high cost resulting from plant operations being curtailed while the pump is being repaired, many users find that the extra money spent to do a more thorough pump repair is money well spent.

Just as there are different levels of completeness or accuracy of pump repair that can affect the longevity of the pump between repair cycles, users should also be aware that there may be different levels of quality in the replacement parts being used. Parts are usually available from the original maker of the pump, but they may also be available from nonOEM parts makers, or some parts may be able to be made in the owner's shop. Not all machine shops have the same levels of accuracy in their machine tools, have the same skill level in machinists, or have available the correct material and latest drawing for the part. This is particularly important in the case of cast parts of irregular shape, such as impellers and volutes, where deviations in dimensions can affect the pump's hydraulic performance as well as its operability and longevity.

As a final general suggestion on pump repair, the pump owner should maintain as thorough a repair history on each pump as possible, complete with the documentation from each repair (see Section VI.B.1 to follow) and repair parts inventory. Accurate records help when making decisions on when repair operations should be done, speed up the repair, reduce the cost of parts and repairs, and may allow a better analysis of the reasons for overly frequent pump repair so that design modifications can be implemented to lengthen the operating time between repairs.

B. Repair Tips

Below are a few important practices to follow during the repair of pumps.

1. Document the Disassembly

The disassembly of a pump to be repaired should be well documented. Components arranged in a particular order on the pump (such as the bowls on a multi-stage vertical turbine pump) should be match-marked or punched with a number to ensure that the pump is reassembled with the components in the same order. Remove any extraneous marks to avoid confusion.

Key dimensions should be measured with micrometers after the components have been disassembled and cleaned. This should include all fits and running clearances, so that these can be compared with the manufacturer's recommended clearances, or with the user's own standards, and replaced or reworked if they fall outside the standard. See further comments on fits and clearances in Section VI.B.5 to follow.

Make sketches or photographs of the components if a drawing is not available, showing clearly where the measurements have been taken and indicating particularly worn or damaged parts.

2. Analyze Disassembled Pump

The disassembled pump should be thoroughly inspected and analyzed, particularly if the user believes that the length of time since the last overhaul is unusually short. A thorough inspection and analysis of the disassembled pump can have the major benefit of reducing the likelihood of pump failure reoccurring so quickly from the same cause.

The inspection of the disassembled pump is an attempt to determine the root cause of the pump failure, if that is not already known, by the nature of the damage to the components. The second question to answer through the analysis of the pump prior to beginning repair operations is whether the runtime of the pump can be economically extended through a design modification or a particular repair operation. Examples of design modifications that might be considered to increase the runtime of the pump include hardfacing or coating of shafts or sleeves at bushing locations or under the packing; changing the material of key wear components such as wear rings, sleeves, or bushings; switching from packing to a mechanical seal or changing seal type or materials; installing heavier-duty bearings; and applying coatings in the casing at areas subject to excessive wear.

3. Bearing Replacement

Bearings are usually replaced during a pump overhaul. Radial and ball bearings commonly used in pumps often fail due to the presence of dirt, moisture, or other debris. Consequently, every effort should be made to keep the bearing and its environment as clean as possible during the repair operation.

Keep new bearings enclosed in their protective wrapping paper and box until they are ready for installation. Keep the bearings covered after they have been unwrapped until they are installed. The mechanic who changes the bearings should have reasonably clean hands and a clean workbench. Carefully inspect the bearing housings for dirt, machine cuttings, etc., and clean the bearing housings with an appropriate solvent or cleaning agent. Thoroughly inspect the shafts on which the bearings are to be mounted, removing any burs or nicks with emery cloth; and check to ensure that the shoulders on which the bearings will rest are square. Finally, the new bearings themselves should be thoroughly inspected for foreign materials or moisture, flat spots or rust on balls or rollers, or any other visible problems.

The tolerances between the inner race of bearings and the shaft, and between the outer race and the bearing housing, are quite tight. Therefore, be very sure that bearings are oriented properly before mounting them on the shaft and in the housing. Many bearings have the shields arranged in such a way that there is a correct way and an incorrect way for them to be oriented. A bearing that is incorrectly mounted and must be pulled usually is out of tolerance and has to be tossed.

If the repair facility permits, it is better to heat bearings with oil or an electric heater prior to mounting them on the shaft. The heating expands the inside diameter of the bearings, allowing them to slip over the shaft easily. If the bearings are not heated, they should be pressed on the shaft very carefully, using a bushing so that the bearings are pressed evenly from all sides and are not allowed to cock.

Finally, make certain when bearings are installed that they are tightly pushed up against the shoulder on the shaft, that they are properly locked in place with the snap rings or lock nuts if there are any, and that they are lubricated properly (or tagged for lubrication as discussed in Section VI.B.9 to follow).

4. Wear Ring Replacement

Recommended new or rebuilt wear ring clearances are discussed in Chapter 4, Section II.A. Good maintenance practice is to replace wear rings or otherwise renew the clearance when the clearance reaches twice the original amount. Alternatively, the decision to renew clearances can be based on a determination through testing of the energy being wasted by having excessive wear ring clearances.

Before removing old wear rings, check and record the worn wear ring clearance and compare with the specification for the original ring clearance. Especially for large pumps or pumps with expensive wear ring materials, it may be possible to replace only one part of the wearing ring (casing ring or impeller ring only), truing up the ring that is going to be used again, and then machining a new mating ring to have the proper ring clearance. This technique requires wear ring components to be purchased or made with undersized inside diameters on casing rings and oversized outside diameters on impeller rings, and requires careful measurement and record keeping. However, it may save the user money in the long run on wear ring replacement.

Before removing old wear rings, remove any set screws that are holding the rings in position. Most old wear rings, which are usually mounted with a press fit in the casing and

Installation, Operation, and Maintenance 495

an interference fit on the impeller, can be simply pulled off. If that does not work, they may need to be machined off. Alternatively, casing rings in certain alloys such as stainless steel can be removed by running a weld bead around the inside of the ring. This causes the ring to shrink, allowing it to be pulled by hand.

After new wear rings have been mounted and pinned in place, the concentricity of the impeller wear ring should be checked by mounting the impeller on the shaft or a mandrel, mounting the shaft in a lathe or on rollers, and checking the runout on the outside diameter of the impeller wear ring with a dial indicator. By performing this operation in a lathe, the ring can be given a final skin cut if it is not running concentric. The ring is especially likely to have a problem with concentricity if it is mounted with set screws, with the ring being likely to dimple at the set screw locations.

5. Guidelines for Fits and Clearances

If the pump manufacturer or owner does not have established criteria which differ from these, the guidelines shown in Table 8.3 may be used for fits and clearances in centrifugal pump repair. These are conservative guidelines for industrial pumps, and can be relaxed on pumps used in lighter-duty services.

6. Always Replace Consumables

All gaskets, O-rings, set screws, springs, lip seals, packing, and lubrication should be replaced any time a pump is repaired. Mechanical seals should be replaced or rebuilt. The dynamic seal faces of many mechanical seals can be re-lapped to restore their flatness and surface finish.

7. Balance Impellers and Couplings

In general, the major rotating components of pumps (impellers and couplings) should be balanced to maintain the pump within acceptable vibration levels. The need for balance, type of balance, and limits of imbalance vary with the weight of

Table 8.3 Fits and Clearances for Centrifugal Pump Repair

Component Description	Fit or Clearance
Ball bearing i.d. to shaft	0.0001–0.0007 in. interference
Ball bearing o.d. to housing	0.0005–0.001 in. clearance; bush housing if > 0.0015 in. oversize
Sleeve to shaft	0.001–0.0015 in. clearance
Impeller to shaft	Metal-to-metal to 0.0005 in. clearance
Wear ring to wear ring	See Chapter 4, Section II.A
Impeller ring to hub and case ring to casing	0.002–0.003 in. interference
Throat bushing to case	0.002–0.003 in. interference
Throat bushing to shaft	0.015–0.020 in. clearance
Coupling to shaft	Metal-to-metal to 0.0005 in. clearance
Deflector to shaft (set screw mtd.)	0.002–0.003 in. clearance
Seal running faces	Flat within three light bands
Alignment fits for case, covers, bearing housings, stuffing box	0.004 in. max. clearance; weld pads and re-machine if over-size

the component and its radius, as well as the operating speed of the pump. Refer to the manufacturer's recommendations for each pump. Many couplings do not require balancing, and smaller impellers may only require a *static balance*. This is done by mounting the impeller on a shaft on rollers, and seeing whether one side of the impeller tends to rotate to the bottom. Material is then removed from this heavier side.

A *dynamic balance* is performed on larger impellers, and some industrial users do a dynamic balance on all impellers of repaired pumps. Dynamic balance is performed using special balancing machines, with the pump mounted on a shaft or mandrel.

8. Check Runout of Assembled Pump

This check of shaft runout on the fully assembled pump is a final assurance that all components are machined accurately, that the shaft is straight, and that the pump has been aligned properly. Mount a dial indicator on the pump bedplate using

Installation, Operation, and Maintenance 497

a clamp attachment or magnetic base, and locate the pin of the indicator on the shaft at the mechanical seal or packing. Turn the shaft by hand at the coupling and note the total runout in one revolution. Shaft runout should be as low as possible. For comparison with a common standard, the API 610 Standard (Chapter 4, Section XIV.C) calls for a maximum shaft runout of 0.002 in. This is a tight standard; so for pumps not built to API 610, this standard can be relaxed to 0.003 to 0.004 in., but 0.002 in. is a worthy goal.

9. Tag Lubrication Status

If the pump or motor requires lubrication prior to being run (and most do), the pump should be tagged to indicate the status of the lubrication (i.e., if the lubrication was done in the shop during the overhaul, and what type of lubricant was used). If the equipment has no lubrication, an incorrect oil level, or requires relubricating prior to start-up, it is especially important to indicate this on a tag to prevent damage to the pump or motor.

10. Cover Openings Prior to Shipment

This prevents debris from getting trapped inside the pump during transit from the repair shop to the field.

Appendix A

Major Suppliers of Pumps in the United States by Product Type

Table A.1 provides a listing of most of the major pump suppliers in the United States, segmented according to the pump types they offer. This list is by no means exhaustive, containing only those manufacturers with which the author is familiar. Readers who are interested in locating information on a particular pump type should be able to use this appendix as a guide, and should be able to locate manufacturers by a Web search, or in a *Thomas Register* or similar index of manufacturers. Note that the appendix lists the manufacturers alphabetically, and that it includes both centrifugal and positive displacement pump suppliers. The list is sorted by the major brand names by which the pump products are most commonly known, and, where applicable, the parent company is shown following the brand name. The types of pumps manufactured by each supplier are indicated in the table, with the product types referring to the following key.

PRODUCT TYPE KEY

A Single stage, single suction, close-coupled
B Single stage, single suction, frame mounted
C Self-priming centrifugal
D Single stage, double suction
E Multi-stage
F Submersible (except vertical turbine)
G Vertical turbine (submersible)
H Vertical turbine (lineshaft)
I Mixed and axial flow
J Sliding vane
K Flexible impeller
L Progressing cavity
M Rotary external and internal gear
N Lobe and circumferential piston
O Multiple screw
P Other rotary
Q Piston/plunger (nonmetering)
R Diaphragm (nonmetering)
S Metering
T ANSI
U API

Table A.1 Major Pump Suppliers

Company	Product Type																				
	A	B	C	D	E	F	G	H	I	J	K	L	M	N	O	P	Q	R	S	T	U
Abel Pumps, Roper Industries	X	X															X	X			
ABS Pumps						X															
A-C Pump, ITT	X	X		X				X	X											X	
Ace Pump Corporation	X	X														X					
Aermotor Pumps, Wicor	X	X	X		X		X	X													X
Afton Pumps, Inc.	X	X						X													X
Aldrich, Flowserve																	X				
Alfa Laval	X	X	X		X									X							
Allweiler Pumps											X				X	X		X			
American Lewa																		X	X		
American Stainless Pumps	X																				
American Turbine Pump							X	X													
Ampco Pumps	X	X	X																		
Ansimag, Inc., Sundyne	X	X																		X	
APV, Invensys	X	X	X																		
Armstrong Pump	X	X																			
Aro Corporation, Ingersoll Rand																	X	X			
Ash Pumps, Weir		X																			
Aurora Pump Company, Pentair	X	X		X	X	X			X												

Table A.1 Major Pump Suppliers (continued)

Company	A	B	C	D	E	F	G	H	I	J	K	L	M	N	O	P	Q	R	S	T	U
Barber-Nichols	X																				
Barnes Pumps, Crane Pumps & Systems						X															
Bell & Gossett, ITT	X	X		X	X																
Blackmer Pump, Dover										X											
Blue White																			X		
Bornemann															X						
Bran + Luebbe, SPX	X	X										X	X				X	X	X		
Buffalo Pumps, Inc.	X	X		X	X																
Burks, Crane Pumps & Systems	X	X	X																		
Butterworth, Pacific Jetting International																	X				
Byron Jackson, Flowserve	X	X		X	X		X	X													X
Camac Industries	X	X																			
Carver Pump Company	X	X	X	X	X																
Cascade Pump Company									X												
Caster Pumps, Sundyne		X																		X	
Cat Pumps Corporation																	X				
Charles S. Lewis, Weir	X																				
Chempump, Teikoku	X																		X	X	
Chicago Pump, Yeomans Chicago	X					X															

Table A.1 Major Pump Suppliers (continued)

Company	Product Type																				
	A	B	C	D	E	F	G	H	I	J	K	L	M	N	O	P	Q	R	S	T	U
Coffin Turbo Pumps	X				X																
Corcoran Company	X	X	X																		
Corken, Idex				X						X											
Cornell Pump Company, Roper Industries	X	X																			
CPC Internalift																X					
Crisafulli Pump Company	X					X			X												
Crown Pumps, Crane Pumps & Systems			X																		
Davis EMU, U.S. Filter						X			X												
Dean Pump, Met-Pro	X	X																			
Delasco																X					
Deming, Crane Pumps & Systems	X	X	X	X	X	X	X	X	X	X								X		X	
Dempster Industries							X	X													
Dickow Pump Company	X	X			X																
Diener Precision Pumps													X						X		
Dorr Oliver, Inc.		X																			
Durco, Flowserve		X																		X	
Ebara International	X		X		X	X															
Edwards Manufacturing, Hypro														X							
Elro, Crane Pumps & Systems																X					

Table A.1 Major Pump Suppliers (continued)

Company	A	B	C	D	E	F	G	H	I	J	K	L	M	N	O	P	Q	R	S	T	U
Essco Pump Division		X				X											X				
F.E. Myers, Pentair	X		X			X															
Fairbanks Morse, Pentair		X	X	X		X	X	X	X												
Filter Pump Industries	X																				
Finish Thompson	X	X																			
Flint & Walling	X		X		X		X														
Flojet Corporation																		X			
Floway Pump Company, Weir							X	X	X												
Flowserve	X	X	X	X	X	X	X	X	X					X						X	X
Fluid Metering, Inc.																			X		
Flux Pumps	X																				
Flygt, ITT						X															
FMC																	X				
Fristam	X	X	X		X									X							
Fybroc Division, Met-Pro		X																		X	
Galigher, Weir	X	X				X															
Gardner-Denver																	X				
Gator Pump		X																			
GIW Industries, KSB		X																			

Table A.1 Major Pump Suppliers (continued)

Company	Product Type																				
	A	B	C	D	E	F	G	H	I	J	K	L	M	N	O	P	Q	R	S	T	U
Godwin Pumps		X	X																		
Gorman-Rupp	X		X		X	X										X					
Goulds Pumps, Inc., ITT	X	X	X	X	X	X	X	X	X											X	X
Graco, Inc.																	X	X			
Granco Pumps																X					
Graymills Corporation	X	X			X																
Grindex						X															
Griswold Pump Company	X	X	X				X	X												X	
Grundfos Pumps Corporation	X	X		X	X		X												X		
Gusher Pump	X	X	X	X																	
Hayward Tyler					X																
Hazelton Pumps, Weir	X	X			X	X															
HOMA Pump						X															
Hotsy Corporation																	X				
Hydromatic, Pentair					X																
Hypro Corporation																	X	X			
IDP, Flowserve	X	X	X	X	X	X	X	X	X											X	X
IMO, Colfax		X											M		O						
Innomag		X																			

Table A.1 Major Pump Suppliers (continued)

Company	A	B	C	D	E	F	G	H	I	J	K	L	M	N	O	P	Q	R	S	T	U
Ikawi Pumps	X															X			X		
Jabsco, ITT	X	X								X	X										
Jacuzzi	X		X		X		X														
Jaeco Pump Company																			X		
Johnston Pump Company, Sulzer Pumps							X	X	X												
Klaus Union		X	X		X											X					X
Komline-Sanderson																	X				
Kontro, Sundyne	X	X	X																		X
Krogh Pump Company	X	X																			
KSB Pumps, Inc.	X	X		X	X	X	X	X	X												
LaBour-Tabor, Peerless		X	X		X															X	
Lancaster Pump	X				X																
Layne & Bowler, Pentair							X	X	X												X
Leistritz Corporation															O						
Lincoln																	X				
Liquiflo	X	X											M							X	
Little Giant	X					X															
LMI, Milton Roy																			X		
Lobee	X	X	X			X							M								

Table A.1 Major Pump Suppliers (continued)

Company	A	B	C	D	E	F	G	H	I	J	K	L	M	N	O	P	Q	R	S	T	U
Lutz Pumps, Inc.	X																				
Luwa													M								
Maag													M								
Magnatex	X	X								X										X	
March Pumps	X																				
Marlow Pumps, ITT	X	X	X			X											X	X			
McDonald A.Y.	X		X	X			X											X			
McFarland																	X	X	X		
Megator Corporation																P					
Micropump Corporation, Idex	X									X			M				X				
Milton Roy																			X		
Milton Roy, Hartell Div.	X																				
Morris, Yeomans Chicago Corporation		X																			
MP Pumps, Tecumseh	X	X	X																		
MTH	X	X		X																	
Multiquip	X		X			X												X			
Nagle Pumps, Inc.	X	X				X															
National Pump Company							X	X													
Neptune																			X		

Table A.1 Major Pump Suppliers (continued)

Company	A	B	C	D	E	F	G	H	I	J	K	L	M	N	O	P	Q	R	S	T	U
Netzsch, Inc.	X											X		X							
Nikkiso Pumps America			X														X		X		
Oberdorfer Pumps, Thomas Industries	X										X	X	X								
Osmonics, G. E.					X																
Pacer Pumps	X		X																		
Paco Pumps, Inc., Sulzer Pumps	X	X		X	X	X															
Patterson Pump Company	X	X		X					X												
Peerless Pump Company	X	X		X	X	X	X	X	X											X	
Pioneer Pumps	X	X	X			X															
Price Pump Company	X	X	X		X													X			
Procon Products										X											
ProMinent																			X		
Prosser, Crane Pumps & Systems						X															
Pulsafeeder, Idex	X				X					X			X						X		
Pumpex						X															
Red Jacket Pumps	X						X														
Robbins & Myers Inc.											X										
Roper Pump Company, Roper Industries											X	X	X	X	X	X					
Roth Pump Co.	X	X			X																

Major Suppliers of Pumps in the United States by Product Type

Table A.1 Major Pump Suppliers (continued)

Company	A	B	C	D	E	F	G	H	I	J	K	L	M	N	O	P	Q	R	S	T	U
Roto Jet, Weir			X																		
Rotor-Tech Inc.													X								
Schwing America																	X				
Scot Pump, Ardox	X	X	X																		
Seepex												X									
Serfilco, Ltd.	X	X	X															X	X		
Sethco Division, Met-Pro	X	X	X																		
Sherwood, Hypro	X	X	X								X		X	X				X			
Sier-Bath, Flowserve													X	X							
Simflo Pumps, Inc.							X	X													
Sine Pump, Sundyne																X					
Sta-Rite Industries, Wicor	X	X	X				X	X													
Sulzer Pumps	X			X	X			X	X												X
Sundyne	X	X																			X
Sykes Pumps Division			X			X															
Taco, inc.	X	X		X																	
Tech-Mag	X	X	X																		
Teikoku	X															X					
Tri-Rotor, Inc.																X					

Table A.1 Major Pump Suppliers (continued)

Company	A	B	C	D	E	F	G	H	I	J	K	L	M	N	O	P	Q	R	S	T	U
Tuthill Pump Company														X	X						
TXT/Texsteam																			X		X
Union Pump Company, David Brown		X			X																
Vanton Pump	X	X	X													X					
Vaughn	X	X	X			X															
Versa Matic, Idex																		X			
Vertiflo	X	X																			
Viking Pump, Inc., Idex													X	X							
Wallace & Tiernan																X			X		
Wanner Engineering																X					
Warman, Weir	X	X																			
Warren Pumps, Inc., Colfax		X													X						
Warren Rupp, Inc., Idex																		X			
Watson Marlow																X			X		
Waukesha Cherry Burrell, SPX	X	X												X							
Wayne Home Equipment	X					X															
Webster Fluid Power													X								
Weil Pump Company	X					X															
Weinman, Crane Pumps & Systems	X	X	X	X		X	X	X													

Table A.1 Major Pump Suppliers (continued)

Company	Product Type																				
	A	B	C	D	E	F	G	H	I	J	K	L	M	N	O	P	Q	R	S	T	U
Wemco, Weir		X				X																
Wheatley GASO, Inc.																	X					
Wilden Pump & Engineering, Dover																		X				
Wilfley & Sons, Inc.		X																		X		
Williams, Milton Roy																				X		
Wilson-Snyder Pumps, Flowserve		X			X													X				
Yamada America, Inc.																						
Yeomans, Yeomans Chicago Corporation		X				X								X								
Zenith Pumps, Parker																				X		
Zoeller Company	X					X																

Appendix B

Conversion Formulae

Formulae used in this book are generally stated in United States Customary System (USCS) units, the system most widely used by the pump industry in the United States. This appendix provides simple conversion formulae for USCS and SI (metric) units. The most common terms mentioned in this book are stated in both units.

Multiply	By	To Obtain
Acres	43,560	Square feet
Acres	4047	Square meters
Acres	1.562×10^3	Square miles
Acres	4840	Square yards
Acre-feet	43,560	Cubic feet
Acre-feet	325,851	Gallons
Acre-feet	1233.48	Cubic meters
Atmospheres	76.0	Cm of mercury
Atmospheres	29.92	Inches of mercury
Atmospheres	33.90	Feet of water
Atmospheres	10,332	Kg/sq. meter
Atmospheres	14.70	Lb/sq. in.
Atmospheres	1.058	Tons/sq. ft
Barrels-oil	42	Gallons-oil
Barrels-beer	31	Gallons-beer

Multiply	By	To Obtain
Barrels-whiskey	45	Gallons-whiskey
Barrels/Day-oil	0.02917	Gallons/min-oil
Bags or sacks-cement	94	Pounds-cement
Board feet	144 sq. in. × 1 in.	Cubic inches
British Thermal Units	0.2520	Kilogram-calories
British Thermal Units	777.6	Foot-lb.
British Thermal Units	3.927×10^4	Horsepower-hrs.
British Thermal Units	107.5	Kilogram-meters
British Thermal Units	2.928×10^4	Kilowatt-hr.
B.T.U./min.	12.96	Foot-lb/sec
B.T.U./min.	0.02356	Horsepower
B.T.U./min.	0.01757	Kilowatts
B.T.U./min.	17.57	Watts
Centares (Centiares)	1	Square meters
Centigrams	0.01	Grams
Centiliters	0.01	Liters
Centimeters	0.3937	Inches
Centimeters	0.01	Meters
Centimeters	10	Millimeters
Centimeters of mercury	0.01316	Atmospheres
Centimeters of mercury	0.4461	Feet of water
Centimeters of mercury	136.0	Kg/sq. meter
Centimeters of mercury	27.85	Lb/sq. ft.
Centimeters of mercury	0.1934	Lb/sq. in.
Centimeters/sec	1.969	Feet/min
Centimeters/sec	0.03281	Feet/sec
Centimeters/sec	0.036	Kilometers/hr
Centimeters/sec	0.6	Meters/min
Centimeters/sec	0.02237	Miles/hr
Centimeters/sec	3.728×10^{-4}	Miles/min
Cms./sec./sec	0.03281	Feet/sec/sec
Cubic centimeters	3.531×10^{-5}	Cubic feet
Cubic centimeters	6.102×10^{-2}	Cubic inches
Cubic centimeters	10^{-6}	Cubic meters
Cubic centimeters	1.308×10^{-6}	Cubic yards
Cubic centimeters	2.642×10^{-4}	Gallons
Cubic centimeters	9.999×10^{-4}	Liters
Cubic centimeters	2.113×10^{-3}	Pints (liq.)
Cubic centimeters	1.057×10^{-3}	Quarts (liq.)
Cubic feet	2.832×10^{-4}	Cubic cms.

Conversion Formulae

Multiply	By	To Obtain
Cubic feet	1728	Cubic inches
Cubic feet	0.02832	Cubic meters
Cubic feet	0.03704	Cubic yards
Cubic feet	7.48052	Gallons
Cubic feet	28.32	Liters
Cubic feet	59.84	Pints (liq.)
Cubic feet	29.92	Quarts (liq.)
Cubic feet/min	472.0	Cubic cms./sec
Cubic feet/min	0.1247	Gallons/sec
Cubic feet/min	0.4719	Liters/sec
Cubic feet/min	62.43	Pounds of water/min.
Cubic feet/sec	0.646317	Millions gal/day
Cubic feet/sec	448.831	Gallons/min.
Cubic inches	16.39	Cubic centimeters
Cubic inches	5.787×10^{-4}	Cubic feet
Cubic inches	1.639×10^{-5}	Cubic meters
Cubic inches	2.143×10^{-5}	Cubic yards
Cubic inches	4.329×10^{-3}	Gallons
Cubic inches	1.639×10^{-2}	Liters
Cubic inches	0.03463	Pints (liq.)
Cubic inches	0.01732	Quarts (liq.)
Cubic meters	10^6	Cubic centimeters
Cubic meters	35.31	Cubic feet
Cubic meters	61023.	Cubic inches
Cubic meters	1.308	Cubic yards
Cubic meters	264.2	Gallons
Cubic meters	999.97	Liters
Cubic meters	2113	Pints (liq.)
Cubic meters	1057	Quarts (liq.)
Cubic Meters/hr	4.40	Gallons/min.
Cubic yards	764,554.86	Cubic centimeters
Cubic yards	27	Cubic feet
Cubic yards	46.656	Cubic inches
Cubic yards	0.7646	Cubic meters
Cubic yards	202.0	Gallons
Cubic yards	764.5	Liters
Cubic yards	1616	Pints (liq.)
Cubic yards	807.9	Quarts (liq.)
Cubic yards/min	0.45	Cubic feet/sec
Cubic yards/min	3.366	Gallons/sec

Multiply	By	To Obtain
Cubic yards/min	12.74	Liters/sec
Decigrams	0.1	Grams
Deciliters	0.1	Liters
Decimeters	0.1	Meters
Degrees (angle)	60	Minutes
Degrees (angle)	0.01745	Radians
Degrees (angle)	3600	Seconds
Degrees/sec	0.01745	Radians/sec
Degrees/sec	0.1667	Revolutions/min
Degrees/sec	0.002778	Revolutions/sec
Dekagrams	10	Grams
Dekaliters	10	Liters
Dekameters	10	Meters
Drams	27.34375	Grains
Drams	0.0625	Ounces
Drams	1.771845	Grams
Fathoms	6	Feet
Feet	30.48	Centimeters
Feet	12	Inches
Feet	0.3048	Meters
Feet	⅓	Yards
Feet of water	0.0295	Atmospheres
Feet of water	0.8826	Inches of mercury
Feet of water	304.8	Kg/sq. meter
Feet of water	62.43	Lb/sq. ft.
Feet of water	0.4335	Lb/sq. inch
Feet/min.	0.5080	Centimeters/sec
Feet/min.	0.01667	Feet/sec
Feet/min.	0.01829	Kilometers/hr
Feet/min.	0.3048	Meters/min
Feet/min.	0.01136	Miles/hr
Feet/sec.	30.48	Centimeters/sec
Feet/sec.	1.097	Kilometers/hr
Feet/sec.	0.5924	Knots
Feet/sec.	18.29	Meters/min
Feet/sec.	0.6818	Miles/hr
Feet/sec.	0.01136	Miles/min
Feet/sec/sec	30.48	Cms./sec/sec
Feet/sec/sec	0.3048	Meters/sec/sec
Foot-pounds	1.286×10^{-3}	British Thermal Units

Conversion Formulae

Multiply	By	To Obtain
Foot-pounds	5.050×10^{-7}	Horsepower-hr
Foot-pounds	3.240×10^{-4}	Kilogram-calories
Foot-pounds	0.1383	Kilogram-meters
Foot-pounds	3.766×10^{-7}	Kilowatt-hours
Foot-pounds/min	2.140×10^{-5}	B.T.U./sec
Foot-pounds/min	0.01667	Foot-pounds/sec
Foot-pounds/min	3.030×10^{-5}	Horsepower
Foot-pounds/min	5.393×10^{-3}	Gm-calories/sec
Foot-pounds/min	2.280×10^{-5}	Kilowatts
Foot-pounds/sec	7.704×10^{-2}	B.T.U./min
Foot-pounds/sec	1.818×10^{-3}	Horsepower
Foot-pounds/sec	1.941×10^{-2}	Kg.-calories/min.
Foot-pounds/sec	1.356×10^{-3}	Kilowatts
Gallons	3785	Cubic centimeters
Gallons	0.1337	Cubic feet
Gallons	231	Cubic inches
Gallons	3.785×10^{-3}	Cubic meters
Gallons	4.951×10^{-3}	Cubic yards
Gallons	3.785	Liters
Gallons	8	Pints (liq.)
Gallons	4	Quarts (liq.)
Gallons-Imperial	1.20095	U.S. gallons
Gallons-U.S.	0.83267	Imperial gallons
Gallons water	8.345	Pounds of water
Gallons/min	2.228×10^{-3}	Cubic feet/sec
Gallons/min	0.06308	Liters/sec
Gallons/min	8.0208	Cu. ft./hr
Grains (troy)	0.06480	Grams
Grains (troy)	0.04167	Pennyweights (troy)
Grains (troy)	2.0833×10^{-3}	Ounces (troy)
Grains/U.S. gal	17.118	Parts/million
Grains/U.S. gal	142.86	Lbs./million gal
Grains/Imp. gal	14.254	Parts/million
Grams	980.7	Dynes
Grams	15.43	Grains
Grams	.001	Kilograms
Grams	1000	Milligrams
Grams	0.03527	Ounces
Grams	0.03215	Ounces (troy)
Grams	2.205×10^{-3}	Pounds

Multiply	By	To Obtain
Grams/cm	5.600×10^{-3}	Pounds/inch
Grams/cu. cm	62.43	Pounds/cubic foot
Grams/cu. cm	0.03613	Pounds/cubic inch
Grams/liter	58.416	Grains/gal.
Grams/liter	8.345	Pounds/1000 gal
Grams/liter	0.06242	Pounds/cubic foot
Grams/liter	1000	Parts/million
Hectares	2.471	Acres
Hectares	1.076×10^5	Square feet
Hectograms	100	Grams
Hectoliters	100	Liters
Hectometers	100	Meters
Hectowatts	100	Watts
Horsepower	42.44	B.T.U./min
Horsepower	33,000	Foot-lb/min
Horsepower	550	Foot-lb/sec
Horsepower	1.014	Horsepower (metric)
Horsepower	10.547	Kg.-calories/min
Horsepower	0.7457	Kilowatts
Horsepower	745.7	Watts
Horsepower (boiler)	33,493	B.T.U./hr
Horsepower (boiler)	9.809	Kilowatts
Horsepower-hours	2546	B.T.U.
Horsepower-hours	1.98×10^6	Foot-lb
Horsepower-hours	641.6	Kilogram-calories
Horsepower-hours	2.737×10^5	Kilogram-meters
Horsepower-hours	0.7457	Kilowatt-hours
Inches	2.540	Centimeters
Inches of mercury	0.03342	Atmospheres
Inches of mercury	1.133	Feet of water
Inches of mercury	345.3	Kg/sq. meter
Inches of mercury	70.73	Lb/sq. ft.
Inches of mercury (32°F)	0.491	Lb/sq. inch
Inches of water	0.002458	Atmospheres
Inches of water	0.07355	Inches of mercury
Inches of water	25.40	Kg/sq. meter
Inches of water	0.578	Ounces/sq. inch
Inches of water	5.202	Lb/sq. foot
Inches of water	0.03613	Lb/sq. inch
Kilograms	980,665	Dynes

Conversion Formulae

Multiply	By	To Obtain
Kilograms	2.205	Lb
Kilograms	1.102×10^{-3}	Tons (short)
Kilograms	10^3	Grams
Kilograms-cal/sec	3.968	B.T.U./sec
Kilograms-cal/sec	3086	Foot-lb/sec
Kilograms-cal/sec	5.6145	Horsepower
Kilograms-cal/sec	4186.7	Watts
Kilogram-cal/min	3085.9	Foot-lb/min
Kilogram-cal/min	0.09351	Horsepower
Kilogram-cal/min	69.733	Watts
Kgs./meter	0.6720	Lb/foot
Kgs./sq. meter	9.678×10^{-5}	Atmospheres
Kgs./sq. meter	3.281×10^{-3}	Feet of water
Kgs./sq. meter	2.896×10^{-3}	Inches of mercury
Kgs./sq. meter	0.2048	Lb/sq. foot
Kgs./sq. meter	1.422×10^{-3}	Lb/sq. inch
Kgs./sq. millimeter	10^6	Kg/sq. meter
Kiloliters	10^3	Liters
Kilometers	10^5	Centimeters
Kilometers	3281	Feet
Kilometers	10^3	Meters
Kilometers	0.6214	Miles
Kilometers	1094	Yards
Kilometers/hr	27.78	Centimeters/sec
Kilometers/hr	54.68	Feet/min
Kilometers/hr	0.9113	Feet/sec
Kilometers/hr	.5399	Knots
Kilometers/hr	16.67	Meters/min
Kilometers/hr	0.6214	Miles/hr
Km/hr/sec	27.78	Cm/sec/sec
Km/hr/sec	0.9113	Ft/sec/sec
Km/hr/sec	0.2778	Meters/sec/sec
Kilowatts	56.907	B.T.U./min
Kilowatts	4.425×10^4	Foot-lbs/min.
Kilowatts	737.6	Foot-lbs/sec.
Kilowatts	1.341	Horsepower
Kilowatts	14.34	Kg-calories/min
Kilowatts	10^3	Watts
Kilowatt-hours	3414.4	B.T.U.
Kilowatt-hours	2.655×10^6	Foot-lb

Multiply	By	To Obtain
Kilowatt-hours	1.341	Horsepower-hr
Kilowatt-hours	860.4	Kilogram-calories
Kilowatt-hours	3.671×10^5	Kilogram-meters
Liters	10^3	Cubic centimeters
Liters	0.03531	Cubic feet
Liters	61.02	Cubic inches
Liters	10^{-3}	Cubic meters
Liters	1.308×10^{-3}	Cubic yards
Liters	0.2642	Gallons
Liters	2.113	Pints (liq.)
Liters	1.057	Quarts (liq.)
Liters/min	5.886×10^{-4}	Cubic ft/sec
Liters/min	4.403×10^{-3}	Gal/sec
Lumber Width (in.) × Thickness (in.)/12	Length (ft)	Board feet
Meters	100	Centimeters
Meters	3.281	Feet
Meters	39.37	Inches
Meters	10^{-3}	Kilometers
Meters	10^3	Millimeters
Meters	1.094	Yards
Meters/min	1.667	Centimeters/sec
Meters/min	3.281	Feet/min
Meters/min	0.05468	Feet/sec
Meters/min	0.06	Kilometers/hr
Meters/min	0.03728	Miles/hr
Meters/sec	196.8	Feet/min
Meters/sec	3.281	Feet/sec
Meters/sec	3.6	Kilometers/hr
Meters/sec	0.06	Kilometers/min
Meters/sec	2.287	Miles/hr
Meters/sec	0.03728	Miles/min
Microns	10^{-6}	Meters
Miles	1.609×10^5	Centimeters
Miles	5280	Feet
Miles	1.609	Kilometers
Miles	1760	Yards
Miles/hr	44.70	Centimeters/sec
Miles/hr	88	Feet/min
Miles/hr	1.467	Feet/sec

Conversion Formulae

Multiply	By	To Obtain
Miles/hr	1.609	Kilometers/hr
Miles/hr	0.8689	Knots
Miles/hr	26.82	Meters/min
Miles/min	2682	Centimeters/sec
Miles/min	88	Feet/sec
Miles/min	1.609	Kilometers/min
Miles/min	60	Miles/hr
Milliers	10^3	Kilograms
Milligrams	10^{-3}	Grams
Millimeters	10^{-3}	Liters
Millimeters	0.1	Centimeters
Millimeters	0.03937	Inches
Milligrams/liter	1	Parts/million
Million gal/day	1.54723	Cubic ft/sec
Miner's inches	1.5	Cubic ft/min
Minutes (angle)	2.909×10^{-4}	Radians
Ounces	16	Drams
Ounces	437.5	Grains
Ounces	0.0625	Pounds
Ounces	28.3495	Grams
Ounces	0.9115	Ounces (troy)
Ounces	2.790×10^{-5}	Tons (long)
Ounces	2.835×10^{-5}	Tons (metric)
Ounces (troy)	480	Grains
Ounces (troy)	20	Pennyweights (troy)
Ounces (troy)	0.08333	Pounds (troy)
Ounces (troy)	31.10348	Grams
Ounces (troy)	1.09714	Ounces (avoir.)
Ounces (fluid)	1.805	Cubic inches
Ounces (fluid)	0.02957	Liters
Ounces/sq. inch	0.0625	Lb/sq. inch
Parts/million	0.0584	Grains/U.S. gal
Parts/million	0.07015	Grains/Imp. gal
Parts/million	8.345	Lbs./million gal
Pennyweights (troy)	24	Grains
Pennyweights (troy)	1.55517	Grams
Pennyweights (troy)	0.05	Ounces (troy)
Pennyweights (troy)	4.1667×10^{-3}	Pounds (troy)
Pounds	16	Ounces
Pounds	256	Drams

Multiply	By	To Obtain
Pounds	7000	Grains
Pounds	0.0005	Tons (short)
Pounds	453.5924	Grams
Pounds	1.21528	Pounds (troy)
Pounds	14.5833	Ounces (troy)
Pounds (troy)	5760	Grains
Pounds (troy)	240	Pennyweights (troy)
Pounds (troy)	12	Ounces (troy)
Pounds (troy)	373.2417	Grams
Pounds (troy)	0.822857	Pounds (avoir.)
Pounds (troy)	13.1657	Ounces (avoir.)
Pounds (troy)	3.6735×10^{-4}	Tons (long)
Pounds (troy)	4.1143×10^{-4}	Tons (short)
Pounds (troy)	3.7324×10^{-4}	Tons (metric)
Pounds of water	0.01602	Cubic feet
Pounds of water	27.68	Cubic inches
Pounds of water	0.1198	Gallons
Pounds of water/min	2.670×10^{-4}	Cubic ft/sec
Pounds/cubic foot	0.01602	Grams/cubic cm
Pounds/cubic foot	16.02	Kgs./cubic meters
Pounds/cubic foot	5.787×10^{-4}	Lbs./cubic inch
Pounds/cubic inch	27.68	Grams/cubic cm
Pounds/cubic inch	2.768×10^{4}	Kg/cubic meter
Pounds/cubic inch	1728	Lb/cubic foot
Pounds/foot	1.488	Kg/meter
Pounds/inch	1152	Grams/cm
Pounds/sq. foot	0.01602	Feet of water
Pounds/sq. foot	4.882	Kg/sq. meter
Pounds/sq. foot	6.944×10^{-3}	Pounds/sq. inch
Pounds/sq. inch	0.06804	Atmospheres
Pounds/sq. inch	2.307	Feet of water
Pounds/sq. inch	2.036	Inches of mercury
Pounds/sq. inch	703.1	Kgs./sq. meter
Quadrants (angle)	90	Degrees
Quadrants (angle)	5400	Minutes
Quadrants (angle)	1.571	Radians
Quarts (dry)	67.20	Cubic inches
Quarts (liq.)	57.75	Cubic inches
Quintal, Argentine	101.28	Pounds
Quintal, Brazil	129.54	Pounds

Conversion Formulae

Multiply	By	To Obtain
Quintal, Castile, Peru	101.43	Pounds
Quintal, Chile	101.41	Pounds
Quintal, Mexico	101.47	Pounds
Quintal, Metric	220.46	Pounds
Quires	25	Sheets
Radians	57.30	Degrees
Radians	3438	Minutes
Radians	0.637	Quadrants
Radians/sec	57.30	Degrees/sec
Radians/sec	0.1592	Revolutions/sec
Radians/sec	9.549	Revolutions/min
Radians/sec./sec	573.0	Revs./min/min
Radians/sec/sec	0.1592	Revs./sec/sec
Reams	500	Sheets
Revolutions	360	Degrees
Revolutions	4	Quadrants
Revolutions	6.283	Radians
Revolutions/min	6	Degrees/sec
Revolutions/min	0.1047	Radians/sec
Revolutions/min	0.01667	Revolutions/sec
Revolutions/min/min	1.745×10^{-3}	Rads./sec/sec
Revolutions/min/min	2.778×10^{-4}	Revs./sec/sec
Revolutions/sec	360	Degrees/sec
Revolutions/sec	6.283	Radians/sec
Revolutions/sec	60	Revolutions/min
Revolutions/sec/sec	6.283	Radians/sec/sec
Revolutions/sec/sec	3600	Revs./min/min
Seconds (angle)	4.848×10^{-6}	Radians
Square centimeters	1.076×10^{-3}	Square feet
Square centimeters	0.1550	Square inches
Square centimeters	10^{-4}	Square meters
Square centimeters	100	Square millimeters
Square feet	2.296×10^{-5}	Acres
Square feet	929.0	Square centimeters
Square feet	144	Square inches
Square feet	0.09290	Square meters
Square feet	3.587×10^{-4}	Square miles
Square feet	1/9	Square yards
1/Sq. ft./gal./min.	8.0208	Overflow rate (ft./hr.)
Square inches	6.452	Square centimeters

Multiply	By	To Obtain
Square inches	6.944×10^{-3}	Square feet
Square inches	645.2	Square millimeters
Square kilometers	247.1	Acres
Square kilometers	10.76×10^6	Square feet
Square kilometers	10^6	Square meters
Square kilometers	0.3861	Square miles
Square kilometers	1.196×10^6	Square yards
Square meters	2.471×10^{-4}	Acres
Square meters	10.76	Square feet
Square meters	3.861×10^{-7}	Square miles
Square meters	1.196	Square yards
Square miles	640	Acres
Square miles	27.88×10^6	Square feet
Square miles	2.590	Square kilometers
Square miles	3.098×10^6	Square yards
Square millimeters	0.01	Square centimeters
Square millimeters	1.550×10^{-3}	Square inches
Square yards	2.066×10^{-4}	Acres
Square yards	9	Square feet
Square yards	0.8361	Square meters
Square yards	3.228×10^{-7}	Square miles
Temp. (°C.) +273	1	Abs. temp. (°C.)
Temp. (°C.) +17.78	1.8	Temp (°F.)
Temp. (°F.) +460	1	Abs. temp (°F.)
Temp. (°F.) –32	5/9	Temp. (°C.)
Tons (long)	1016	Kilograms
Tons (long)	2240	Pounds
Tons (long)	1.12000	Tons (short)
Tons (metric)	10^3	Kilograms
Tons (metric)	2205	Pounds
Tons (short)	2000	Pounds
Tons (short)	32,000	Ounces
Tons (short)	907.1843	Kilograms
Tons (short)	2430.56	Pounds (troy)
Tons (short)	0.89287	Tons (long)
Tons (short)	29166.66	Ounces (troy)
Tons (short)	0.90718	Tons (metric)
Tons of water/24 hrs.	83.333	Pounds water/hr.
Tons of water/24 hrs.	0.16643	Gallons/min.
Tons of water/24 hrs.	1.3349	Cu. ft./hr.

Conversion Formulae

Multiply	By	To Obtain
Watts	0.05686	B.T.U./min.
Watts	44.25	Foot-lbs./min.
Watts	0.7376	Foot-lbs./sec.
Watts	1.341×10^{-3}	Horsepower
Watts	0.01434	Kg.-calories/min.
Watts	10^{-3}	Kilowatts
Watt-hours	3.414	B.T.U.
Watt-hours	2655	Foot-lbs.
Watt-hours	1.341×10^{-3}	Horsepower-hrs.
Watt-hours	0.8604	Kilogram-calories
Watt-hours	367.1	Kilogram-meters
Watt-hours	10^{-3}	Kilowatt-hours
Yards	91.44	Centimeters
Yards	3	Feet
Yards	36	Inches
Yards	0.9144	Meters

References

1. Karassik, I. J., Krutzsch, W. C., Fraser, W. H., and Messina, J. P., *Pump Handbook,* Third Edition, McGraw-Hill, New York, 2000.

2. *Hydraulic Institute Standards,* Hydraulic Institute, Parsippany, NJ, 2003.

3. Stepanoff, A. J., *Centrifugal and Axial Flow Pumps,* Second Edition, Krieger Publishing, Melbourne, FL, 1992.

4. Lobanoff, V. and Ross, R. R., *Centrifugal Pumps: Design and Application,* Second Edition, Gulf Publishing, Houston, TX, 1992.

5. *Engineering Data Book,* Second Edition, Hydraulic Institute, Parsippany, NJ, 1991.

6. Chen, C. C., "Cope with Dissolved Gases in Pump Calculations," *Chemical Processing,* October 1993.

7. ASME B73.1 (ANSI B73.1), *Specification for Horizontal End Suction Centrifugal Pumps for Chemical Process,* 2001.

8. *API Standard 610, Centrifugal Pumps for Petroleum, Petrochemical, and Natural Gas Industries,* Third Edition, American Petroleum Institute, Washington, D. C., 2004.

9. Piotrowski, J., *Shaft Alignment Handbook,* Marcel Dekker, New York, 1995.

Index

A

Acceleration head, 114–115
Adjustable speed drive, 374
Affinity laws, 122–127, 379, 381
Air entrainment, 197–200, 490
Air operated diaphragm pump, 44–46
Alignment, 287–293, 447–448, 452–462
Angle, 447–448
Angular misalignment, 447
ANSI, 113, 282–284
API, 219, 220, 284–286
Axial flow pump, 118–121, 277
Axially split case, 245–248, 250–253
Axial thrust, 13, 30, 196, 222–230, 248, 275, 465

B

Back pull out, 284
Balance drum, 230
Barrel pump, 255
Barske impeller, 444
Bearing, 11–14, 31–33, 197, 217–224, 227–240, 357–360, 446, 477
Bearing frame, 113, 237, 284
Bedplate, 155, 237, 285, 454–456
Benchmark, 204, 462, 480, 488
Bernoulli equation, 10
Best efficiency point (BEP), 11–12, 83–89, 116–122, 157, 364, 377, 381, 463–468
Bi-wing lobe pump, 35–36
Brake horsepower (BHP), 81–87, 100, 119–123, 156, 468
Break away, 90
Buffer liquid, 348–351

C

Canned motor pump, 359–362
Capacity, 10, 16, 18, 54, 73, 85, 88, 116, 121, 152, 157, 225
Casing, 7–14, 216–219, 235, 238, 243, 245–251
Cavitation, 9, 89–114, 127, 415, 431
Centrifugal, 1, 2, 5–11, 15–18, 54–56, 73–75, 85, 118, 122, 152, 157
Ceramic coating, 435
C-face motor, 236–237, 312–313
Check valve, 38, 41, 49, 78, 197, 463
Circumferential piston pump, 35–36
Clearances, 17, 29–32, 82, 204, 219–221, 279, 373, 495–496
Close-coupled pump, 233–237
Closed loop system, 57, 67, 128
Computer software, 58, 93, 185–196
Constant level oiler, 239, 240
Control system, 18

527

Corrosion 91, 134, 426–432
Corrosive liquid, 353, 430
Coupling, 217, 235, 237, 287–291, 447–448, 454–456
Cross-over, 13
Cutwater, 10, 12, 14
Cylinder, 7, 38–40, 330, 333

D

Dead head, 145
Deflection, shaft, 11, 12, 154, 465
Demonstration CD, 80, 101, 127, 183
Density, 2, 55, 189
Diaphragm pump, 41–47
Diffuser, 11–14, 55, 197, 250, 251, 254–256, 269, 445
Diffusion, 2, 10, 55
Dilatant liquid, 162
Direct acting, 38
Double acting, 38, 39
Double barrel pump, 255
Double case pump, 255, 256
Double suction, 105, 116, 121, 196, 223–227, 245–249
Double volute, 12, 14, 238, 243, 248
Drum pump, 259, 260, 261
Dry pit configuration, 257
Duplex, 39, 256, 257
Duty cycle, 54, 114, 143, 230, 280, 366
Dynamic balance, 92, 496

E

Economic analysis, 53
Efficiency, 12, 15, 80–89, 116, 126, 157, 203–204, 217, 230, 296, 322–324
Elastomeric, 289, 328, 439–441
Electric motor (see also Motor, electric), 72, 291–326
End suction pump, 233–237

Energy
 conservation, 363–422
 kinetic, 10
 potential, 10
 transformations, 1–2
External gear pump, 29–33

F

Field testing, 203–212
Fits and clearances, 495, 496
Fittings, head loss in, 58, 66, 68, 69
Flexible coupling 287–293
Flexible impeller pump, 25–26
Flexible tube pump, 26
Foot mounted motor, 237
Foot valve, 198, 214, 461
Frame mounted pump, 237–240
Friction head, 56–66, 128, 133, 152
Fudge factor, 152–153, 394
Full emission, 445

G

Gauge pressure, 3, 57, 70, 212
Gear pump, 29–33
Gland, 329, 335–337, 341
Gland quench, 337
Grinder pump, 263
Grout, 454, 456

H

Hard-face, 267
Head
 acceleration, 114–115, 401
 friction, 56–66, 128, 133, 152
 pressure, 56, 66, 67, 70
 shutoff, 119, 120, 151, 279
 static, 56–57, 70, 100, 103, 128, 133, 378–380
 system, 56, 73, 127–139
 total, 10, 54, 56, 133, 209–211
 velocity, 3, 56, 70–71, 210

Index

Head-capacity curve, 10, 56, 73–75, 119–121, 135, 279
High speed pump, 441, 444–446
Horizontally split case, 224, 245–249
Horsepower, 19, 80–89, 120–121, 156, 211, 307
Hose pump, 26–27
H-Q curve (see also Head-capacity curve), 10, 56, 73–75, 119–121, 135, 279
Hydraulic performance, 90, 218, 273, 480, 481

I

Impeller
  balancing, 92, 496
  Barske, 444
  between bearings, 7
  closed, 215–223
  double suction, 116, 121, 197, 223–227, 248
  filing of vanes, 230–232
  large eye, 225, 226, 253, 256, 467
  nonclog, 17, 232, 233, 243, 256, 262
  open, 215–223, 227, 229, 284
  overhung, 7, 246
  recessed, 17, 233, 266, 267
  semi-open, 223
  shroud, 118, 216, 223, 466
  single suction, 223–227
  solids handling, 232–233
  stages, 7, 250, 251, 269, 364
  trimming, 73, 111, 367
  vortex, 17, 232–234
Inducer, 225–226, 445
Inline pump, 240–242, 246
Internal gear pump, 33
ISO, 286–287
Iso curve, 86, 100, 111

J

Jet pump, 7, 313

K

Kinetic pump, 5

L

Lantern ring, 337–338
Lapping, 341
Leakage joint, 216–219
Life-cycle costs, 395–421
Lobe pump (see also Rotary lobe pump), 33–36
Low flow damage, 468–477
Lubrication, 46, 239, 289, 293, 358, 446, 477–478

M

Magnetic bearing, 446
Magnetic drive pump, 354–359
Magnetic flowmeter, 205
Maintenance, 186, 197, 339, 367, 370, 404, 477, 489
Manometer, 211
Mass flowmeter, 205
Mating ring, 340, 343, 348, 351
Mechanical seal
  advantages, 338–339
  balanced, 346–347
  description, 339–343
  double, 347–349
  gas lubricated, 351–352
  inside, 343–345
  outside, 345–346
  single, 343–347
  tandem, 349–351
  unbalanced, 346–347
Metering pump, 18, 42–44
Miniature positive displacement pump, 47–49
Minimum flow, 11, 73, 463–477

Mixed flow pump, 118–121, 233
Motor
 electric, 72, 291–326
 enclosure, 300–302
 frame size, 296, 303–314
 hazardous location, 302
 insulation, 88, 297, 303, 307, 319, 320
 open drip proof, 300, 305
 single phase, 211, 298, 314–319
 totally enclosed air over, 301
 totally enclosed fan cooled, 301
 totally enclosed non-ventilated, 301, 306
Multiple branch system, 136–139
Multiple screw pump, 36–38
Multi-stage pump, 13, 156, 222, 250–256

N

Natural frequency, 158, 274, 485–488
Net positive suction head (see also NPSH), 89–116
Newtonian liquid, 162, 163
Noise, 197, 372, 404, 448, 468, 490
Nonmetallic pump, 32, 49, 259, 283, 432–435
Nonoverloading motor, 88, 153
Nozzle, 205, 206, 284
NPSH
 calculation of, 98–101
 definition of, 89–98
 dissolved gases and, 113
 examples, 101–102, 106–109
 for reciprocating pumps, 114–116
 measuring, 212
 reduction chart, 112
 safe margin, 109–114
 testing, 109–110
 variation with pump speed, 127
 viscosity and, 183

O

Offset, 292, 447, 448
Oil mist lubrication, 446–448
Open drip proof motor, 300, 305
Orifice, 206, 373, 470, 471
O-ring, 328–333, 341, 435–441
Overhung design, 7, 246
Oversizing pumps, 152–155

P

Packing, 229, 249, 313, 333–338, 453, 478–479
Packing gland (see Gland)
Paddle wheel, 206
Parallel misalignment, 288, 447
Parallel pumping, 139–146
Partial emission, 444–445
Performance curve, 71–76
Performance envelope, 74, 86
Performance test, 83–86, 109
Peripheral pump, 7, 278
Peristaltic pump, 26–27
Piping, head loss in, 58–66
Piping design and layout, 186–200
Piston pump, 38–40
Pitot tube, 206–207
Plastic components, 32, 432–435
Plastic liquid, 162
Plunger pump, 39–41
Positive displacement, 5–7, 14–49
Power (see Horsepower)
Prerotation, 465, 468
Pressure
 absolute, 3, 67, 70, 98
 barometric, 67, 212
 gauge, 3, 57, 70, 212
 head, 56, 66, 67, 70
Primary ring, 340, 343, 346, 348
Priming, 8, 15, 16, 242–245, 460–461
Progressing cavity pump, 27–29
Propeller pump, 118, 277
Pseudo-plastic liquid, 162
Pulsation, 15, 34, 41, 46, 486–488

Index

Pulsation dampener, 41, 116
Pump
    air operated diaphragm, 44–46
    axial flow, 118–121, 277
    barrel, 255
    bi-wing lobe, 35–36
    canned motor, 359–362
    centrifugal, 1, 2, 5–11, 15–18, 54–56, 73–75, 85, 118, 122, 152, 157
    circumferential piston, 35–36
    classification of, 5–7
    close-coupled, 233–237
    definition of, 1–2
    diaphragm, 41–47
    double barrel, 255
    double case, 255, 256
    drum, 259, 260, 261
    end suction, 233–237
    external gear, 29–33
    flexible impeller, 25–26
    flexible tube, 26
    frame mounted, 237–240
    gear, 29–33
    grinder, 263
    high speed, 441, 444–446
    hose, 26–27
    inline, 240–242, 246
    installation, 281–282, 400–402, 411, 452–462
    internal gear, 33
    jet, 7, 313
    lobe, 33–36
    magnetic drive, 354–359
    metering, 18, 42–44
    miniature positive displacement, 47–49
    mixed flow, 118–121, 233
    multiple screw, 36–38
    multi-stage, 13, 156, 222, 250–256
    non-metallic, 32, 49, 259, 283, 432–435
    peripheral, 7, 278
    peristaltic, 26–27
    piston, 38–40
    plunger, 39–41
    positive displacement, 5–7, 14–49
    priming, 8, 15, 16, 242–245, 460–461
    progressing cavity, 27–29
    propeller, 118, 277
    radial flow, 228–121
    reciprocating, 17–20, 38–49, 114–116
    regenerative turbine, 7, 278–279
    ring-joint, 255
    ring-section, 255
    rotary, 7, 22, 24–38
    rotary lobe, 33–36
    screw, 36–38
    sealless, 15, 43, 259, 352–362
    self-priming, 8, 15, 16, 242–245
    sinusoidal rotor, 25
    size designation, 74, 76
    sizing, 56–80
    sliding vane, 24
    slurry, 222, 223, 264–267
    split case, 245–256
    start-up, 402, 452–462
    submersible, 260–264, 270–272
    three-screw, 36–38
    two-screw, 36–38
    vertical column, 256–260
    vertical turbine, 268–277
    wobble plate, 46
Pump-out vanes, 218, 227, 229, 266

R

Radial bearing loads, 11–14, 32, 197, 248, 465
Radial flow pump, 118–121
Radially split case, 249, 250, 254–256
Reciprocating pump, 17–20, 38–49, 114–116
Recirculation, 122, 204, 466–467
Regenerative turbine pump, 7, 278–279

Relief valve, 19, 279, 473–475
Repair, 404–406, 489–497
Resistance coefficient, 66–68
Resonance, 274, 486, 488
Rim and face alignment, 447, 456
Ring-joint pump, 255
Ring-section pump, 255
Rotary lobe pump, 33–36
Rotary pump, 7, 22, 24–38
Rotor, 24–38
Runout flow, 100, 121, 275

S

Saturated liquid, 107
Screw pump, 36–38
Seal (see Mechanical seal)
Seal head, 340
Sealless pump, 15, 43, 259, 252–362
Seal ring face, 340
Seconds Saybolt Universal (see SSU)
Segmental wedge, 207
Self-priming, 8, 15, 15, 242–245
Separation, 466–467
Series pumping, 146–152
Serrated rings, 221
Service factor, 88, 294, 295, 298–299, 302–303
Shaft
  critical speed, 158
  runout, 460, 490, 496–497
  sealing (see Mechanical seal)
Shear, 15, 17, 162, 289
Shutoff head, 119, 120, 151, 279
Simplex, 39
Single volute, 8, 10–14, 235, 238, 251, 257
Sinusoidal rotor pump, 25
Sleeve, 238, 248, 249, 258, 267, 273, 284, 334–343
Sliding vane pump, 24
Slip, 11, 18, 72, 82, 299, 382
Slurry pump, 222, 223, 264–267
Software (see Computer software)

Solids, 17, 18, 32, 34, 216, 229, 232–233
Solids handling impeller, 232–233
Specifications, 279–287
Specific gravity, 4, 55, 56, 67, 82, 87
Specific speed, 116–122, 157
Speed, 18, 71–73, 122–127, 155–159
Split case pump, 245–256
SSU, 162–178
Stages, 7, 250, 251, 269, 364
Start-up, 402, 452–462
Static head, 56–57, 128, 133, 378–380
Stator, 27–29, 72, 315, 359, 383
Stilt-mounted base, 454, 456
Stuffing box, 42, 217, 229, 333–338
Submergence, 199–200
Submersible pump, 260–264, 270–272
Suction bell, 269
Suction head, 57
Suction lift, 20, 57
Suction specific speed, 121–122, 156, 225–226, 467
Sump design, 200–202
System design, 185–196
System head curve, 127–139

T

TH, 10, 54, 56, 76–80, 209–211
Thermoplastic, 433, 434
Thermoset, 433
Three-screw pump, 36–38
Throttling, 84, 134, 154, 373–382
Thrust, axial, 30, 218–224, 227–230, 465
Thrust balancing, 227–230
Thrust bearing, 218, 224, 227, 248, 275, 465
Timing gears, 31–36
Total head, 10, 54, 56, 133, 209–211
Triplex, 39
Troubleshooting, 452, 485–489

Index

Turbine
 meter, 207
 regenerative, 7, 278–279
 steam, 1, 382, 456
 vertical, 268–277
Two-phase flow, 15
Two-screw pump, 36–38

U

Ultrasonic flowmeter, 207–208
Units, 3–5, 54–56, 80–81, 116–117, 511–523

V

Vacuum, 66–67, 108, 110, 209–211, 244
Vacuum gauge, 209–210
Valves, head loss in, 66–69
Vane, 9, 12, 24, 89–92, 119, 230–232
Vapor pressure, 93–96
Variable frequency drive (VFD), 319–320, 383–395
Variable speed, 373–395

Velocity, 1–3, 9–10, 17, 55, 58–59, 70
Velocity head, 3, 56, 70–71, 210
Vent, 8, 239, 242, 460
Vent and drain, 344–345
Venturi, 205, 206, 208
Vertical column pump, 256–60
Vertical turbine pump, 268–277
Vibration, 91, 158, 274, 468, 476, 482–488
Viscosity, 15–18, 58, 162–184
Volumetric measurement, 208
Volute
 double, 12, 14, 238, 243, 248
 single, 8, 10–14, 235, 238, 251, 257
Vortex, 199–201
Vortex flowmeter, 208–209

W

Water horsepower, 80, 82, 84
Wear ring, 218–221, 228, 249, 286, 494–495
Wire drawing, 254
Wire-to-water horsepower, 82, 84
Wobble plate pump, 46